JAPANESE URBAN SYSTEM

The GeoJournal Library

Volume 56

The titles published in this series are listed at the end of this volume.

Japanese Urban System

by

YUJI MURAYAMA
Institute of Geoscience,
University of Tsukuba, Japan

with collaborations by

Takashi Inoue

College of Economics,
Aoyama Gakuin University, Japan

and

Yuichi Hashimoto

Faculty of Letters,
Hokkaido University, Japan

KLUWER ACADEMIC PUBLISHERS
DORDRECHT / BOSTON / LONDON

Library of Congress Cataloging-in-Publication Data

Murayama, Yuji, 1953-
Japanese urban system / Yuji Murayama.
p. cm.
ISBN 0-7923-6600-X (hc. : alk. paper)
1. Cities and towns--Japan. 2. Sociology, Urban--Japan. I. Title.

HT147.J3 M85 2000
307.76'0952--dc21

00-062197

ISBN 0-7923-6600-X

Published by Kluwer Academic Publishers,
P.O. Box 17, 3300 AA Dordrecht, The Netherlands.

Sold and distributed in North, Central and South America
by Kluwer Academic Publishers,
101 Philip Drive, Norwell, MA 02061, U.S.A.

In all other countries, sold and distributed
by Kluwer Academic Publishers,
P.O. Box 322, 3300 AH Dordrecht, The Netherlands.

Printed on acid-free paper

Printed in the Netherlands.

Preface

This book is an attempt to explain Japanese regional structure and associated dynamism in terms of urban systems. It is extremely effective to use the urban systems approach to explain the regional changes in today's Japan, which is undergoing changes wrought by economic globalization and the information revolution. This is because the transformation into a service economy has become the key component of the economic activities of cities, linkages are being mutually strengthened, and regional development is being determined by the interdependency of cities.

Readers hoping to gain an understanding of the *regional geography of Japan* may feel that the structure and content of this book are lacking something. However, it is not the intention of this book to systematically paint a total geographical image of Japan within the context of East Asia. Instead, by focusing on urban systems theory, it might be possible to theorize about the factors related to the changing geography of Japan, such as the growth and decline processes of Japanese urban systems, the strengthening of ties among cities and associated factors, and the expansion of socioeconomic exchanges with cities overseas, from a perspective that is different from the conventional approach.

This book is the result of urban systems research that I have been conducting over the past 15 years. I would like to thank many people who have been involved in this research for their help in bringing it to fruition. First and foremost, I would like to thank professor emeritus Takashi Okuno of the University of Tsukuba for providing me with leadership and guidance while I was both an undergraduate and a graduate student. From him I learned the rudiments of urban systems research and quantitative geography. I am also grateful for the constant scholarly stimulus of Shozo Yamamoto, Nobuo Takahashi, Isao Saito, Akira Tabayashi and Akira Tezuka, all professors at the University of Tsukuba. For teaching me about the concepts of urban systems and analytical framework, I am indebted to J.W. Simmons and L.S. Bourne, both professors in the Geography Department at the University of Toronto. While I was pursuing graduate studies at that university in 1979 and 1980, they provided me with a wealth of suggestions and assistance to help me on my way in urban systems research. They also encouraged me to participate in the Commission on Urban Systems sponsored by the International Geographical Union. Through this commission I was able to meet leading geographers from around the world. My recognition of the substantial differences between the formation processes of Japanese urban systems and those of the European and American systems stems in large part from my membership in this commission and my discussions and debates with its other members.

Dr. Du Guoqing at the University of Tsukuba skillfully shaped my numerous diskettes into a final manuscript. Cartographic expertise under tight constraints was contributed by Mr. Kazuto Miyasaka at the University of Tsukuba. I am very grateful for their assistance. Special thanks must be extended to Mr. Kevin Smith, Ms. Emi Kainuma and Mr. Shingo Satoh for their critical reading of the manuscript.

This book is a reedited collection of various academic papers on urban systems. All of the individual papers contained herein have been amended and/or revised. Yuji Murayama (1994) City systems research – outcomes and issues, *The Human Geography*, **46**, 396-417 (in Japanese, Chapter 1). Yuji Murayama (1982) Diffusion channels of innovation in the urban systems: the case of Lions Clubs, *The Annals of the Tohoku Geographical Association*, **34**, 224-235 (in Japanese, Chapter 2). Yuji Murayama (1996) Development process of the East Asian city-system, NLI Research Institute – BAPPENAS Joint Research Publication, 1-53 (Chapter 3). Yuji Murayama (1996) Regional structure of information flow in Japan, *Tsukuba Studies in Human Geography*, **20**, 173-184 (in Japanese, Chapter 4). Yuji Murayama (1984) Regional structure of Japanese commodity flows: an approach of geographical field theory, *Annals of the Japan Association of Economic Geographers*, **30**, 95-111 (in Japanese, Chapter 5). Yuji Murayama (1994) The impact of railways on accessibility in the Japanese urban system, *Journal of Transport Geography*, **2**, 87-100 (Chapter 6). Yuji Murayama (1990) Space-time analysis of internal migration in postwar Japan, *Tsukuba Studies in Human Geography*, **14**, 169-188 (in Japanese, Chapter 7). Yuji Murayama and Yuichi Hashimoto (1995) The spatial structure of population in the Nagoya Metropolitan Area by age and gender, *Science Reports of the Institute of Geoscience, University of Tsukuba, Section A*, **16**, 69-89. (Chapter 8. Note: The latter part of this paper is based on Yuichi Hashimoto and Yuji Murayama (1991) Regional characteristics of the migration by sex and age classes in Aichi Prefecture, *Ann. Tohoku Geogr. Assoc.* **43**, 122-138, in Japanese). Yuji Murayama (1999) The decline of the central commercial district and future prospects in small cities in Japan, Kiyoshi Kobayashi, Yasutaka Matsuo and Kenji Tsutsumi eds., *Local Knowledge and Innovation: Enhancing the Sustainability of Non-Metropolitan Regions*, MARG, Kyoto University, 143-149 (Chapter 9). Yuji Murayama, Takashi Inoue and Yuichi Hashimoto (1997) Spatial chain patterns of intra-urban migration, *Geographia Polonica*, **69**, 135-152 (Chapter 10). Yuji Murayama (1994) The process of land use conversion in the Tokyo Metropolitan Area as shown by Markov chain model, *Annual Report of the Institute of Geoscience, University of Tsukuba*, **20**, 5-10 (Chapter 11). Yuji Murayama (1997) Paris and Tokyo: A statistical comparison of urban spatial structure, *Science Reports of the Institute of Geoscience, University of Tsukuba, Section A*, **18**, 37-56 (Chapter 12).

Yuji Murayama

CONTENTS

List of figures

List of tables

List of photos

PART I

INTRODUCTION

OVERVIEW

The 20th century was the *Age of Urbanization*. As the economic structure shifted from agriculture to manufacturing, there was an exodus of population from rural areas into cities. In the advanced nations of Europe and North America, urbanization began to accelerate around the beginning of the century, and cities were continuously joining the ranks of the million population class.

In the latter half of the century, industrialization was rapidly spreading to developing countries, and urban populations there grew at a rapid pace. The concentration of population into the capital cities was particularly noteworthy, and many high primacy "megacities" appeared. In contrast, the movement of population into the large cities in the advanced countries waned, and houses were being replaced with office buildings. The service economy stimulated the growth of core management functions, and, with the accumulation of capital investment, cities came to have enormous economic power. Thus, the 20th century saw the concentration of both people and money into the cities, and urban functions increased in both the advanced and developing worlds.

The 20th century also witnessed the *Transportation Revolution*. Revolutionary modes of transportation, such as subways, automobiles and airplanes, appeared one after another, greatly increasing not only people's mobility but also facilitating the transport of large volumes of goods. Transportation costs dramatically declined. The world was shrinking. It was no longer unusual to see the "elite businessmen" of multinational corporations making weekly round-trips between New York and Tokyo, and daily trips between Tokyo and Seoul by traders became commonplace.

By late century, many cutting-edge innovations for information transmission such as the Internet had appeared, instantly eliminating the friction of distance that had existed in the business world. This increased accessibility strengthened the mutual dependency of urban systems and accelerated the division of cities into functions and hierarchies. In other words, the growth or decline of cities was being decided not by their own economic power and stock, but by their social and economic relationships with other cities.

Thus was born a new research approach called the urban systems theory, which aimed for a deeper understanding of this phenomenon. "Urban systems" is a comprehensive term that refers to the interdependence of groups of cities arising from the movement of goods, people, information and capital. Today, when cities are more dependent on one another than ever before, it is extremely useful to take such a holistic approach in order to understand the regional structure of modern society and its dynamism. The growth of cities and the division of functions are altering their existing relationships with one another, which is feeding on itself, leading to new growth and division of functions. Urban systems research seeks to gain a quantitative understanding of the dynamic interdependence, and to understand the mechanisms of growth and change in the systems. If these mechanisms can be identified, we will be able to know how urban systems as a unit undergo transformation, and may thus be able to predict which cities will grow and which will decline. It may also become possible to suggest measures for maintaining, managing and/or regulating urban systems that are in harmony, and to make concrete proposals for stimulating growth in only a selected few urban systems while ensuring a stable transition for the remaining ones.

The theoretical framework for urban systems theory was fine-tuned as quantitative geography becomes more entrenched. In the 1960s and 1970s, an ambitious approach was taken in corroborative and theoretical research to examine urban systems in terms of their distribution patterns, rank sizes, functional division, urban dimension, diffusion process, spatial interaction and economic fluctuation. By the latter half of the 1970s, research had expanded to include comparisons of urban systems among nations, future predictions, appropriate management, and policies designed to harmonize urban systems. The late 1980s saw an increasing effort to consolidate geographical research with that of related disciplines such as political science, sociology, and economics. Recently, research on international urban systems, which are ranked above national systems, has been gaining attention.

Given the above trends, the Introduction in Part I will systematically arrange the theoretical and corroborative research related to urban systems that has been conducted so far, examine the effectiveness of methodologies and analytical methods, and look at future topics.

In terms of spatial scale, we can classify urban systems into three levels: international, national and regional. The regional systems are subsystems of the national systems, which in turn are subsystems of the international systems. This book will focus on national systems, which are the function as the most closed, most politically unified systems, to develop a theoretical framework for ranking Japanese systems within the international urban system.

Part II takes a close look at international urban systems. First, Chapter 2 examines how innovations, having spatial and temporal order, are disseminated from the international system through the national system to the regional system. The Lions Club, which has an active mission worldwide to contribute to the good of society, is used as an index.

The results of the analyses show that the dissemination of innovation in urban systems has the following features: 1) There is a time lag in dissemination in its route from international to national to regional systems. In the case of the Lions Club, the peak at the international level was in the mid 1950s, the national (Japanese) peak was around 1960, and the regional (in this case, Yamanashi Prefecture) peak was in the mid-1960s. 2) When an innovation spreads through an urban system, there are various effects at work, including the hierarchical effect, the neighboring effect, and the random effect. In the case of the Lions Club, the strongest effect at the regional level is the hierarchical effect; at the national level it is both the hierarchical and neighboring effects; and at the international level it is the random effect. 3) The dissemination of innovation at the regional and national levels progresses through cities that have strong social and economic ties. However, in the international urban system, political systems and historical circumstances have a greater effect.

Chapter 3 looks at the formation process and associated mechanism of the East Asian international urban system. Since the latter half of the 1970s, Japanese manufacturers have been expanding overseas, leading to robust economic growth, especially in East Asia. In other words, this overseas expansion was a driving force behind the economic development of East Asia, which formed an economic sphere that rivaled those of Europe and North America.

Given this background, various analyses were conducted in this chapter, especially by focusing on changes in intercity linkages. These analyses resulted in the following conclusions: 1) Since the 1980s, industrialization in East Asia has led to rapid growth of urban areas, and many "megacities" have appeared. Megacities attract not only manufacturing industries and population, but also a variety of service industries, and medium and small cities in the country are incorporated into the economic sphere of the megacity. Even within megacities, large cities that have a particularly dense concentration of economic core management functions are strengthening economic ties with large cities in other countries, contributing to the spatial organization of international urban systems in East Asia.

2) In the 1980s, the primate city of each country in East Asia coordinated the economic activities of the other cities in the country. These primate cities also established weak linkages with large cities in developed countries and capital cities of neighboring countries, forming a hierarchical urban system. However, today these primate cities are strengthening other linkages, and the second- and third-ranking cities in their countries are strengthening ties with cities in other lands, creating a level-network type of intercity relationships.

3) The large cities in Japan, whether they serve large or relatively small regions, have become nodes in the East Asian international urban system, and are growing as important focal cities. However, Tokyo, which had long reigned as a city at the very top of the East Asian hierarchy, has been receiving intensive competition from Singapore and Hong Kong, and its dominance has relatively declined.

Part III looks at the national urban system of Japan. In Chapter 4, telephone traffic, which is an integral part of the daily affairs of individuals, businesses and governments, is used as an index to characterize the nodes and functional regions of Japan in terms of information flow. According to the analysis, these nodes and functional regions show a clear multi-level hierarchical structure. Alone at the top is Tokyo, which dominates the rest of the cities in Japan. At the second-highest rung on the hierarchy are the wide-hinterland cities of Osaka, Nagoya, Fukuoka, Sapporo and Sendai. Osaka reigns over the Kinki region, Nagoya over the Chubu region, Fukuoka dominates Kyushu, Sapporo is supreme over Hokkaido, and Sendai is the dominant city in the Tohoku region. Third on the hierarchy are the prefectural capitals, which oversee the activities of the other cities in their respective prefectures.

Chapter 5 deals with the movement of 30 major categories of goods in Japan, including farm products, industrial goods, goods delivered by courier services, and waste products. Using factor analysis, the main distribution patterns were derived, and characterized as follows:

The vast majority of manufactured goods move within the Pacific belt, especially among the three major metropolitan areas. While the movement of primary and light industrial goods occurs throughout the country, it is centered on Hokkaido, Tokyo, Aichi, Osaka, and Fukuoka. Agricultural products and related goods are mostly imported, so the hinterlands of large trading ports act as distribution points for these goods. Chemical products, whose main center of production is western Japan, are transported relatively long distances, in contrast to miscellaneous goods, which are usually shipped only short distances and for which the cores of each of these regions form a closed system of linkages.

Chapter 6 examines the developmental process of Japan's urban system in terms of improvements in accessibility. The most important factor that has contributed to modern Japan's economic development is the building of the railways. The development of a nationwide railway network helped to facilitate and quicken the mass movement of goods and people over long distances. Network analysis employing graph theory revealed the following points: From the mid-18th century until about 1960, intercity differences in accessibility were constantly shrinking. However, after about the mid-1960s, the opening of Shinkansen "bullet train" lines created an accessibility differential between the cities lying along the lines, and those that were not. As a result, there was a relative increase in the locational potential of cities along the Shinkansen lines, which augmented their nodality.

Chapter 7 deals with population movements since the mid-1950s, and how spatial movement patterns changed within Japan. Quantitative analyses using origin-destination matrices showed that trends in population movements were closely related with economic trends. During the period of high economic growth (mid-1950s to early 1970s), there were massive population movements toward cities like Tokyo and Osaka, but during the period of relative stagnation (early 1970s to early 1980s), this trend ceased. During the current period of stable economic growth (early 1980s to date) as well, major population movements have generally been non-existent, except for noteworthy movement from relatively rural areas into Tokyo. However economic decline has been occurring in Osaka, which has lost its ability to attract population from a large area.

Part IV, the *Regional Urban System*, takes up the case of Aichi Prefecture in Chapter 8, revealing the spatial characteristics of population movement patterns. The core of this regional system is Nagoya, one of the three largest metropolitan areas in Japan. Over a long period of time, people have been moving out of the central city into surrounding areas. Analyses have shown that most of the people leaving central Nagoya are families who are looking to purchase a home, thus the proportion of single and elderly residents in the city center has been increasing. This phenomenon is a universal feature of all regional urban systems in Japan.

Part V examines the spatial features of the elements that make up urban areas, that is, cities as nodes. Chapter 9 looks at a small outlying city, Hitachiota, Chapter 10 discusses a medium-sized city (Yokkaichi in Mie Prefecture), and Chapter 11 takes up the investigation of a major city, Tokyo. Chapter 12 compares the spatial changes in the urban structure of Tokyo, which is growing as a world city, with the changes that are occurring in Paris.

Given the above discussions, the *Conclusion* in Part V clarifies the formation process of Japanese urban systems and points out the characteristics of the country's modern urban system. In brief, the formation of the Japanese national urban system had been supported by the development of railways from the latter 19th century until about 1920, when the basic infrastructure was completed. During this period the two highest cities in the hierarchy were Tokyo and Osaka, which formed a bipolar system. This structure lasted about 50 years, until the end of the period of high economic growth in the early 1970s. After that, Osaka went into decline, while Tokyo grew at a rapid pace, eventually becoming the primate city, that is, the only city at the top of the Japanese hierarchy.

In the future, the Japanese national urban system will undoubtedly change from the current hierarchical pyramid into a non-hierarchical network. However, this process is very slow, and it appears at present that the hierarchical system is still entrenched. If the promotion of power sharing with provincial areas destroys the centralized power structure that has supported the hierarchy, as well as the corporate territorial management system that came into existence as a result of government policies, the transition to a non-hierarchical network can be expected to accelerate.

CHAPTER 1

STUDY OF URBAN SYSTEMS: OUTCOMES AND ISSUES

1.1. Introduction

Urban system is a comprehensive collection of cities which are interdependent through economic fluctuations, diffusion and exchange of information, and flow of goods, capital and people (Pred, 1977). Cities which are parts of such a system do not grow, stagnate, and decline independent of one another; instead, they change in response to growth or decline in other cities. The study of urban systems is not concerned with the spatial expansion of the cities or its internal structure; rather, it emphasizes interurban relationships by viewing cities as points in a network. Recently the study of urban systems has become interested in the modification of interurban dependencies brought about by the expansion of economic activities over a wide area and the intensification of transportation and information networks. Since openness, or the dependency of cities on the outside, is increasing, we cannot gain a clear understanding of the functionality and growth of cities without thinking these systems as a combination of social and economic factors.

According to Vining (1949), who first made the connection between urban systems and urban growth, the term "urban system" was already in use before 1940. However, it was not until the publication of the work of Duncan *et al.* (1960) that the national urban system of the United States was described in terms of the economic role of individual cities and their interdependent relationships (Bourne and Simmons, 1978). Berry (1964) made geographers aware of urban systems by explaining urban population size distributions and the hierarchical urban structure using general system theory. Since Berry's study, research on urban systems, which was influenced by the development of quantitative geography, has been rapidly progressing.

With the establishment of analytical frameworks using multivariate statistics in the 1960s and 1970s, exhaustive research was conducted on rank size, functional classification, dimensional analysis, nodal and functional structures, diffusion processes, spatial interactions, and the wide ranging effects of economic fluctuation. In the late 1970s, research was expanded to include international comparisons of urban systems, forecasts for the future, and control, management and even policy to maintain urban systems in equilibrium. In the late 1980s, interdisciplinary research in urban systems increased as ideas from related disciplines such as political science, sociology, and economics were incorporated into geographical research. More recently, research concerning international and global urban systems (which supersede a national urban system of cities) has been gaining attention. The theoretical conceptualization has evolved as well.[1]

By using graph theory concepts, it is possible to indicate the complex, interdependent urban systems with a topological graph comprising only the two elements of node (city) and linkage (interdependency). Node refers to size, regional influence, function and location, while linkage can be used to express vectors for volume, length and direction of flows (Murayama, 1982a). Research into urban systems began with a static analysis which emphasized the explanation of spatial order of nodes, eventually including analysis of interurban relationships which incorporated linkage attributes. This in turn led to a search for spatial processes — explaining the mechanisms for growth and intensification in urban systems in terms of the mutual relationships of these two elements: the effect of node variation on linkages, and the effect of linkage variations on nodes.

There are many collections and summaries of urban systems studies (Wärneryd, 1968; Berry and Horton eds., 1970; Bourne, 1975; Hansen, 1978; Joo, 1981; Hino, 1981; Kawashima and Korcelli, 1982; Morikawa, 1982; Bourne *et al.*, 1984; Marshall, 1989). The study of urban systems encompasses a wide range of disciplines, including not only geography but also economics, politics, sociology, transportation engineering, urban planning, and other fields. Even within geography, there are many different approaches and definitions. Although this review focuses on the development of methodologies in urban systems and the growth of corroborative research, it is limited to a selective cross-section of the extensive literature.

This chapter reviewing methodological development of urban systems studies consists of four parts. First, researches emphasizing the spatial order of nodes are examined, and then those incorporating linkages are discussed. Thirdly, we pay attention to global urban systems research which is very active due to the internationalization of economic activity. A final attempt is to discuss problems associated with analytical operations.

1.2. Research Emphasizing the Spatial Order of Nodes

This section will look at how spatial order is given to the distribution, hierarchy and functions of nodes as a result of interaction among cities; however, in this literature, these interactions themselves are not clearly understood and must be approached as a "black box".

1.2.1. DISTRIBUTION PATTERNS

There exists a certain regularity to the distribution patterns of nodes, i.e., location, density, arrangement, and urban hierarchy. From the 1950s well into the 1970s, a wealth of research considered the relationship between spatial order and Central Place Theory (CPT).[2] Particular attention was given to three elements (scale, number, spacing) of central places (cities), in order to determine the extent to which the predicted structure of the central places coincided with the actual distribution of cities. The extent to which theoretical and actual distribution differed was the subject of considerable debate, and methods were developed to measure this difference quantitatively (Dacey, 1960). The index of centrality provided the basis for determining the urban hierarchy, and point pattern analysis (nearest neighbor and grid methods) was used to gain an understanding of nodal distribution patterns, i.e., uniform, random, aggregate (Boots and Getis, 1988).

Corroborative research on nodal distribution patterns suggested that 1) systems of nodes show a random distribution during the initial stages of their development, but as interurban dependencies increase, nodes grow as they compete against one another; thus distribution heads toward equilibrium, and 2) cities which are not developed economically are small in size, and close together; however, as urban economies develop, both their size and distance from other cities increases, and small cities disappear.

The scope of distribution pattern research was limited to small areas or regions; little research was conducted at the national level. Urban distribution patterns in national urban systems were studied by regional scientists and spatial economists more than by geographers. As a result, the search for the optimal number and size of cities within urban systems was concerned with the advantages of economies of scale and optimal resource allocation (Kanamoto, 1981).

1.2.2. DISTRIBUTION OF CITY SIZE

There is a regularity to city size (represented by population, income, or number of workers) and its hierarchy, which was first pointed out by Auerbach (1913). This regularity was generalized into the principle of city size hierarchies by Zipf (1949), who argued that the equilibrium between uniformity (the force acting to concentrate population into a few large cities) and dispersion (the force acting to distribute population to small- and medium-sized cities) determined the curve of rank-size (hierarchy).

Zipf's ideas were developed further by Berry (1961), using probability concepts from system theory. According to Berry's Entropy Maximization Model, a primate (leading) city emerges during the stable state of nodes (Jefferson, 1939), while the rank-size principle (log normal distribution) appears when there is disorder (maximum entropy). When economic development occurs and interurban linkages are strengthened (many forces are at work), there is gradual movement from primacy to intermediacy to rank-size type. Furthermore, the transition from rank-size patterns to primacy reinforces the interdependence among cities. Berry explained this as being the result of the work of strong unifying forces. Since then, numerous theoretical models have been proposed for city size distributions, and evaluated by applying them to actual nodal distribution patterns (Richardson, 1973; Kohsaka, 1978; Carroll, 1982).

At the spatial scale of national urban systems, analysis has been conducted under the two broad themes: 1) How do city size distribution patterns differ from one country to another, and what causes these differences? 2) What kind of relationship exists between changes in the city size curve and the growth of a national urban system? The conclusions drawn from corroborative research around the world have been summarized by Carroll (1982): 1) When a strong interdependency is concentrated in a certain link, the city size distribution is primate; when interdependency is diffused in all links, the curve approaches rank-size. 2) A closed economic system with high internal interdependence supports rank-size patterns; openness (i.e., external dependency) supports primacy. 3) When the population and area are small, and the social, political, and economic structures are relatively simple with little urbanization, primacy is more likely. 4) Political systems reflects city size distribution. For example, socialism systems are more primate, while federal systems (e.g., Germany and Canada) approach rank-size. The more government intervention, the greater the trend is toward primacy. 5) City size distribution reflects the balance between social equilibrium and disequilibrium. Primacy reaches its peak during the intermediate stage of economic development, but urban systems then move toward rank-size patterns.

In recent years, primacy patterns in developing countries have caught the attention not only of geographers, but also of sociologists, political scientists, and economists (Fujimaki, 1984). The continuing development of primacy is caused by the incorporation of primate cities into the global urban system, as economic linkages to cities in developed countries are strengthened. The spatial interaction of national urban systems of developing countries is weak, so primate cities become growth poles due to the influence of the major industrialized countries. The secondary level of cities cannot easily receive the benefits of this development. Since domestic transportation and information networks are underdeveloped, the flow of goods is not smooth there; since administrative organizations have not been established to coordinate the nation's land resources, new technologies and growth-inducing policies are not effectively transferred from the primate city to secondary cities. Therefore, economic disparity between primate and secondary cities in developing countries is increasing.

There are both positive and negative ways of viewing the impact of primate cities on urban systems (Sheppard, 1982). Those who emphasize the demerits cite the following points: 1) Primate cities hinder the efficient use of resources, 2) primate cities promote interactions with foreign countries while reducing domestic interactions, thus hampering economic growth, and 3) primacy creates regional social inequity. On the other hand, proponents of this system counter that these problems are nothing more than a temporary condition, that concentration of capital into the primate city is, in fact, efficient due to the economic effects of scale and agglomeration. Moreover, they claim that primacy leads to the specialization of knowledge and the accumulation of capital and human resources, while acting as the driving force behind the expansion of transportation networks and the spread of innovation.

City size distribution is effective for comparisons between nations, and also for time series analyses within a country and the growth process of a national urban system. For example, Strickland and Aiken (1984) tested a model to measure the stability and transformation of urban systems by considering the German urban system. They found that an urban system which has formed does not easily change, and that some tremendous force is needed to extricate it from the primacy pattern.

The following problems have come to light in the application of city size models to urban systems research. First, there may be a problem involved in analyzing a nation as an entire urban system. Second, there may be a need to incorporate spatial distribution into the model. Third, urban population may not be the best measure of size for urban systems research (Wilson, 1969).

Concerning above point 1), Suzuki (1981) developed a useful theory on the city size distribution (the decomposability of the primacy principle) in national and lower-order regional urban systems. The construction of models considering the "point pattern distribution" is expected to solve the second weakness. For the third, researches could use, for example, the cumulative amount of capital and information, or the concentration of core administrative functions and numbers of elite workers as measures.

1.2.3. THE URBAN DIMENSION

During the period under economic development, the specialization of cities and division of labor among cities progresses, reinforcing interurban dependencies. Some cities — usually the smaller one — specialize in certain economic functions, while other cities — the larger and more central — take on diversified roles. In this way, the division of the functional roles of cities progresses. Urban dimensional analysis seeks to understand the basic functional characteristics of the urban system as a whole, i.e., the urban dimension. This aggregates measure of urban activity. Using the fundamental dimension as a base, the relative functional differences of each city can be identified. Furthermore, studies over time reveal the extent to which transfigurational processes and the division of urban functionality have progressed.

The "urban function classifications" have long been promoted by Ogburn (1937), Harris (1943), and Nelson (1955), among others. Their early research, using single indices, mainly involves the classification of cities and does not consider urban functions which are created as a result of interdependencies. The full-scale incorporation of system concepts into urban dimension analysis reduced the multitude of urban attributes into a few significant factors. The identification of the urban dimension has been the result of factor analysis, which first came into use in the 1960s (Moser and Scott, 1961). During the initial stage, only the so-called "urban economic functions" of demographics, production labor, number of offices, and retail prices were used as variables. However, soon a wide range of variables, such as culture, living environment, social traits, transportation, and individual attributes (a result of factorial ecological studies of inner cities), were incorporated as indices, providing a comprehensive perspective on the urban dimension. Naturally, the value of urban dimension analysis is not merely that it is an index which can distinguish urban functional divisions (Yamaguchi, 1973).

The urban dimension has often been identified at the national level. This has been grounded on the hypothesis that urban dimension is emerged through the interactions of national attributes such as history, ethnicity, industry and policies (Berry, 1969). In the advanced industrialized countries, factors such as social class, economic standing, (ethnic) cultural foundation, and city size have been derived from common research data. In developing countries, the variables are mixed together since urban functions have not been divided. Corroborative research has reported that factors such as social class, economic standing and cultural foundation, which are clearly separated in the developed countries, are difficult to interpret as a distinct dimension in the developing countries (Sung, 1977).

It has been argued that there is a relationship between fluctuations in the urban dimension, and urbanization and economic development in a country.[3] For example, in his 1951 and 1961 comparisons of the urban dimension in Canada, King (1966) explained on a region-by-region basis the degree to which independent urban systems in the country had been incorporated into the national system. On the other hand, in his studies of the evolution of the Japanese urban dimension between 1950 and 1970, Hino (1977) found that the urban system in post-war Japan was relatively stable from 1950-1960, but entered a state of instability from about 1960 to 1970. In his analysis of the national urban system in the Republic of Korea, Joo (1982) revealed considerable changes in national urban dimension since the 1960s. Many of these studies have concluded that there is no stable transition of the urban dimension when seen in a time series. However, a limited number of studies have pointed out its stability. For example, in his analysis of the urban system of Chile, a developing country, Berry (1969) found no long-term change in the urban dimension. This suggests that countries experiencing marked urbanization and economic growth "rearrange" the urban dimension over time, while in stagnant countries with unstable economic growth, the urban dimension is unchanged. In any event, fluctuations in the urban dimension are closely related to urbanization and economic development. Furthermore, by noting the dynamics of cities within urban systems, we can see a trend that city size affects fluctuation patterns, i.e., the larger the city is and the more diversified its functions are, the more stable its transition will be. Besides, the more pronounced economic growth is in a country, the more advanced will the interurban division of functions be.

The following three points should be incorporated into future analysis: 1) The degree of variations in interurban functions may be related to the degree and pattern of functional interdependencies among cities.[4] 2) The factors which cause the diversity in these functions must be identified. 3) The way how differences in urban functions shape social structure and growth processes must be clarified.

This section has summarized the three approaches (distribution patterns, city size distribution, urban dimension) used to explain the spatial order of urban nodes. Each of these approaches treats interurban relationships as a black box, and have argued that the spatial order of cities is a result of mutual dependency. There are, however, limitations in the reliance on studies of nodes to explain the mechanisms involved in the creation of urban systems. The analysis of linkages is important, and there is a need to identify the forces which connect cities and how these forces work. In this respect, the three approaches above have contributed little to urban system research.

1.3. Research Incorporating Linkages (Interurban Relationships)

1.3.1. INTERURBAN FLOW AND NODAL FUNCTIONAL STRUCTURES

The study of linkages began with analyses of transportation networks such as roads, railways, and air routes. Quantitative methods, which were developed to measure the contiguity and potential of transport nodes and the connectivity of transportation networks, were applied to urban system research (Haggett and Choley, 1969) as "network analysis." When interurban origin-destination (O-D) matrices became readily available at the national and regional levels, researchers studied them in order to understand nodal and functional structures and spatial interactions. The identification of nodal and functional regions enabled a few strong links of interdependency to be culled from the possible $n \times n$ links (n = number of cities) (Simmons, 1978). Numerous quantitative methods, such as maximum flow, graph theory, factor analysis, Markov chains, and cluster analysis, were developed and applied (Morikawa, 1978). The newly-introduced patterns and structures of nodal and functional regions were supplemented by research on the integration and nodality (i.e., measurements using entropy and directivity) of both subregions and regions as a whole. Furthermore, investigations of time-series changes in nodal and functional regions explicitly evaluated systems concepts (Murayama, 1992).

Indices should be selected which can most accurately identify interurban linkages as data in the O-D matrix. Past studies have utilized communications flow (telephone, telegraph, telex, the postal system, newspapers, magazines), population flow (migration, commuting), transportation (tourism, movement of goods), and the flow of investment capital as indices. Although it has been said that the flow of investment capital, the driving force of the economy, is the most appropriate to be used as a comprehensive index, data needed for general application is rarely made available to the public.[5]

Characteristics of nodal and functional structures which have been elucidated from empirical research include: 1) Nodal and functional regions often correspond to administrative boundaries (e.g., state and county borders). 2) As spatial organization progresses over time, independent nodal regions decrease, and unity within a system strengthens (i.e., subsystem integration progresses).[6] 3) During the course of development, the urban hierarchy tends to be divided into upper and lower orders, with a middle order conspicuously absent. Small cities are theoretically subordinate to medium cities, which are in turn subordinate to large cities. However, indirect control by the larger cities, which is supported by the position of these cities at the apex of the urban hierarchy, moves toward direct control in which both small and medium cities are directly subordinate to some large cities.[7] 4) With the decline of vertical linkages, horizontal linkages are reinforced as cities at the same level of the urban hierarchy strengthen their interlinkages.[8] This trend is particularly evident among high-ranking cities.

The study of spatial interactions constructs models to estimate interurban flows within an urban system, based on their size and distance from one another. This field lays the groundwork for understanding future nodal and functional structure, for if a detailed model of spatial interactions could be created, it would become possible to make an accurate estimate of the O-D matrix, thus opening the way for making analogies, albeit indirectly, of future nodal and functional structure. Among the most important studies leading in that direction is that of Nam (1981) who estimated the Korean O-D matrix from 1977 person-trips and identified the likely nodal region in 1991.

Since the late 1970s, "map pattern problems" in spatial interactions have been gaining attention from the standpoint of city distribution. The effect of nodal location on interurban flow has been confirmed, and sophisticated models of interactions (competing destination models, etc.) which incorporate nodal location (as represented by accessibility, etc.) have been proposed. In map pattern problems, distribution pattern analysis, which had been vigorously promoted in the 1960s, once again got into in the spotlight.

Spatial interaction can be also linked to studies of urban dimension (function). Berry (1968) developed his General Field Theory in which spatial interactions and spatial structure affect one another. The location of cities, and the functions that each city possesses, regulate the interurban mutual dependencies on the one hand while these mutual dependencies promote the spatial structure, i.e., the functional division of cities and the growth of each of these cities on the other hand. When seen in a time series, these two factors cause a transfiguration of urban systems (i.e., growth) as they co-vary (Murayama, 1990). Whether this General Field Theory is merely hypothetical, or whether it actually applies to the real world, will be proven only after much more corroborative research has been completed.

1.3.2. INTERURBAN LINKAGES FROM THE PERSPECTIVE OF BUSINESS ACTIVITIES

The previous section has discussed the actual (physical) interactions or flow which occurs between cities, and interactional research which tried to identify interurban relationships. This section will examine studies of the creation of interurban flow, that is, from the aspect of the interurban linkages provided by the economic stimulus. Businesses play an important role as economic stimulators, especially the boardrooms (headquarters) of large corporations. With the economic power of these large corporations, we may well be able to say that they control national urban systems since they develop their corporate activities with an eye on national markets. Corporate headquarters (the source of economic core management functions) coordinate branch activities through their directives, transmission of information, and provision of goods, labor, and capital. Branches, meanwhile, control subordinate organizations such as local offices and production facilities (Nishihara, 1991). The spatial behavior of corporations promotes both the division and linkage of urban functions and supports an urban hierarchy. From this perspective, researchers have tried to identify the development of urban systems derived from spatial behavior of corporations such as the development of office network.

Interurban relationships are identified from the locations of corporate headquarters and cities which have branches.[9] Using the numbers of companies and employees as indices, an interurban O-D matrix of the headquarters-branch office relationship can be constructed. Separate O-D matrices can be created for each type of production or business based on the assumption that the formation of interurban links differs according to the type of company involved.[10] Using this data matrix as a starting point, we can proceed to search for an understanding of the strength of interurban connections and the domination by large cities.[11] Specifically, the urban hierarchy is investigated from the perspectives of the spatial location, organization, routes for transmitting decisions, business territory, branch location, etc., of corporate offices.

This approach, which has looked at urban systems through the development of corporate offices, has produced a wealth of data, though there are still some problems. First of all, it is assumed that in the operations of corporations, there is a strict hierarchical dominant-subordinate relationship in which the corporate headquarters co-ordinate the branches while the branches control subordinate offices; this ignores the horizontal relationships between branches, and direct connections between headquarters and subordinate offices. Secondly, since the strength of interurban linkages is estimated from the numbers of companies and employees based on headquarter-branch relationships, we cannot consider the qualitative linkage of transmission of ideas or control. The latter can be incorporated by using such measure as the volume of communications (business contact) and decision-making, capital flow between headquarters and branches, numbers of exchange personnel, etc.

In recent years, interurban networks (Green, 1981; Green and Semple, 1981; Rice and Semple, 1993) of interlocking directorates[12] and the volume of interurban business communications have been used in an attempt to unmask interurban linkages. For example, Wheeler and Mitchelson (1989) used express mail O-D data from 48 U.S. cities to show corporate transmission of information (most special delivery service is used by business). They found that: 1) The transmission of information is more important than its reception; the greater the volume of information that a city transmits, the more important will be its role in economic activities. Large cities function as an information control base and transmit much more information than they receive. 2) In contrast to information transmission which is limited to a few large cities, information reception is often dispersed to small and medium cities. Furthermore, most of the information being transmitted from these smaller cities is going to large cities, with little of it being transmitted to other smaller cities. Therefore, small and medium cities are highly subordinate to large cities, and have weak interlinkages among themselves. 3) The flow of information, which differs from the flow of people and materials, shows little attenuation over distance. In particular, large cities show strong linkages even though there may be considerable distances between them; the effect of "distance decay" is negligible. 4) The situation in which a few large cities control many small and medium cities materializes gradually over many years.

Like corporate interurban linkages, interurban linkages between government and administrative organizations (e.g., national capital, state capitals, local agencies) are also important factors in the development of urban systems. Territory of governmental agencies roughly corresponds to corporate territory, and the granting of various patents and licenses for economic activity is proof that government agencies wield tremendous influence over corporate activities. Future research should attempt to sort out the relationships between corporations and government agencies by investigating the extent to which the latter regulate urban systems and how their behavior affects which parts of the interurban linkages (Hino, 1981).

1.3.3. INTERURBAN DIFFUSION OF INNOVATION

The interurban diffusion of innovation (the introduction of new technologies, organizations, ideas, devices, etc.) is closely related to urban systems. They are mutually regulating, as urban systems control diffusion channels while diffusion channels influence urban systems (Pedersen, 1970; Cohen, 1972; Hudson, 1972; Pred, 1977; Brown, 1981; Sugiura, 1987). Urban systems research which deals with innovation must meet at least two preconditions. One is the impact of growth-inducing innovations upon the formation of urban systems. Past research has reported on new technologies such as satellite hook-ups and supercomputers as being used in everything from the establishment of clubs and organizations (e.g., Lions and Rotary clubs) to radio and television stations to shopping centers, supermarkets, electric companies, newspaper companies, banks, and municipal offices. The most important point to bear in mind is that the characteristics of these new innovations, such as the extent to which they are becoming socially and regionally diffused, the economic effects of diffusion, and whether non-economic factors are at work in this diffusion, should be clearly understood before analysis can begin.

The second precondition is knowing beforehand the route for spatial diffusion. If we cannot characterize those cities transmitting and receiving innovation, in other words, if we do not know from which cities innovation is emanating, we cannot find the relationship between urban system process and structure. There are few indices which can specify the route of diffusion. Murayama (1982b)'s use of Lions Club[13] data in his analysis of interurban diffusion is one example of a natural index for diffusion routes.[14]

There are two general themes in the research on the interurban diffusion of innovation. The first is the time lag involved in the diffusion of innovation and the route that it takes to pass through the urban system. The second focuses on how the continuous introduction of innovations into the urban system causes changes in these systems over the long term. In the first theme, researchers describe sequences of changes in an urban system, when diffusion routes are already known. As substitutes for interurban linkages, past research has used central place systems (Hudson, 1969), nodal and functional structure (Murayama, 1982c), or MDS (multi-dimensional scaling) to reconstruct the interactive (i.e., contiguous) space (Sugiura, 1982). In any event, this information should be easy to process during analysis; furthermore, it must be able to describe accurately actual interurban linkages.

We can roughly divide innovations into those used by households (individuals and families) and those used by business and government (corporations and administrative agencies, etc.); the latter which are also called urban innovations due to their use in city units, the driving force of urban development, are of most interest to urban systems research. A business application gives a competitive advantage to early adopters, but this type of diffusion does not trickle down to the lowest orders of the urban hierarchy; it is characterized by stagnation at a certain economic threshold. Given that the transmission of innovations follows strong linkages of interurban functions, the stronger the mutual dependencies are between cities, the faster the rate of diffusion is. Numerous corroborative studies have shown that, when seen from a time dimension, the diffusion of innovation is S-shaped (i.e., the cumulative application is an S-shaped curve); introduction during the initial stage is slow, then gradually accelerates until it becomes sluggish again in the final stage.

City size and distance are the two major factors explaining the spatial diffusion of innovation. They are called, respectively, the hierarchical (i.e., city size) effect and the proximity (i.e., distance) effect. The hierarchical effect is usually stronger than the proximity effect within the urban system.

As it has previously been mentioned in 1.3.2, the spatial behavior of large corporations has attracted the attention of diffusion researchers because of its impact on the growth of urban systems. In particular, there is a wealth of research focusing on the process of creating new branches and local offices which accompanies business expansion; the expansion and division of territory; routes for transmitting directives and information, and the speed at which they are transmitted; and "permeation channels" for technological innovations, management goals, and directives. The negative effects of branch closures and reductions in commercial activities (the "priority order of withdrawal") has also come under consideration. In his detailed study of the spatial behavior of large corporations in the United States, Pred (1977) introduced a new model (the so-called "Pred Model"). According to Pred, the diffusion of innovation within the corporate organization is not only hierarchical but includes a considerable amount of lateral diffusion between cities of the same size, upward diffusion from small to large cities, and mutual diffusion among large cities (Pred, 1976).

Semple (1985) sparked interest in the relationship between site development of large corporate office networks and fluctuations in urban systems with the introduction of his quaternary place theory, a 6-stage development model.

Recently, global locational development, such as the location of multinational corporate offices, has been attracting attention, with analytical research expanding to include international and global urban systems. However, we cannot say that the extent to which urban systems influence the diffusion of innovation and the ways in which its "feedback" causes the transfiguration of urban systems (i.e., the mechanisms of cycle and cumulative effect) are clearly understood (Clark, 1980; Jones, 1983). The understanding of cyclical and accumulative mechanisms of urban systems could provide the raw material needed for formulating effective policies for growth centers.

1.3.4. INTERURBAN SPREAD OF ECONOMIC FLUCTUATION

One school of geographical thought has attempted to explain the processes by which economic fluctuation spreads among cities. A theme is closely related to spatial diffusion research. Specifically, it seeks to identify the transmission routes and leading and latent economic cycles using wages, employment, unemployment, etc., as indices.[15] These factors correlated to the spatial location, urban functions, and urban hierarchy of urban systems. Vining (1945) and others first identified the spread of such regional economic fluctuations in the field of regional economics but it was King *et al.* (1969) who first stressed the importance from a spatial perspective. Using these studies as a starting point, corroborative research in the 1970s and 1980s began to investigate the interurban wave of economic fluctuation, focusing on the national urban systems of advanced industrialized nations such as the United States, West Germany, Great Britain, Canada, and Australia. This research concentrated on the following three themes: 1) the identification of cyclical elements which influence economic fluctuations, and the degree of this influence, 2) the leading-latent relationship in regional economic fluctuations, and 3) the interurban transmission routes of economic imbalance (Kohsaka, 1984; Martin and Spence, 1981). In identifying the latent chain relationship, various types of multivariate analyses, such as factor analysis (applied to T-type factor analysis in which the region is the rows and the time is the columns of the matrix), multi-regression, correlation analysis, spectral analysis, and the Box-Jenkins model have been used.

The relationship between city systems and the interurban wave of economic fluctuation, which has been identified by past research (Jeffrey and Webb, 1972; Jeffrey, 1974; King, 1976; Asami, 1980; Pigozzi, 1980) can be summarized in the following seven points: 1) Because of the diversity of their industries, large cities enjoy relative economic stability, in contrast to small and medium cities with their smaller industrial bases. 2) A hierarchical sequence diffuses economic fluctuations from large cities to small and medium cities. 3) The time lag involved in economic fluctuation is closely related to the industrial structure of a city; there is quick response to these fluctuations in cities with related industrial agglomerations. 4) Economic fluctuations of large cities correspond to those at the national level, but there is often an uneven dynamic lag in the smaller cities, and there is more lag than in the large cities. 5) Economic fluctuation travels routes corresponding to strong interurban linkages. 6) The strengthening or integration of urban systems weakens interurban differences in economic fluctuation. 7) Mid-term economic fluctuations (3-4 years) are leading in cities specializing in industries with a strong manufacturing specialization, while they are latent in cities specializing in core management functions and the service sector.

1.4. Research on International and Global Urban System

1.4.1. BACKGROUND OF DEVELOPMENT

Viewed on a spatial scale, urban systems can be classified into four levels: regional, national, international, and global. A completely closed system is found only in the unique global urban system; all others are open at the upper levels. Although comparative studies of national urban systems do not have a long history, this section will discuss the direction of research in this field by introducing studies from fields related to geography.

Research in international urban systems looks at two or more countries while the entire world is the object of research involving a global or world urban system. International systems research is looking at areas which are becoming economic blocks, such as North America and the European Union, making continents analytical units and studying cross-national interurban relationships, and examining the influence of national borders, degree of unity, openness and closure of national urban systems as a subsystem of this. Of particular note is European research, which has been rapidly growing in recent years due to the implementation of EC integration policies.[16]

Geographical research involving international and global urban systems has been influenced by theories which have been proposed in political science, economics, sociology, and other disciplines (Friedmann and Wolff, 1982). These include New Dependency Theory/Metropolis-Satellite Theory (Frank, 1967), Interdependency Theory (Morse,1976), World Systems Theory (Wallerstein, 1979), New International Division of Labor (Fröebel *et al.*, 1980), and its offshoot Global Spatial Restructuring,[17] and World City Theory (Friedmann, 1986).

1.4.2. WORLD CITIES AS SYSTEM ELEMENTS: CHARACTERSTICS AND FUNCTIONS

The rapid development of transportation and communication technologies, the activities of multinational corporations, and the internationalization of capital flow[18] have worked to strengthen interdependencies of cities on a global scale (Brunn and Leinbach, 1991). Presently, at least the primate city of each country is directly incorporated into international or even global urban system.

The structural elements of global urban system are the large cities, including the primate city of each country, and those cities which maintain strong interdependent relationships with many other cities have been termed as "world cities".[19] The latters are defined as cities which possess economic influence which supersedes national boundaries, and act as interconnection nodes (Machimura, 1989) for global and international urban systems which have a "global control capability" (Sassen-Koob, 1984).

In world cities, international banking and investment companies, and accounting, brokerage, and insurance firms, as well as complementary specialized, high-level service functions such as legal offices, advertising firms, etc., are growing rapidly. These functions are concentrated in the central cities which hitherto have been experiencing decline,[20] that is, they are reversing the process of deurbanization. They have also been characterized by the widening polarization of social classes. In other words, core management functions and international capital of the multinational corporations accumulate here, resulting in the growth of a multinational elite class, while at the same time, low-tech blue-collar workers employed in these corporations, and low-paying service industries supporting these corporate elite, are also found (Sassen, 1988; Machimura, 1992). We do not know whether this type of internal structure in world cities is simply the result of a city's position within the national urban system, or whether this condition results from the incorporation of a primate city as a central node into the global urban system through its strong interdependencies with large cities overseas. The latter perspective has led to the emergence of a new school which explains this phenomenon through economic interdependencies with foreign cities.

We must not forget, however, that world cities possess an enormous capacity to manage themselves, and that we may be overemphasizing interdependencies. The growth of world cities is the mutual effect of the internal strengths of the cities (including the power that the reigning national system possesses) and interdependencies with foreign cities. These two factors are separated during analysis and used to estimate the differences in influence each has in global urbanization. This is one theme in the study of global urban system.[21]

1.4.3. CLASSIFICATION OF WORLD CITIES

Since the intensification of global urban system research in the late 1980s, researches have tried to characterize the components of this system, (i.e., the major cities of the world) and to rank their capacity for growth. The discussion in section 1.2 reminds us that such studies are less interested in interurban relationships (linkages) than in the ordering of nodes. It could be called the international urban dimension (section 1.2.3).[22] Using the data from 20 socio-economic attributes to conduct urban dimensional analysis on 30 major world cities, Yamaguchi (1988) revealed three significant dimensions: 1) population structure (Factor 1); 2) city centrality in the world economy (Factor 2); and an industrial profile concerning the stages of economic growth (Factor 3). Yamaguchi used three dimensions to classify the 30 cities into 7 groups.

Of many approaches for classifying world cities, one which has been attracting attention focuses on the activities of multinational corporations. The "multinationals" are the key elements of the global urban system. When enormous transportation and communication networks are formed that can handle the flow of services, capital, and information on a global scale, headquarters, decision-making organizations of multinational corporations, begin to concentrate in certain core nodes.

The significant studies dealing with the multinational corporation in the global scale were carried out by Feagin and Smith (1987) and Rimmer (1986). Using the number of locations for headquarters of multi-national corporations, they attempted to classify a hierarchy for world cities. Friedmann (1986) has classified 30 world cities based on locational conditions and urban functions (primary level or secondary level). At the primary level there were 4 cities, including New York, London, Tokyo, and Paris, and there were 9 cities at the secondary level (e.g., Brussels, Milan). Two cities are local in the semi-periphery of the primary level (Sao Paulo and Singapore) and 10 cities (including many capitals) in the semi-periphery of the secondary level. Friedmann referred to "spheres of influence" (i.e., areas where multinational corporations and international organizations conduct their activities), and divided the global urban system into three subsystems: Europe, America, and Asia.

This urban dimension research has brought to light the rapid polarization of social classes, the interurban functional divisions, and the primary functions of the world's large cities. The distinctions of world cities, from the perspective of urban systems, are the result of interurban competition and intensifying interdependencies among cities on a global scale.

1.4.4. INTERURBAN RELATIONSHIPS

The above research concerning growth and functional division of world cities as structural elements of a global urban system produces valuable information, but analysis of global interurban dependencies themselves, which believed to be the main factors causing fluctuations in the world, is only now getting under way. The cause of this late start is the difficulty in obtaining interurban O-D data for world cities, which places severe limitations on corroborative research. For example, there is no comprehensive data set for the flow of information, labor, or capital for all regions of the world. Working with limited material, investigations have only recently begun to consider the flow of goods in world trade (Conkling and McConnell, 1981), the flow of people like air passengers (Forsström and Lorentzon, 1991), the branch network of international banks (Meyer, 1986), the relationships between the headquarters of multinational corporations and their branches, the interurban networks of international interlocking directorates, and the interurban diffusion of clubs and organizations.

The important element of interurban flow is, above any other thing, information. It is no exaggeration to say that the force which has worked to create global urban system is the development of electronic communication technology and associated networks, which have enabled the unprecedented, smooth transmission of directives on a global scale. This has strengthened linkages among major world cities and has removed national and regional borders as an obstacle to economic activity (Hepworth, 1989). Seen from that perspective, it would appear that focusing on the flow of information would be the ideal way to gain a precise understanding of interurban linkages. Unfortunately, however, the relevant flow data, such as international telephone calls, facsimile transmissions, telex communications, personal computer networks, etc., have not been made available to the public in O-D matrix form. Instead, analysis of interurban relationships has involved indices of cost and information transmission capacity of networks, and the accessibility and connectivity of networks actually in place such as leased telephone lines, cables (including undersea), and fiber optics.[23] Hereafter, we should make special reference to the actual volume and content of flow, and the qualitative conditions of linkages must be clarified (Langdale, 1989; Hottes, 1992). In order for this to occur, however, data must be made available to the public.

1.5. Problems Associated with Analytical Operations in Research

The underlying concept of urban systems is their recognition as collection of mutually-dependent cities. However, the specific concept of urban systems depends on the approach taken by the researchers. The following problems must be resolved during analytical operations:

1) Selection of cities.
The analyst must determine the threshold size of cities in the system which depends on the spatial scale of the analysis. For example, in the analysis of regional urban systems, smaller cities of 20,000-30,000 population may become system elements, while international and global systems analysis deals with cities with a population of at least one million. Traditionally, cities have been selected on the basis of their size (e.g., population and the cumulative economic volume), as well as centrality and nodality, but as a matter of principle, the strength of interdependencies with other cities should be determined. Since the collective body of interdependent cities is treated as a system, cities which are small in size but which have a significant impact on the economies of other cities (e.g., small cities which are home to corporate or political headquarters) should be included. While a large city with weak linkages to other cities such as a "bedroom town" can be excluded. The "centrality scale" in nodal structure and the proportion of basic to non-basic functions (the B/N ratio), which can indirectly measure the strength of external dependencies, are believed to be effective indices. Finally, the spatial boundary of a city should extend beyond the administrative unit (municipalities) or even the built-up area. It should be the sphere of influence which serves the city.

2) Selecting the number of cities.
The number of cities to be selected depends on the purpose of the research. In theory, accuracy should increase as the number of cities increase; unfortunately, this also makes the analysis more complicated. For example, if the number of cities is increased from 10 to 20, the number of relationships which must be considered (including the cities themselves, i.e., intra-urban linkages) quadruples, from 100 to 400. There is some merit, then, in choosing the minimum number of cities which can depict the overall structure of the urban system.

3) The problem of openness.
Urban systems which are subjected to analysis have generally been treated as closed systems. However, with the exception of global urban system, all urban systems, whether they are regional, national, or international, are open to other systems of the same order at the spatial level; these are treated as subsystems when seen from higher orders. Furthermore, when there is a high degree of openness, each city within the study system will be affected economically by other urban systems; so that it is often difficult to understand observed changes. National urban systems with a high degree of openness should be considered a subsystem of the higher international hierarchy. For example, the American national urban system has strong linkages with neighboring countries, and highly open national urban systems such as that of Canada should probably be considered a subsystem of the international urban system covering North

America (Murayama, 1984). In any case, regardless of the spatial level, the openness of urban systems is increasing year by year, with closed systems becoming theoretically implausible.

1.6. Topics for Future Research

The author would like to suggest three topics for future research which are derived from above-mentioned discussions.

First, and most important, is the improvement of quality of O-D data which can enable us to gain a precise understanding of interurban relationships. At present, there is little data showing interurban linkages, especially at the international and global levels. Almost no interurban O-D time series data, such as that on airline passenger and airmail flow compiled by the ICAO, is available. Data on communications flow, such as facsimile transmissions and telephone calls, should be made publicly available by the relevant international organizations.[24] In the meantime, while data are still not easily obtainable, researchers must develop what limited data is available. For example, if we examine the location of Japanese-owned businesses at the global level, the construction of interurban O-D becomes possible through the use of the Japan Overseas Investment Data File and Japan Overseas Investments by Company, since the locations of corporate headquarters and their branches are a matter of public record. While there are few data for the initial stage of the formation of urban systems, in their analyses of the growth of the American urban system of the 19th Century, Pred (1973) and Goheen (1990) noted the dates in which newspaper articles of major incidents appeared in order to estimate the speed of information transmission and the routes that it took during that period. It is important that we make maximum use of what little data remains to us, in historical documents and other materials, to obtain data for interurban linkages.[25]

Secondly, although many researchers have treated interurban relationships independently of one another, it is essential that we begin to consolidate the data from these studies. As the reader has noted in this paper, the nodal and functional structures, spatial interactions, the interurban linkages of business activities, the diffusion of innovation, and the wave of economic fluctuation all combine to form an urban system. How strong are the functions of one city compared to those of another, and at what level are they linked? In which sectors do these linkages play a role in interurban dependencies as a whole? The consolidation of past research data, and the tackling of challenges posed by these questions, are ways in which research on urban systems can be made more productive. At the same time, it is becoming more important that interurban linkages should not be perceived merely as a collection of urban interactions; instead, they should be identified on a function-by-function basis. Furthermore, the author would like to see analysis of urban systems proceed not only at the interurban level, but to be extended to internal systems within the cities themselves. The interrelationships of functions within these internal systems, that is, the extent to which one function is related to another, as well as the way these functions are linked to functions in other cities, should be understood; ideally, an interurban-interfunctional dependency matrix such as a chart of interurban input-output relationships should be created.

Thirdly, the growth and functional specialization of cities are changing the existing interurban relationships. The feedback from this fluctuation encourages the growth and functional specialization of cities. The ultimate goal for city systems research is to gain a quantitative understanding of this dynamic interrelationship and obtain a firm grasp of the mechanisms behind the growth and fluctuation of the urban system. If these mechanisms can be brought to light, we can tackle the question of how to transform the urban system as a whole, and be able to predict beforehand those systems which are growing and those which are headed toward decline. It will also become possible to present methods for maintaining, managing, and even controlling urban systems in harmony; to develop policies which target specific cities for growth; and to transform the remaining urban systems in a stable manner. Instead of simply looking back at the past, geographers should consider that the study of urban systems can provide the materials and perspectives for them to direct their attention toward future forecasts and policy issues.

Notes

1. An important impetus for urban system research was the formation in 1976 of the National Settlement System by the Commission on Urban Systems in Transition (Chair- K. Dziewonski of Poland) under the auspices of the IGU(Bourne, *et al.* eds. 1984).

2. This section will not attempt to deal with the Central Place Theory which looks at the core structure as an interrelated spatial system of the order of core functions and the level of the central place. Central Place is discussed in detail in Usui (1979).

3. The following technical problem occurs in making time series comparisons of the urban dimension: Even if the numbers of analyzed cities and variables are maintained constant, the variable composition of the factors will undergo slight changes from year to year since factor analysis is conducted separately. This impedes accurate comparisons. To solve this problem, we must 1) derive an inter-factor "correspondence coefficient" for each time period and use the degree of similarity of factor compositions to make comparisons, 2) derive the rate of change for each variable of the time periods we are comparing, and 3) conduct our factor analysis by arranging the matrices for each time period from top to bottom (Murayama, 1993).

4. In urban systems, core management functions are significant even within the other urban functions; in fact, some studies classify cities according to interurban differences in these core management functions (Harper, 1987).

5. Among the limited studies considering nodal and functional structures as raw material for O-D capital flow data is Conzen's masterful work (Conzen, 1975, 1977).

6. For example, a Canadian urban systems analysis which used air passenger flow showed that a unitary system was formed between 1961, when there were four subsystems, and 1976, when Toronto appeared at the top of the hierarchy (Murayama, 1982).

7. This point is evident in studies of historical geography. For example, in Kohno's (1990) study of the Nagano regional urban system from the early Meiji Era to the middle of the Showa Era, each city was gradually incorporated into a system with the dominant (primary central place) Nagano at the apex of the hierarchy; however, the hierarchy from second-order cities on down was not clear. In addition, Ishimaru (1993), who considered recent changes in the Japanese urban system from the locations of service industries, indicated that striking differences appeared between high-order and low-order cities, while medium-order

cities began to lose clearly-defined patterns.

8. For example, Bourne *et al.* (1983) reported that in Europe and the United States, city systems tended to spatially diffuse, with differences between high-order and low-order cities diminishing and the hierarchical structure weakening. However, Morikawa (1985), who looked at population movement in the Japanese national urban system, did not recognize this type of trend in Japan; rather, he pointed out that the urban hierarchical structure was becoming stronger. Clearly, conditions differ from country to country.

9. In addition to headquarters-branch relationships, the following indices have been used: Interurban links from the perspective of parent company-subsidiary relationships (Wheeler, 1986), interurban locational movement of head offices (Fujita, 1984), and locational movement of head offices of amalgamated companies (Semple *et al.*, 1985).

10. For example, Nishihara analyzed the way in which spatial patterns of interurban relationships differ according to the type of industry, business or office (Nishihara, 1991).

11. The analysis of office networks and interurban linkages of headquarters-branches for individual companies includes many types of businesses such as banks and trust companies, life insurance companies, trading companies, legal offices, newspaper companies, communications bureaus, manufacturing companies, department stores and supermarkets. In urban systems research, these activities can be consolidated to derive general characteristics of interurban linkages.

12. In the interlocking directorate network, various relationships exist between countries, such as parent company-subsidiary relationships, other connections based upon ownership, personal linkages, and so on. (Hosoi, 1993).

13. The Lions Club is an organization comprised of a white-collar membership which seeks friendship, international mutual understanding, and social volunteerism. It provides labor and financial support for new branches.

14. The establishment of new (affiliate) clubs is the work of existing (parent) clubs, accomplished by recruiting new members through business colleagues. Socio-economic linkages reflect relationships between the cities in which affiliate and parent clubs are located.

15. Using Product Cycle Theory (Vernon, 1966), several studies have shown how advanced technology is transferred from developed to developing countries.

16. In 1993 alone, at least two journals issued special editions devoted exclusively to European urban systems: Vol.30, No.6 of *Urban Studies* and Vol.84, No.4 of *Tijdschrift voor Econ. en Soc. Geografie.*

17. Machimura (1992) used this term within the framework of the *New International Division of Labor* to refer to the spatial dimension.

18. According to Machimura (1992), national intervention is in decline, giving way to cooperative relationships between the public and private sectors with the freedom for capital investment increasing through markets.

19. Papers which have overviewed past research data on world cities include Friedmann and Wolff (1982).

20. Takahashi (1989) considered the regional perspective in his analysis of the agglomeration of foreign-owned financial institutions in downtown Tokyo.

21. For example, *Economic Base Theory* states that the higher the ratio of basic activity to non-basic activity in a city, the stronger the interdependencies with other cities are.

22. Research which looks at national units instead of city units has been conducted since the 1960s, especially in the field of political science (Russett, 1967).

23. For example, KDD (Kokusai Densin Denwa) releases some of its communications volume for individual

countries, but does not do so on a city-by-city basis.

24. American Telephone and Telegraph (AT&T) has released data on the volume of international calls being made from points within the United States to cities overseas.

25. Japanese research in this area includes the Sugiura (1978) 's, case study of Okage-mairi in the Edo Period. He noted the discrepancies between methods for transmitting information and from this information diffusion process built an analogy of the regional system of that time.

References

Asami, Y. (1980) Leads and lags of economic fluctuations in terms of new job offers: the case study of the northwestern part of the Tokyo metropolitan area, *Geographical Review of Japan*, **53** (5), 329-344. (J with E)

Auerbach, F. (1913) Das Gesetz der Bevolkerungskonzentration, *Petermann's Geographische Mitteilungen* **59**, 73-76.

Berry, B.J.L. (1961) City size distributions and economic development, *Economic Development and Cultural Change* **9**, 673-687.

Berry, B.J.L. (1964) Cities as systems within systems of cities, *Papers of the Regional Science Association* **13**, 147-162.

Berry, B.J.L. (1968) Interdependency of spatial structure and spatial behavior: a general field theory formulation, *Papers of the Regional Science Association* **21**, 205-227.

Berry, B.J.L. (1969) Relationships between regional economic development and the urban system: the case of Chili, *Tijdschrift voor Economische en Sociale Geografie* **60**, 283-307.

Berry, B.J.L. and Horton, F. E. eds. (1970) *Geographic Perspectives on Urban Systems*, Prentice-hall, Englewood Cliffs, New Jersey, 564p.

Boots, B.N. and Getis, A. (1988) *Point Pattern Analysis*, SAGE Publications, Newbury Park, 93p.

Bourne, L.S. (1975) *Urban Systems: Strategies for Regulation: A Comparison of Policies in Britan, Sweden, Australia, and Canada*, Clarendon Press, Oxford, 264p.

Bourne, L.S. and Simmons, J.W. eds. (1978) *Systems of Cities, Readings on Structure, Growth, and Policy*, Oxford University Press, New York, 565p.

Bourne, L.S., Korcelli, P. and Wärneryd, O. (1983) Emerging spatial configurations of urban systems: a review of comparative experience, *Geographia Polonica* **47**, 85-99.

Bourne, L.S., Sinclair, R. and Dziewonski, K. eds. (1984) *Urbanization and Settlement System: International Perspectives*, Oxford University Press, New York, 475p.

Brown, L.A. (1981) *Innovation Diffusion: A New Perspective*, Methuen, London, 345p.

Brunn, S.D. and Leinbach T.R. eds. (1991) *Collapsing Space and Time: Geographic Aspects of Communication and Information*, Harper Collins Academic, London, 404p.

Carroll, G.R. (1982) National city-size distributions: what do we know after 67 years of research?, *Progress in Human Geography* **6**, 1-43.

Clark, G.L. (1980) Critical problems of geographical unemployment models, *Progress in Human Geography* **4**, 157-180.

Cohen, Y.S. (1972) Diffusion of an innovation in an urban system, *The University of Chicago, Department of Geography, Research Paper* **140**, 136p.

Conkling, E.C. and McConnell, J.E. (1981) Toward an integrated approach to the geography of international trade, *Professional Geographer* **33** (1), 16-25.

Conzen, M.P. (1975) Capital flows and the developing urban hierarchy: state bank capital in Wisconsin, 1854-1895, *Economic Geography* **51** (4), 321-337.

Conzen, M.P. (1977) The maturing urban system in the United States, 1840-1910, *Annals of the Association of American Geographers* **67** (1), 88-108.

Dacey, M.F. (1960) Analysis of central place and point patterns by a nearest neighbor method, *Lund Studies in Geography* **B** (2), 55-75.

Duncan, O.D. *et al.* (1960) *Metropolis and Region*, Johns Hopkins University Press, Baltimore, 587p.

Feagin, J.R. and Smith, M.P. (1987) Cities and the new international division of labor: an overview, in Smith and Feagin eds., *The Capitalist City*, Basil Blackwell, Oxford, 3-34.

Forsström, A. and Lorentzon, S. (1991) Global development of communication: a frame for the pattern of localization in a small industrialized country, in Brunn and Leinbach eds. (1991), 87-107.

Fröebel, F., Heinrichs, J. and Kreye, O. (1980) *The New International Division of Labour: Structural Unemployment in Industrialized Countries and Industrialization in Developing Countries*, Cambridge University Press, Cambridge, 407p.

Frank, A.G. (1967) *Capitalism and Underdevelopment in Latin America: Historical Studies of Chile and Brazil*, Monthly Review Press, New York, 298p.

Friedmann, J. (1986) The world city hypothesis, *Development and Change* **17** (1), 69-83.

Friedmann, J. and Wolff, G. (1982) World city formation: an agenda for research and action, *International Journal of Urban and Regional Research* **6** (3), 309-344.

Fujimaki, M. (1984) Reconsidering city size distribution in underdeveloped countries, *Tenri University Journal* **141**, 184-213. (J)

Fujita, N. (1984) Formation of a hierarchical regional system in post-war Japan, *Memoirs of the Institute of Cultural Sciences*, Meiji Univ. **23**, 1-32. (J)

Goheen, P.G. (1990) The impact of the telegraph on the newspaper in mid-nineteenth century British North America, *Urban Geography* **11** (2), 107-129.

Green, M.B. (1981) Regional preferences for interlocking directorates among the largest American corporations, *Environment & Planning* **A 13**, 829-839.

Green, M.B. and Semple, R.K. (1981) The corporate interlocking directorate as an urban spatial information network, *Urban Geography* **2**, 148-160.

Haggett, P. and Choley, R.J. (1969) *Network Analysis in Geography*, Edward Arnold, London, 348p.

Hansen, N.M. (1978) *Human Settlement System: International Perspectives on Structure, Change and Public Policy*, Ballinger Publishing Company, Cambridge, Mass., 297p.

Harper, R.A. (1987) A functional classification of management centers of the United States, *Urban Geography* **8** (6), 540-549.

Harris, C.D. (1943) A functional classification of cities in the United States, *Geographical Review* **33**, 86-99.

Henderson, J.V. (1974) The sizes and types of cities, *American Economic Review* **64**, 640-657.

Hepworth, M.(1989) *Geography of the Information Economy*, Belhaven Press, London, 258p.

Hino, M. (1977) Fundamental dimensions of the Japanese urban system in the years of 1950, 1960 and 1970, *Geographical Review of Japan* **50** (6), 335-353. (J with E)

Hino, M. (1981) Critical review of the studies of system of cities - with particular reference to the spatial structure formed by the activities of big enterprises, *The Human Geography* **33** (2), 133-153. (J with E)

Hosoi, K. (1993) The intersocial network theory: analysis of intersocial relationships by the interlocking directorates system, in Nakagawa, N. ed., *International Industrial Theory: Introduction to Global Industrial Theory*, Mineruva Shobo, Kyoto, 111-135. (J)

Hottes, K. (1992) Submarine cables in our times: a report on competition between seacables and satellites, paper presented at the IGU Symposium on the Geography of Telecommunications and Communications, Washington, 15p.

Hudson, J.C. (1969) Diffusion in a central place system, *Geographical Analysis* **1**, 45-58.

Hudson, J.C. (1972) *Geographical Diffusion Theory*, Studies in Geography, University of Northwestern Press, Evanston, 179p.

Ishimaru, T. (1993) The location of service employment and the Japanese urban system, *Geographical Sciences* **48** (3), 142-150. (J with E)

Jefferson, M. (1939) The law of the primate city, *Geographical Review* **29**, 226-232.

Jeffrey, D. (1974) Regional fluctuations in unemployment within the U.S. urban economic system: a study of the spatial impact of short term economic change, *Economic Geography* **50**, 111-123.

Jeffrey, D., and Webb, D.(1972) Economic fluctuations in the Australian regional system, 1955-70, *Australian Geographical Studies* **10**, 141-160.

Jones, D.W. (1983) Mechanisms for geographical transmission of economic fluctuations, *Annals of the Association of American Geographers* **73**(1), 35-50.

Joo, K. (1981) A perspective view on urban system, *Annals of the Tohoku Geographical Association* **33** (2), 119-133. (J with E)

Joo, K. (1982) Urbanization and urban system of Korea:1960-1980, *Geographical Review of Japan* **55** (1), 1-20. (J with E)

Kanamoto, Y. (1981) Development of locational theory, *Studies in Regional Science* **12**, 257-263. (J)

Kawashima, T. and Korcelli, P. eds. (1982) *Human Settlement Systems: Spatial Patterns and Trends*, IIASA, Laxenburg, 210p.

King, L.J. (1966) Cross-sectional analysis of Canadian urban dimensions, 1951 and 1961, *Canadian Geographer* **10**, 205-224.

King, L.J. (1976) Alternatives to a positive economic geography, *Annals of the Association of American Geographers* **66**, 293-308.

King, L., Casetti, E. and Jeffrey, D. (1969) Economic impulses in a regional system of cities: a study of spatial interaction, *Regional Studies* **3**, 213-218.

Kohno, K. (1990) Transformation of the central place system in the Nagano Basin from the early Meiji era to the post-war period, *Geographical Review of Japan, Series* **A 63** (1), 1-28. (J with E)

Kohsaka, H. (1978) A dynamic analysis of city size distribution — a case study of Niigata Prefecture —, *Geographical Review of Japan* **51** (3), 223-234. (J with E)

Kohsaka, H. (1984) *Analysis of Regional Economy*, Kohbundo, Tokyo, 232p.(J)

Langdale, J.V. (1989) The geography of international business telecommunications: the role of leased networks, *Annals of the Association of American Geographers* **79** (4), 501-522.

Machimura, T. (1989) Transnationalization of Japanese corporations and its regional impact, *Tsukuba Journal of Sociology* **14**, 102-119. (J)

Machimura, T. (1992) Global social theory and urban social theory, in Suzuki, H. ed., *Decoding the Modern City*, Mineruva Shobo, Kyoto, 60-87. (J)

Marshall, J.U. (1989) *The Structure of Urban Systems*, University of Toronto Press, Toronto, 394p.

Martin, R.L. and Spence, N.A. (1981) Economic geography, in Wrigley, N. and Bennett, R.J. eds., *Quantitative Geography*, Routledge & Kegan Paul, London, 322-334.

Meyer, D.R. (1986) The world system of cities: relations between international financial metropolises and South American cities, *Social Forces* **64** (3), 553-581.

Morikawa, H. (1978) A study on the techniques of nodal and functional regionalization: a case study of telephone calls in the Chugoku district, *The Human Geography* **30** (1), 17-38. (J with E)

Morikawa, H. (1982) Concepts and research trends of urban system, in Tanabe, K. ed., *Urban System in Japan — a Geographical Study —*, Kokon-shoin, Tokyo, pp.1-26. (J)

Morikawa, H. (1985) The Japanese urban system based on the analysis of internal migration, *The Human Geography* **37** (1), 20-38. (J with E)

Morse, E.L. (1976) *Modernization and the Transformation of International Relations*, Free Press, New York, 203p.

Moser, C.A. and Scott, W. (1961) *British Towns: A Statistical Study of Their Social and Economic Differences*, Oliver and Boyd, Edinburgh, 169p.

Murayama, Y. (1982a) Re-examination of the Nystuen-Dacey model in terms of nodal regionalization, *Geographical Sciences* **37** (2), 73-84. (J with E)

Murayama, Y. (1982b) Diffusion channels of innovation in the urban systems: the case of Lions Clubs, *Annals of the Tohoku Geographical Association*, **34** (4), 224-235. (J with E)

Murayama, Y. (1982c) Canadian urban system and its evolution process in terms of air-passenger flows, *Geographical Review of Japan* **55** (6), 380-402.

Murayama, Y. (1984) International city-system in North America: an approach from air-passenger flows, *Annual Report of the Institute of Geoscience, University of Tsukuba* **10**, 8-13.

Murayama, Y. (1990) Regional structure of commodity flows in Japan: an application of dynamic geographical field theory, *Science Report of the Institute of Geoscience, University of Tsukuba, Section A* **11**, 79-114.

Murayama, Y. (1992) Method for delineating traffic regions, *Mobility* **87**, 102-110. (J)

Murayama, Y. (1993) *Regional Analysis (revised edition)*, Kokon-shoin, Tokyo, 105p. (J)

Nam, Y. (1981) Prediction of the structure of functional regions within Seoul City: as an attempt for prediction of the future, *The Human Geography*, **33** (6), 508-525. (J with E)

Nelson, H. J. (1955) A service classification of American cities, *Economic Geography* **31**, 189-210.

Nishihara, J. (1991) The development of intra-corporate branch networks and the Japanese urban system, *Geographical Review of Japan* **64**A (1), 1-25. (J with E)

Ogburn, W. (1937) *Social Characteristics of Cities: A Basis for New Interpretations of the Role of the City in American life*, International City Manager's Association, Chicago, 69p.

Pedersen, P.O. (1970) Innovation diffusion within and between national urban systems, *Geographical Analysis* **2** (3), 203-254.

Pigozzi, B.W.M. (1980) Interurban linkages through polynomially constrained distributed lags, *Geographical Analysis* **12** (4), 340-352.

Pred, A.R. (1973) *Urban Growth and the Circulation of Information: The United States Systems of Cities, 1790-1840*, Harvard University Press, Cambridge, Mass, 348p.

Pred, A.R. (1976) The interurban transmission of growth in advanced economies: empirical findings versus regional-planning assumptions, *Regional Studies* **10**, 151-171.

Pred, A.R. (1977) *City-Systems in Advanced Economies*, Hutchinson, London, 256p.

Rice, M.D. and Semple, R.K. (1993) Spatial interlocking directorates in the Canadian urban system, 1971-1989. *Urban Geography* **14**, 375-396.

Richardson, H.W. (1973) Theory of the distribution of city sizes: review and prospects, *Regional Studies* **7**, 239-251.

Rimmer, P.J. (1986) Japan's world cities: Tokyo, Osaka, Nagoya or Tokaido megalopolis? *Development and*

Change **17** (1), 121-123.

Russett, B. M. (1967) *International Regions and the International System: A Study in Political Ecology*, Rand McNally, Chicago, 252p.

Sassen, S. (1988) *The Mobility of Labor and Capital*, Cambridge University Press, Cambridge, 224p.

Sassen-Koob, S. (1984) The new labor demand in global cities, in Smith, M.P. ed., *Cities in Transformation*, Sage, Beverly Hills, 139-172.

Semple, R.K. (1985) Toward a quaternary place theory, *Urban Geography* **6** (4), 285-296.

Semple, R.K., Green, M.B. and Martz, D.J.F. (1985) Perspectives on corporate headquarters relocation in the United States, *Urban Geography* **6**, 370-391.

Sheppard, E. (1982) City size distributions and spatial economic change, *International Regional Science Review* **7** (2), 127-151.

Simmons, J.W. (1978) The organization of the urban system, in Bourne and Simmons (1978), 61-69.

Strickland, D. and Aiken, M. (1984) Corporate influence and the German urban system: headquarters location of German industrial corporations, 1950-1982, *Economic Geography* **60** (1), 38-54.

Sugiura, Y. (1978) Spatial diffusion of "Okagemairi" in the year of 1771, *Geographical Review of Japan* **51** (8), 621-642. (J with E)

Sugiura, Y. (1982) A spatial diffusion of electric supply companies through the system of cities in Japan, 1887-1898, *Geographical Review of Japan*, **55** (9), 634-655. (J with E)

Sugiura, Y. (1987) Spatial diffusion of electricity supply companies through a system of cities in Japan, 1887-1898, *Geographical Reports of Tokyo Metropolitan University* **22**, 67-84.

Sung, J. (1977) The urban system in Korean cities, *Geographical Review of Japan* **50** (7), 381-401.

Suzuki, K. (1981) On the decomposability of the Zipf's rank-size rule, *Studies in Regional Science* **12**, 35-52. (J with E)

Takahashi, N. (1989) Agglomeration of foreign financial institutions in Tokyo, *Tsukuba Studies in Human Geography* **13**, 73-94. (J with E)

Usui, T. (1979) The orders of central functions and the hierarchical central place system, *The Human Geography* **31** (6), 481-506. (J with E)

Vernon, R. (1966) International investment and international trade in the product cycle, *Quarterly Journal of Economics* **80**, 190-207.

Vining, R. (1945) Regional variation in cyclical fluctuations viewed as a frequency distribution, *Econometrica* **13** (3), 183-213.

Vining, R. (1949) The region as an economic entity and certain variations to be observed in the study of systems of region, *Papers and Proceedings of the American Economic Association* **39**, 89-104.

Wallerstein, I.M. (1979) *The Capitalist World-Economy: Essays*, Cambridge University Press, Cambridge, 305p.

Wärneryd, O. (1968) *Interdependence in Urban Systems*, Regionkonsult Antiebolag, Göteborg.

Wheeler, J.O. (1986) Corporate spatial links with financial institutions: the role of the metropolitan hierarchy. *Annals of the Association of American Geographers* **76**, 262-274.

Wheeler, J.O. and Mitchelson, R.L. (1989) Information flows among major metropolitan areas in the United States, *Annals of the Association of American Geographers* **79** (4), 523-543.

Wilson, A.G. (1969) Notes on some concepts in social physics, *Papers Proc. Reg. Sci. Assoc.* **22**, 159-193.

Yamaguchi, T. (1973) Some procedures for classifying Japanese cities, *The Proceedings of the Department of Humanities, College of General Education, University of Tokyo* **56**, 73-94. (J with E)

Yamaguchi, T. (1988) Recent changes in Japan's urban system, *The Proceedings of the Department of Humanities, College of Arts and Sciences, University of Tokyo* **88** (Series of Human Geography 10), 71-84.

Zipf, G. K. (1949) *Human Behavior and the Principle of Least Effort: An Introduction to Human Ecology*, Addison Wesley, Cambridge, Mass., 366p.

(J = in Japanese; J with E = in Japanese with English abstract)

PART II

INTERNATIONAL URBAN SYSTEM

CHAPTER 2

DIFFUSION OF INNOVATION

2.1. Introduction

The study of urban systems accompanied the beginning of quantitative geography in the 1960s, and is quickly becoming one of the discipline's most rapidly developing fields (Murayama 1982a). From concepts of the 1960s that emphasized an explanation of static spatial order in one time period such as King's urban dimension (1966), Berry's distribution patterns of urban systems (1961), and Nystuen and Dacey's (1961) functional and nodal structures of urban systems, this field developed into an explanation of the dynamic spatial order of urban systems (Sugiura 1978), evolving to the point where, like this study, it shares common ground with diffusion studies emphasizing spatial processes. In other words, it began to evolve into spatial processes studies (spatial analysis) integrating spatial and temporal components. Representative of such studies are the rate and channels of diffusion of information and innovation in urban systems (Pederson, 1970), the growth of urban systems brought about by diffusion itself (Pred, 1980, pp.119-141), and the way that cyclic economic fluctuations spread among cities (Jeffery and Webb, 1972).

Against the background of such research trends in urban systems, this study will provide a substantial insight into how innovation, possessing spatial and temporal order, is diffused within an urban system. To do this, special emphasis will be placed on diffusion channels, about which very little substantial data has appeared in spatial diffusion studies to date. In the approach from spatial diffusion vis-à-vis urban systems, it is essential that there be an explanation of diffusion channels which demonstrate the interdependent relationships of cities.

The resolution level of urban systems can be dealt with in terms of a hierarchy composed of international, national, and regional levels, as shown in Figure 2.1. The diffusion of innovation in international urban systems begins mainly in the major cities of each country, such as national capitals; then, with such cities as cores, empirical evidence shows that innovation is diffused to the national level. This study will focus on the national system of Japan, approaching it within the context of the spatial diffusion process of the international urban system. There will also be an analysis of the kind of impact this has had on urban systems in the hinterlands ("sphere of influence") of major metropolitan areas. Special attention will be given to functional differences in diffusion patterns at these three levels.

In the analysis, there will be particular emphasis on two topological characteristics of urban systems, that is, the nodes delineated by the cities, and the interurban relationships of mutual dependency that are shown by the linkages. Nodes will be considered to have attributes of scale, while linkages will be approached as vectors having direction and length.

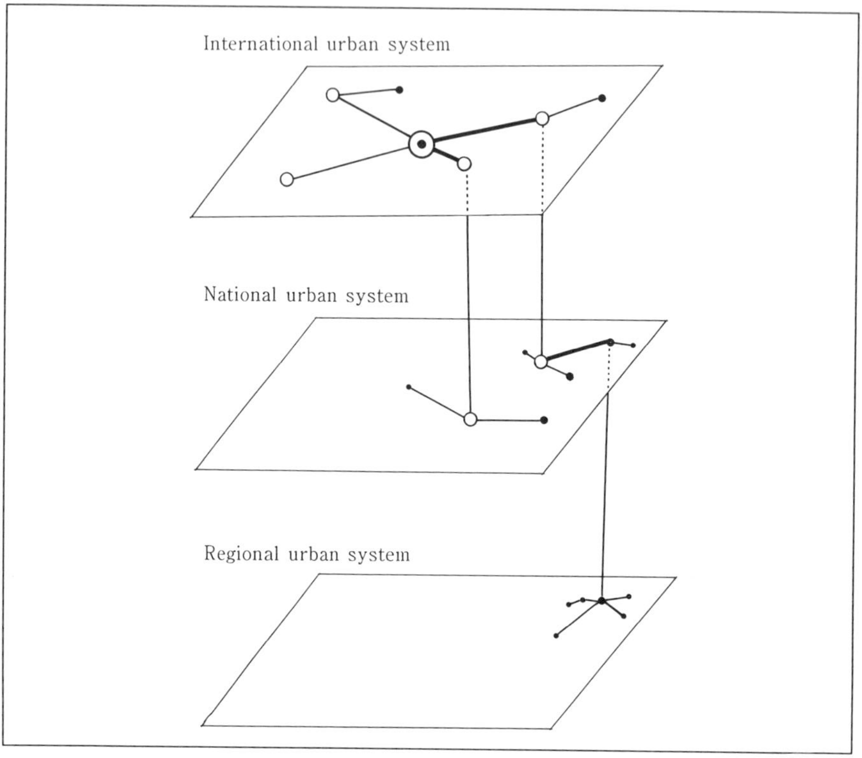

Figure 2.1. Spatial hierarchy of urban systems

2.2. Index of Analysis

Assuming that diffusion permeates the entire region, that the diffusion channels are already known, and that the diffusion itself is ubiquitous, etc., the *Lions Club* was used as an index.

The main purposes of Lions Clubs, which are organizations made up of business people, are to nurture friendship among members, foster international good-will, and make a contribution to society in general. By making donations of labor and money, they seek to contribute to the local community through a host of activities that include social welfare (services for the blind and speech-impaired), international good-will and youth exchanges, educational assistance, environmental protection, and public safety, to name a few. Membership is, in principal, set at one member per type of business, and is managed such that members do not become biased for or against any type of business. The standard club has 40-80 members (with a nationwide average of 55.8), with meetings held at regular intervals (twice per month) to report on the goals and status of activities and to promote mutual understanding among members.

The Lions Club was founded in Chicago (Illinois), USA in 1913 by a young businessman named Melvin Jones. Seven years later, the first international club was started in Windsor (Ontario), Canada, and, after another 7 years, the organization had become established in Mexico. After World War II, it began to take root in Europe, starting with the first European club in Stockholm, Sweden in 1948. Since then, it has spread around the world, including Japan, where the first club was established in 1952. As of July 1, 1981 there were 34,913 Lions Clubs worldwide with a total membership of 1,315,566. At that time, there were 2,508 clubs in Japan with a total of 139,960 members, ranking second in the world only behind the United States (see Table 2.1).[2]

Table 2.1. Number of Lions Clubs in major countries

	Number of clubs		Number of clubs
1. United States	14,732	12. (West) Germany	610
2. Japan	2,508	13. Italy	550
3. India	1,783	14. Sweden	517
4. Brazil	1,746	15. Argentina	460
5. Canada	1,702	16. Norway	417
6. Australia	1,224	17. New Zealand	385
7. France	776	18. Philippines	370
8. South Korea	730	19. Colombia	256
9. Finland	728	20. Taiwan	239
10. Mexico	705	Other	3,779
11. United Kingdom	696	Total	34,913

Note: As of July 1, 1981.

New clubs are established as follows. First, an "extension committee" is formed in an existing club to consider the terms for the new club, such as period of establishment, number of members, and location of office. Then, each member of the existing club works together to recruit new members through friends, associates, and business contacts to start the process of organization building. After a certain number of new members have been recruited, the extension committee submits an application on behalf on the new club to *Lions International*, i.e., the world headquarters in Chicago. If the Chicago headquarters gives its approval, the new club can be certified. To provide this new club with guidance in operations and activities, several existing members are normally transferred to the new group. The club which has done the work on behalf of the new club is called a "sponsor club;" together, the two clubs strengthen their ties through club activities. Soon, after the new club has become well established, it usually becomes the sponsor for yet another new club (which becomes the "grandchild" of the first club). In this way, the organization successively grows. It is an honor for a club to have spawned many "children" and "grandchildren," and, by helping to form new clubs, each existing club lends its support to the growth of the organization (*Lions Magazine* (Japanese edition), Interview with Lions Club officers).

As mentioned previously, the establishment of new clubs is done by existing members who recruit new members through friends, associates, and business contacts. We can assume that the cities in which the new and old clubs are respectively located have well-developed social and economic chain-typed relationships.

2.3. The Process of Spatial Diffusion of Lions Clubs

2.3.1 INTERNATIONAL URBAN SYSTEM

Figure 2.2 is a cartographic representation of the growth and diffusion patterns of Lions Clubs for 4 periods in the international urban system. Only the first city in a country to gain a club is represented. Therefore, there is only 1 analytical node for each country. However, for the most precise analysis, countries leading in the club's growth and diffusion such as the United States, and countries with several major cities should actually have many cities included in the analysis; however, since this would cause problems in the selection of analytical nodes and in the analytical procedure, each country was limited to one node. Of course, there have been no Lions Clubs in communist countries where individual economic freedoms are restricted.

In general, we can see that over the years the club has expanded from North America to Central America, then to South America, Europe, Asia, and finally, Africa. During the period 1917-1945, there were clubs in only 12 countries, all of which were on the American continent, but from 1946-55, the number had grown by 40 countries, most of which were in South America and Europe. From 1956-65, another 46 countries, mostly from Asia, Africa, and the Caribbean joined the ranks. Finally, between 1965 and 1981, another 19 countries gained clubs for the first time, meaning in effect that there was now almost no country in the Free World that did not have at least one Lions Club.

Clubs tend to increase more rapidly in countries with higher populations, although they also show rapid growth in the relatively low-populated nations of Central America, which are close to the United States. We can therefore conclude that there are both a hierarchical effect and a proximity effect (Hudson, 1972) at work here. This means that in the relation between annual growth of Lions Clubs and population scale in the international urban system, as can be clearly seen in Figure 2.3, the correlation coefficient, at 0.339, is very low (although significant at the 5% level), suggesting that the diffusion of the organization does not necessarily occur based on the order imposed by a population hierarchy.

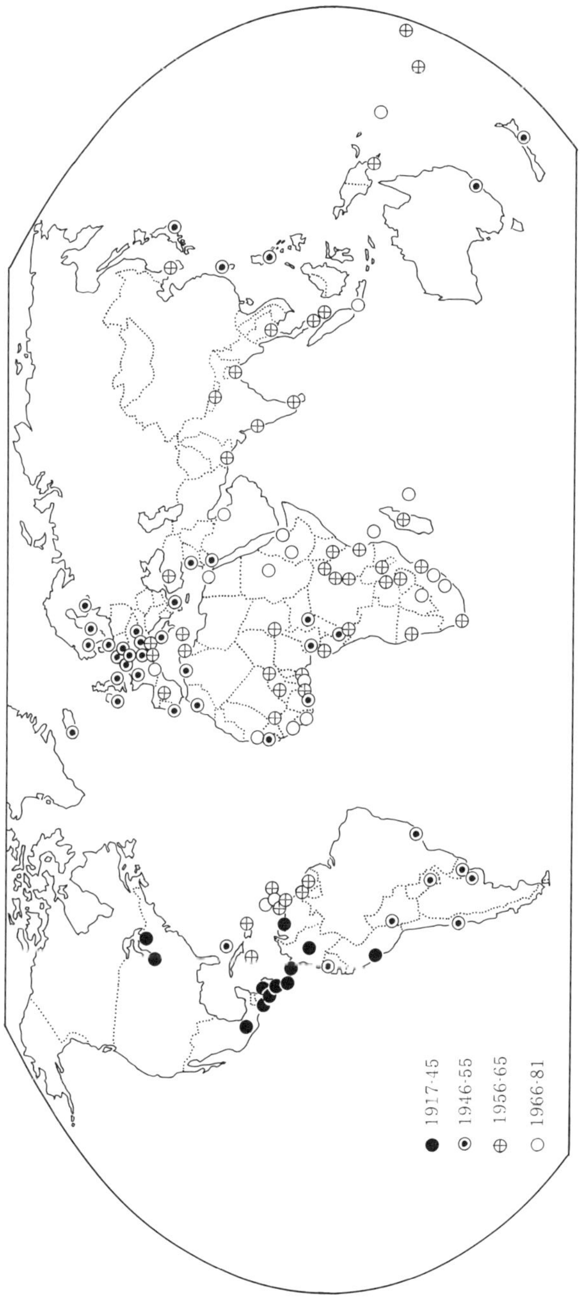

Figure 2.2. Lions Club diffusion patterns in the international urban system

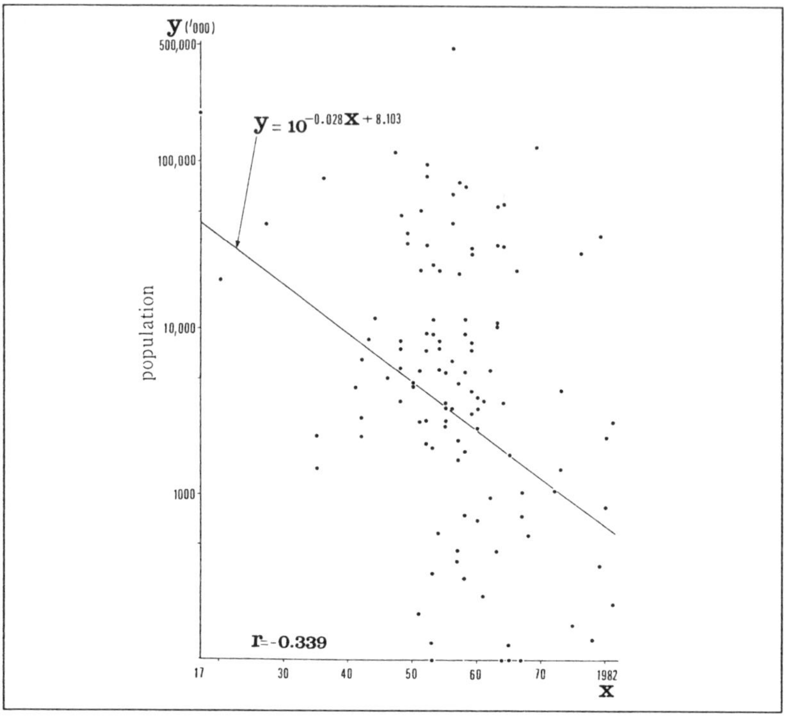

Figure 2.3. Relationship between population scale and yearly diffusion of Lions Clubs in the international urban system

Let's turn our focus from the distribution of nodes (cities) in the diffusion pattern to links, giving attention to diffusion channels (Figure 2.4). The United States, where the Lions Club began, has established affiliated clubs in 25 countries, especially in Central America. The first club established in South America was in Colombia in 1936, followed by Peru in 1944. In both cases, the sponsor was from the United States. Peru served as the core for expansion in South America, becoming the direct sponsor for Bolivia and Chile (both in 1948), and the indirect sponsor for 4 other countries in the same region.

In Europe, the United States in 1948 sponsored the countries of Sweden, Switzerland, and France, all 3 of which went on to serve as the cores for sponsoring clubs in neighboring countries. The core countries for growth in northern Africa were France and Morocco, which France had previously sponsored, while in southern Africa the cores were Great Britain and its "child," the Republic of South Africa.

No regular patterns have been recognized in Asia, which is characterized by direct sponsorship from the United States, Great Britain, and other major countries. Accordingly, the diffusion links themselves are of a much longer distance in Asia than elsewhere.

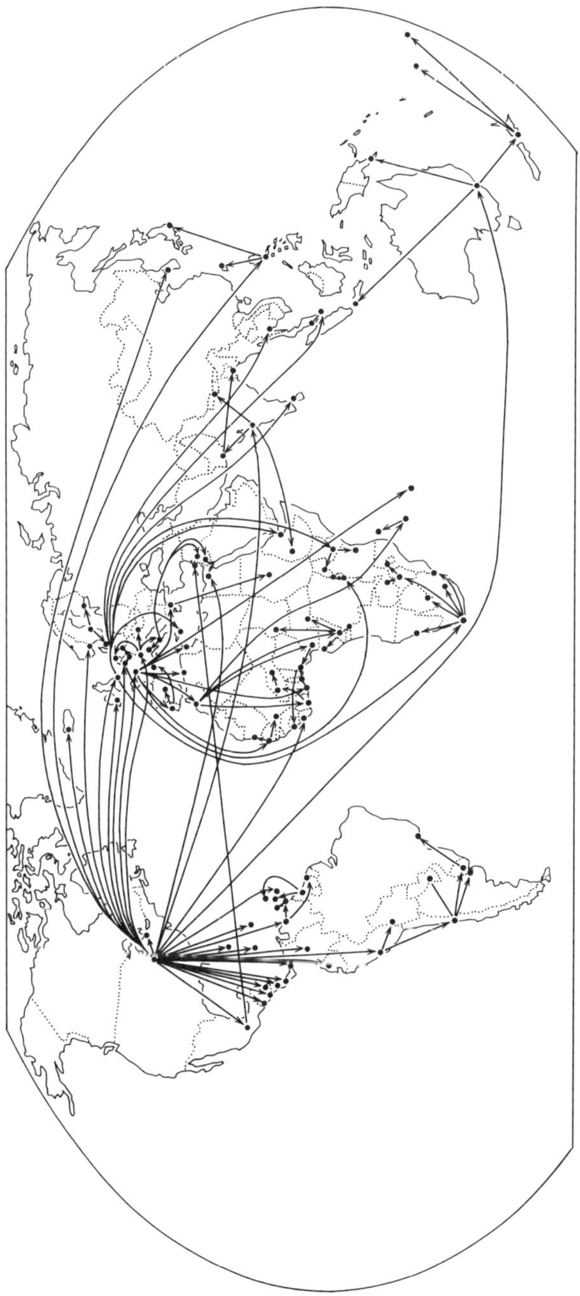

Figure 2.4. Diffusion of Lions Club channels in the international urban system

The cores for diffusion in Oceania are the United States and its child, Australia, which established its first club in 1947.

Generally speaking, diffusion in each continent has been centered on the major country or countries of the continent, successively spreading to neighboring countries. However, there are also cases in which there is no geographic positional relationship, in which former colonial powers such as the United States, Great Britain and France directly sponsor clubs in their former colonies or territories, and, in many cases, the diffusion channel is from less populated countries to a country with higher population density. Thus we can surmise that there has been a great effect from political organizations and/or historical situation. One of the most typical examples of this strong historical situation is the Manila club of Philippines, which sponsored the Tokyo club of Japan.

2.3.2 NATIONAL URBAN SYSTEM

Analytical nodes comprise 208 cities with a DID (densely-inhabited district) population of at least 30,000 in 1965.[3] Figure 2.5 shows the number of members and member density of 47 prefectural capitals. Membership was largest in Tokyo, Osaka, Yokohama, and other metropolitan areas. Member density was highest in Takamatsu (28 per 10,000 population), followed by Akita and Saga. The lowest density was found in Naha, followed by Chiba and Urawa. Generally, membership density was highest in smaller regional cities and lowest in the major metropolitan areas and their suburbs. This is because major cities had a considerable number of large factories and retailers, as well as government offices, while the suburbs were bedroom communities. Therefore, these areas had a relatively low proportion of potential Lions Club members.

Figure 2.6 is a cartographic representation of diffusion patterns of Lions Clubs in the 208 nodes in this study in 4 different periods. During the first period (March 1952- June 1956), clubs were established in Tokyo and Yokohama (1952), and in Kobe, Osaka, Matsuyama and Kyoto (1953). Three more cities were added to the list in 1954, and 8 more the following year. During this period, most clubs were concentrated in the southern and western parts of the country, with Sapporo (est. 1956) as the only club existing north of the Tohoku district. It should also be noted that during the first period, no city had more than one club.

In the second period (July 1956- June 1961), there was some diffusion in the northeastern part of Japan, but at the same population scale, there was an overwhelmingly high proportion of diffusion in the south and west, where there were 4 cities that each had at least 5 clubs (i.e., Tokyo, Osaka, Hiroshima, and Kitakyushu). This period is also characterized by the fact that clubs had not yet expanded into the suburban areas of Tokyo, which strongly maintained their identities as bedroom communities. Given that there were relatively few prospective members in this area, and that these suburbs were located in close proximity to the central city, we can postulate that it was comparatively easy to join existing clubs based in Tokyo.

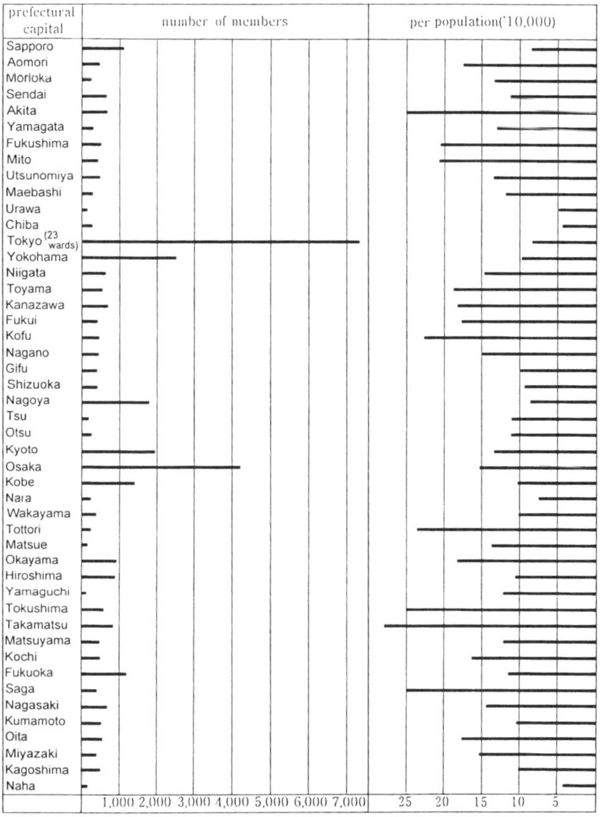

Figure 2.5. Lions Club membership and density in prefectural capitals

During the third period (July 1961- June 1971), new clubs were established in at least 202 cities (excluding 5 cities in the Tokyo area and Tanabe in Wakayama Prefecture), effectively marking the end of diffusion in the Japanese national urban system. In this period, there were three cities (Tokyo, Osaka and Yokohama) with at least 17 clubs. This era also saw the differentials in the number of clubs among cities, and the beginning of a hierarchy based on city scale.

The fourth period (July 1971- June 1981) saw diffusion mainly within the same city, rather than from one city to another, with a well-established urban hierarchy based on the number of local clubs. In June 1981, there were 130 clubs in Tokyo and 59 in Osaka.

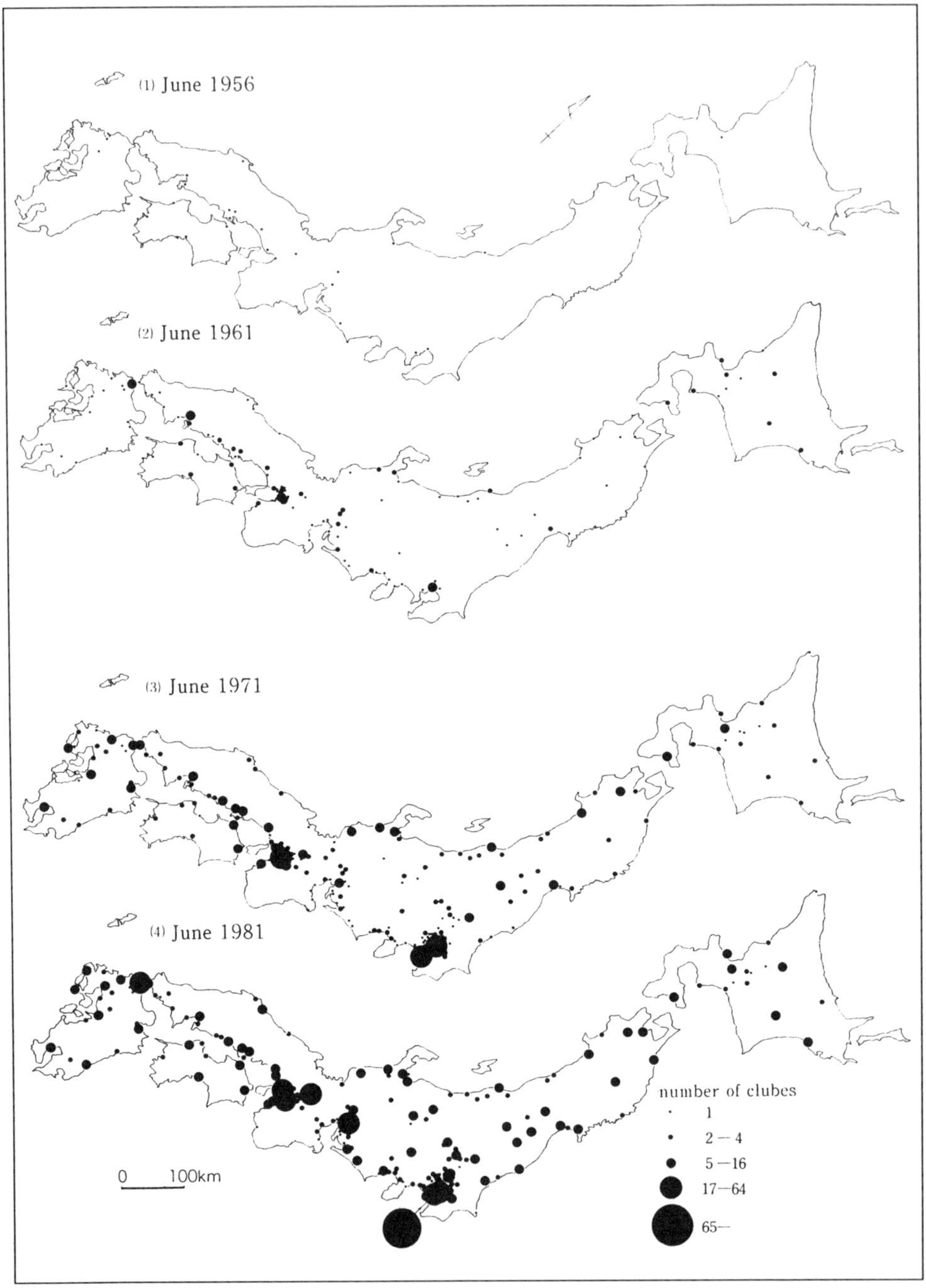

Figure 2.6. Lions Club diffusion patterns in the Japanese national urban system

Figure 2.7 shows the relationship between the year-to-year diffusion of Lions Clubs and population in the 208 cities. Compared with the cities in the international urban system (Figure 2.3), the correlation coefficient, at 0.520 (significant at the 5% level) was high, suggesting that club development was progressing based on some degree of an urban hierarchy. Nevertheless, this coefficient was not exceptionally high, indicating that the proximity effect and random effect (Hudson, 1972) were having a major influence here.

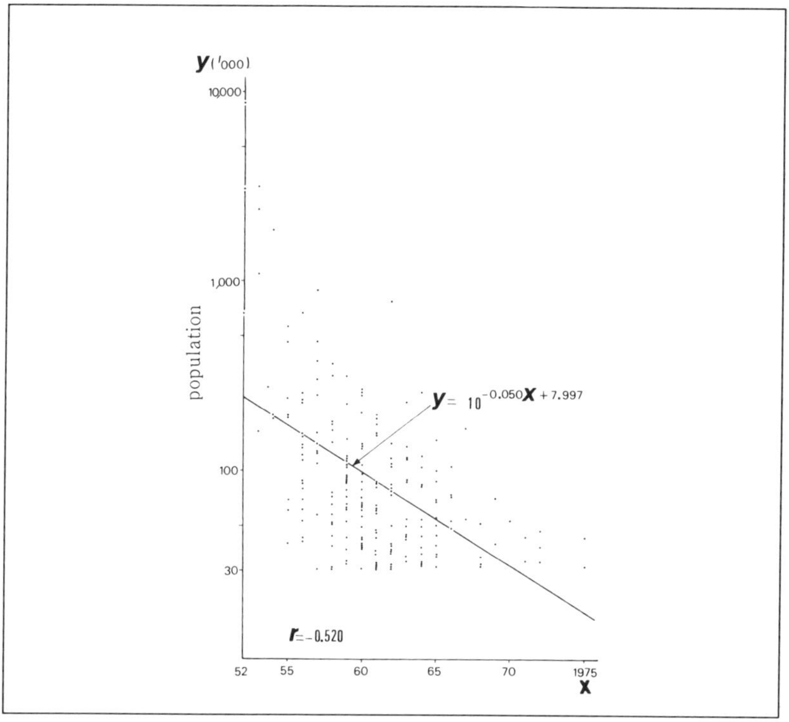

Figure 2.7. Relationship between population scale and yearly diffusion of Lions Clubs in the Japanese national urban system

There were 4 levels of this hierarchy based on population scale: 250,000 and over (23 cities), 100,000-250,000 (53), 50,000-100,000 (67), and 30,000-50,000 (65 cities). Superimposing a logistic curve over the diffusion pattern (Figure 2.8) we can see differences in diffusion periods based on population, but the time lag is not so great. For example, looking along the y-axis where x=1960, we can see that the period of diffusion was complete for the cities at the top of the hierarchy (population over 250,000), while it was just beginning for the smallest cities. Yet, 4 curves overlap, showing that there was diffusion at all levels of the hierarchy. This can be considered to be one factor responsible for the decline in the correlation coefficient in Figure 2.8. It should also be noted that the rate of diffusion was the most sluggish at the lowest level of the hierarchy, and the most rapid in the 50,000-100,000 class.

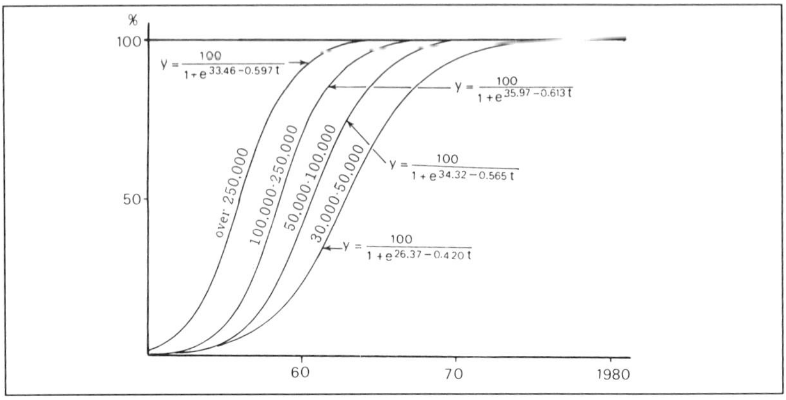

Figure 2.8. Logistic curve for diffusion in the Japanese national urban system (per level of population hierarchy)

Next, Figure 2.9 shows a logistic curve superimposed on a map of Japan divided into 4 zones[5] to consider regional differences in the period and rate of club diffusion. Expansion was completed in the shortest time in the Kyushu and Shikoku regions, and in the longest time in the Kanto, Hokuriku, and Chubu regions. At the same time, the rate of diffusion was fastest in the Hokkaido and Tohoku regions, and the most sluggish in the Kanto, Hokuriku, and Chubu regions. Despite the fact that the diffusion in the Hokkaido and Tohoku regions began much later than in the other regions, it quickly accelerated, riding the wave of the high economic growth of the early 1960s. On the other hand, despite the fact that cities in the Kanto region such as Tokyo, Yokohama, and Kawasaki had been the leaders in nationwide diffusion, diffusion in the Tokyo suburbs was proceeding at a relative snail's pace.

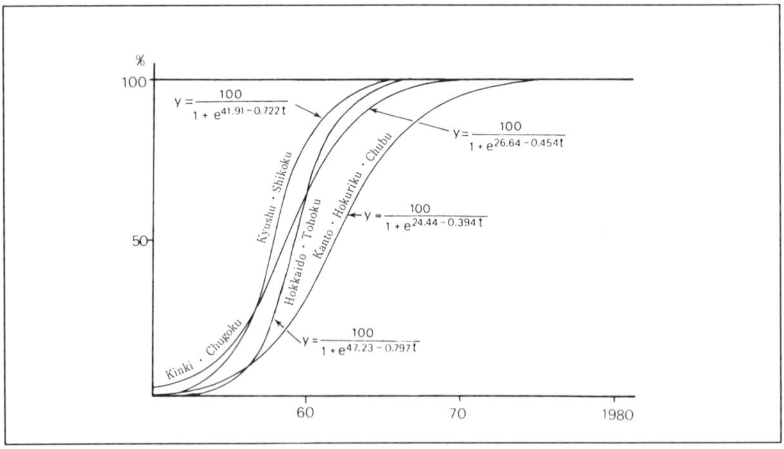

Figure 2.9. Logistic curve for diffusion in the Japanese national urban system (per region)

Given the preceding considerations, we can clearly see that the innovation called Lions Clubs spread within the Japanese national urban system with a certain degree of temporal and spatial order. Given that, why did the system take this particular route in its diffusion? Let us for the time focus on diffusion channels (Figure 2.10). The Tokyo club, which was sponsored by the Manila club, also helped establish the Yokohama club in 1952 and the Kobe club the following year. Kobe then helped establish the Osaka, Matsuyama and Kyoto clubs the same year (1953). These clubs became the core of diffusion, fostering "child," then "grandchild" clubs so that by 1975, the final link in the chain, the Higashi Murayama club, had been established (see Figure 2.10). Core cities for diffusion were Sapporo (for the Hokkaido region), Sendai (for Tohoku region), Tokyo (Kanto region), Niigata (Hokuriku region), Nagoya (Chubu region), Kobe and Osaka (Kansai region), Matsuyama (Shikoku region), and Fukuoka (for Kyushu region).

Looking at the topological structure of diffusion channels, we can see, as indicated in Figure 2.10, that the straight-line distance between links gradually shortened over time. The average distance between links in the 4 periods shown in Figure 2.6 was 129.9 km among the 23 cities in the first period, 43.4 km (94 cities) in the second period, 25.6 km (85 cities) in the third period, and 5.5 km (6 cities) in the fourth period. Such shrinking distances of diffusion links is a combination of the hierarchical and proximity effects, suggesting that diffusion in Figure 2.1 was steadily penetrating into the regional level.

In the general diffusion patterns shown in Figure 2.10, we can recognize a certain spatial order, but why did this order come into being? Let us hypothesize for this analysis that the diffusion channels of the innovation so called Lions Clubs is affected by the strength or weakness of interdependencies between cities. The reason for this is that, as previously mentioned, the establishment of new clubs results from the recruitment of new members through the friends, associates, and/or business contacts of existing members. The respective cities in which the new and existing clubs are located can be assumed to have strong social and economic linkages. Therefore, vehicular traffic flow as an index for the mutual dependency of two cities shall be used to verify the hypothesis.

In the analysis of interdependencies between cities, vehicular traffic flow is valid and appropriate as a general index (Okuno, 1972). An O-D chart of the "Survey of National Road Transportation" (1971), which is conducted every 3 years by the Ministry of Construction, was used as the database. With the restriction that no city can obtain a linkage from a less populated city (larger cities cannot be subordinate to smaller cities), linkages were derived as vectors for maximum movement in each city.[6] Therefore, excluding the terminal node of Tokyo, each of the 207 other cities was determined to have one destination link.

The derived nodal structure of vehicular traffic flow (Figure 2.11) shows patterns that are quite similar to those of the diffusion channels of Figure 2.10. Of the 207 links in Figure 2.10, more than half (112) coincide with the maximum traffic flow links of

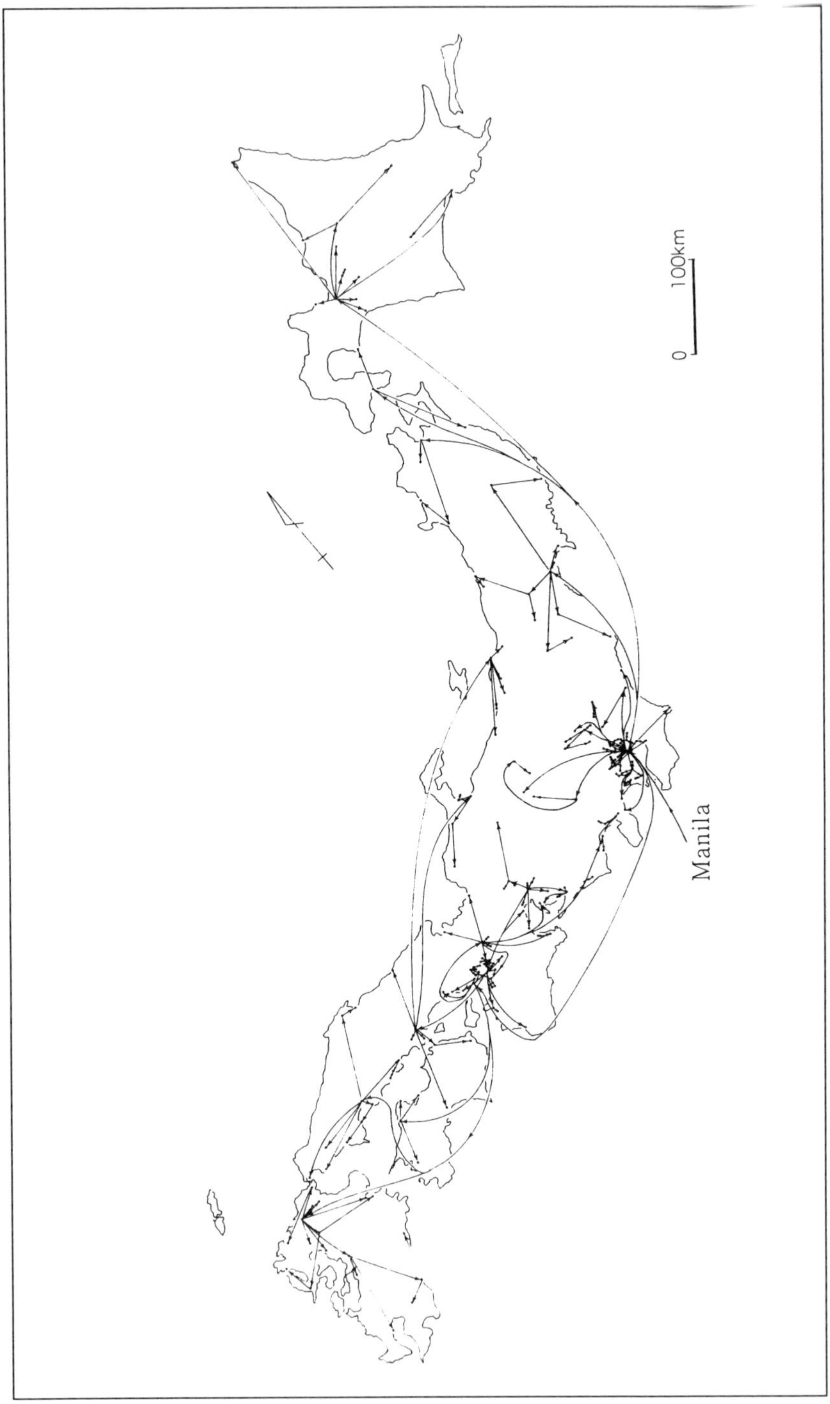

Figure 2.10. Diffusion channels of Lions Clubs in the Japanese national urban system

Figure 2.11. The Japanese national urban system: nodal structure based on traffic flow

Figure 2.11. In addition, 53 of the 95 links that do not coincide with maximum traffic flow do coincide with the second greatest traffic flow. In addition, there were just 8 links which could only correspond with the 4th greatest traffic flow or lower.[7] From this analysis, we can probably say that in the Japanese national urban system, the innovation (Lions Clubs) was diffused chiefly through strong social and economic links between cities.

2.3.3. REGIONAL URBAN SYSTEM

Let us consider Kofu, one of the nodes in the Japanese urban system, for analysis at the regional level. We will call the area in which Kofu-based clubs have had either a direct or indirect hand in sponsoring new clubs the "Kofu sphere of influence." This area is within the bounds of the regional (Yamanashi) urban system. In this region there are 22 cities or core areas (nodes) with clubs. Figure 2.12 shows the relationship between diffusion year and population in these 22 nodes. In comparison with Figure 2.3 (international urban system) and Figure 2.7 (Japanese national urban system), the slope

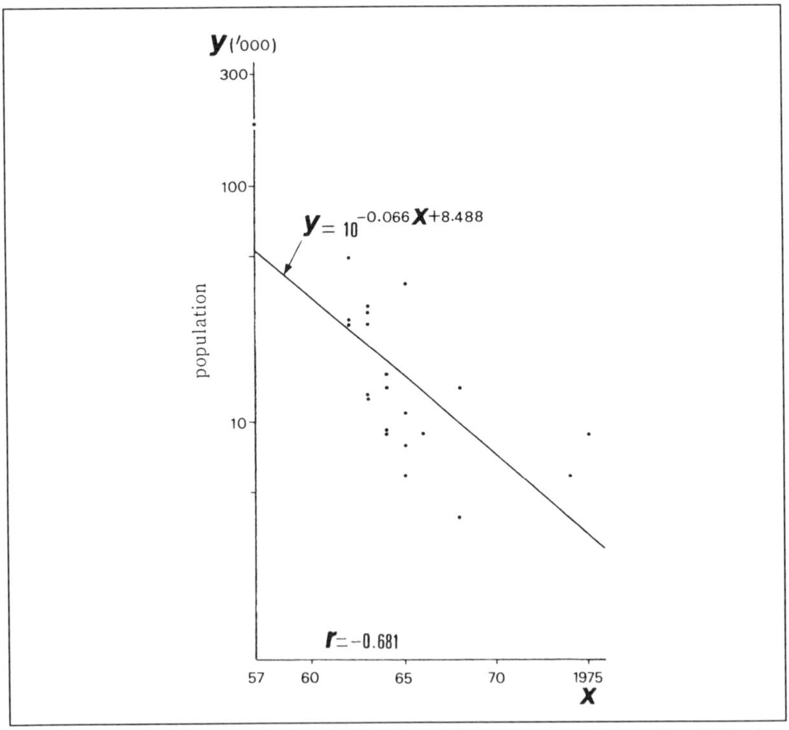

Figure 2.12. Relationship between population scale and yearly diffusion of Lions Clubs in the Yamanashi regional urban system

of the regression line in Figure 2.12 is steepest; Therefore, the diffusion to all these cities took only 18 years, suggesting that growth in the regional (Yamanashi) urban system occurred at a rapid pace. The correlation coefficient, at -0.681 (degree of freedom correction of the correlation coefficient is -0.661, significant at 5%) was the highest of any second-tier system, indicating that diffusion proceeded along a population hierarchy.

A cartographic representation of Lions Club diffusion channels (Figure 2.13) shows us that the Kofu club, which was sponsored by the Tokyo club in 1957, served as the regional core. In 1962, under the sponsorship of the Kofu club, new clubs were established in Fuji-Yoshida (July 15), Shioyama (November 1), and Nirasaki (December 6). After that, the Fuji-Yoshida club established 2 child and 3 grandchild clubs, Shioyama sponsored 3 child clubs, and Nirasaki took one child and one grandchild under its wing.

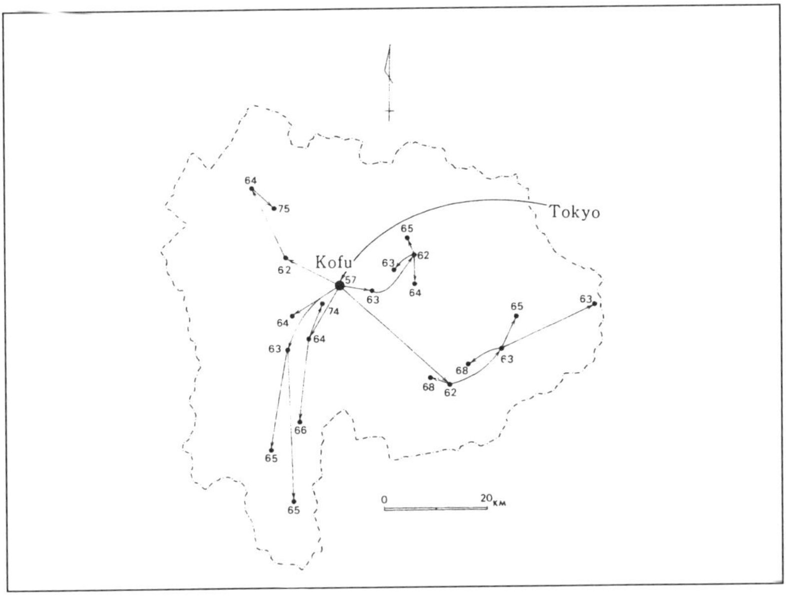

Figure 2.13. Diffusion channels of Lions Clubs in the Yamanashi regional urban system

Note: Numbers in the diagram signify year of establishment (57 = 1957).

Figure 2.14 is a cartographic representation of the nodal structure of vehicular traffic flow using that same indices and methodology as in the preceding section. Here, there are 10 links which do not coincide with the diffusion channel links shown in Figure 2.13 (which are represented by the dotted lines in the figure). Generally speaking, lower directionality of maximum traffic flow links shows nonconformity with diffusion channels.[8] The values of directionality of Shimobe, Minobu, and Nambu in southern Yamanashi Prefecture are all low, at 18, 18, and 24, respectively. These are competitive nodes in the nodal structure (Murayama, 1982b), one of the main factors why traffic volume links do not coincide with diffusion links. Conversely, in interurban links with a high degree of mutual dependency it is likely that diffusion roughly proceeded along these links.

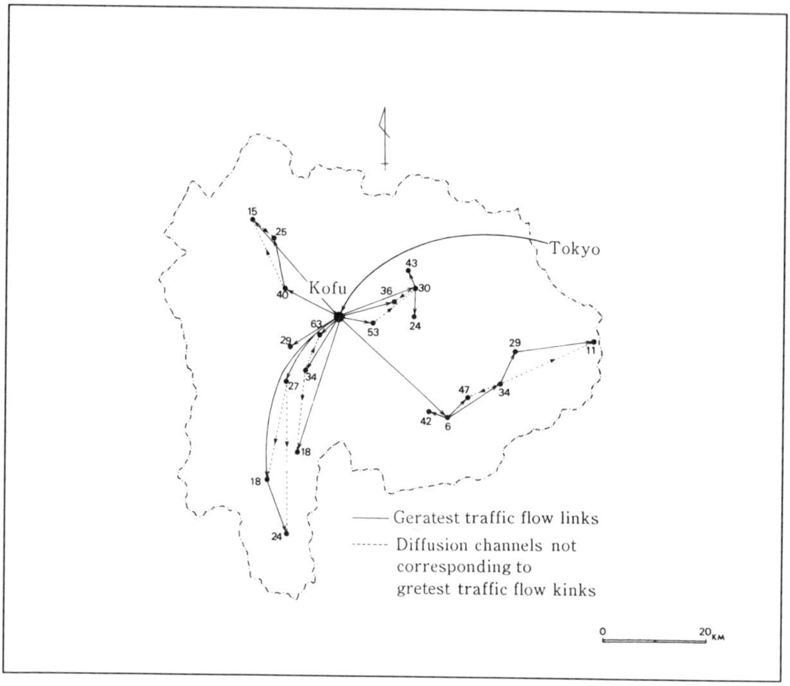

Figure 2.14. The Yamanashi regional urban system: nodal structure based on traffic flow

Note: Numbers in the diagram signify degree of directionality toward node.

However, we must also consider the Rotary Club, which has competing purposes and activities. Unlike national and international urban systems, populations of nodes of regional urban systems are relatively low, the number of qualified prospective members is small, and there may already be a different club that has been established in the area. This would lead us to expect different diffusion patterns.

The dotted lines in Figure 2.15 show diffusion channels of the Rotary Club, while the solid lines show the same for the Lions Club. There appears to be a slight relationship of the lag in year of establishment. It would be interesting as a future research topic to examine how these two competing clubs expand as they wield influence over each other in regional urban systems.

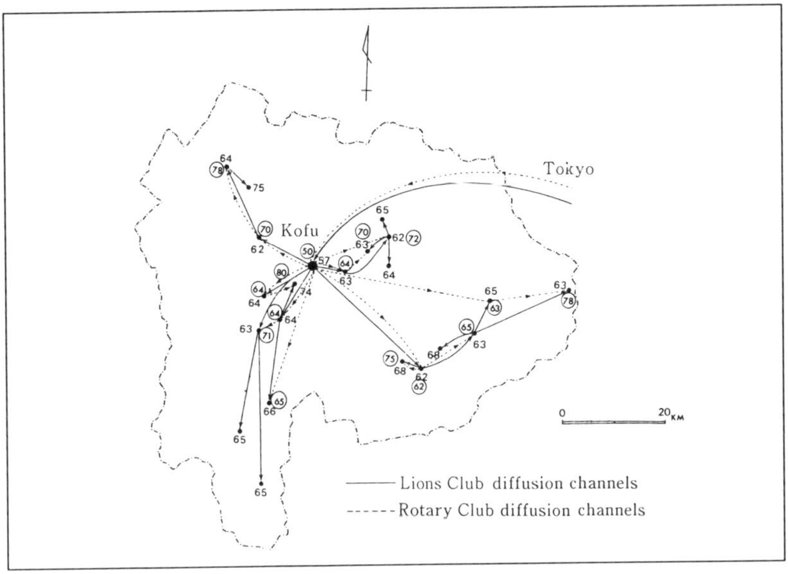

Figure 2.15. Diffusion channels of Rotary Clubs in the Yamanashi regional urban system

Note: Circled numbers in the diagram signify year of Rotary Club establishment; plain numbers show year of Lions Club establishment.

2.4. Conclusion

Using Lions Clubs as an index, this chapter has considered how innovation is diffused in international, national (Japanese) and regional (Yamanashi) urban systems as spatial and temporal order takes shape. The functional differences in process of spatial diffusion at these three levels can be summarized as follows:

1) The peak of Lions Club diffusion in the international urban system lasted from around 1950 to sometime in the 1960s, while it occurred some time around 1960 in the Japanese national system and from 1962-65 in the Yamanashi regional system. Hence, we can see a distinct time lag between these three levels, with growth and diffusion progressing from the international system to the regional (Yamanashi) system.

2) The correlation coefficient in the regression analysis between establishment year and population scale was highest for the regional urban system, and lowest for the international urban system. This indicates that as the hierarchical level decreased, so did the influence of the random and proximity effects, and diffusion proceeded along a population hierarchy.

3) Expansion channel links in the national (Japanese) and regional (Yamanashi) urban

systems examined in this study coincided in large part with the maximum traffic flow links. In the national and regional urban systems, diffusion occurred along strong social and economic linkages between cities, in contrast to the international urban system, which was more affected by political and historical ties between nations.

Notes

1. In conventional usage, "innovation" refers to new developments, a meaning which has been extended to include changes in lifestyles and aspects of the community (Ishimizu, 1976, p.170). In this study, however, the term has simply been used to refer to the appearance of a new thing or organization in a community. On a side note, the spatial diffusion of the Rotary Club has been taken up in such as works as Gould (1974).

2. Committee for Lions Magazine, Japanese version (1981): Lions Annual for 1981 (November Special Edition- 30th Anniversary Issue), p.110.

3. The diffusion into these 208 cities began in 1952 and culminated in 1975. The population data considered here was from 1965, roughly the mid-way point.

4. Committee for Lions Magazine, Japanese version (1977): Special 25th Anniversary Edition of the Japan Lions Club.

5. At the time of this study, there were 25 cities with *Lions Club* in the Kyushu/Shikoku region; 53 in the Kinki/Chugoku region; 94 in the Kanto/Hokuriku/Chubu region; and 36 in the Hokkaido/Tohoku region.

6. For example, if the vector can be drawn from City B to City A, City B has the higher population; thus, City B will account for the largest share of the movement going from City A to cities of a larger population.

7. The formation of these diffusion channels is concentrated in the initial stage (period 1) of diffusion in the national urban system, although there have been many exceptions. For example, both the Niigata and Toyama clubs were sponsored by the Okayama club, which was very active at the time in assisting with relief from disasters, which had been afflicting the former two cities on occasion.

8. The directionality of city (core area) i can be derived from $\frac{b_{ij}}{\sum_{j}^{m} b_{ij}} \times 100$. Here, b_{ij} is the factor value of the ith row in the jth column of the O-D matrix, while m is the number of nodes (=22).

References

Berry, B.J.L. (1961) City size distributions and economic development, *Economic Development and Cultural Change* **9**, 573-587.

Gould, P. (1974) Diffusion in an urban hierarchy: the case of Rotary Clubs, *Economic Geography* **50**, 333-340.

Hudson, J.C. (1972) Geographical diffusion theory, *Northwestern University, Studies in Geography* **19**, 179p.

Ishimizu, T. (1976) *An Outline of Quantitative Geography*, Kokon Shoin, Tokyo, 242p. (J)

Jeffrey, D. and Webb, D.J. (1972) Economic fluctuations in the Australian regional system, *Australian Geographical Studies* **10**, 141-160.

King, L.T. (1966) Cross-sectional analysis of Canadian urban dimensions: 1951 and 1961, *Canadian Geographer* **10**, 205-224.

Murayama, Y. (1982a) Canadian urban system and its evolution process in terms of air-passenger flows, *Geographical Review of Japan* **55**, 380-402.

Murayama, Y. (1982b) Re-examination of the Nystuen-Dacey model, *Geographical Sciences* **37**, 73-84. (J)

Nystuen, J.D. and Dacey, M.F. (1961) A graph theory interpretation of nodal regions, *Papers and Proceedings of the Regional Science Association* **7**, 29-42.

Okuno, T. (1972) Linkage system of the Chukyo region in terms of automobile traffic flow, Itoh, G. ed., *Chukyo Region*, Taimeido, Tokyo, pp.235-245. (J)

Pederson, P.O. (1970) Innovation diffusion within and between national urban systems, *Geographical Analysis* 2, 203-253.

Pred, A. (1980) *Urban Growth and City-Systems in the United States, 1840-1860*, Harvard University Press, Cambridge, 282p.

Sugiura, Y. (1978) Innovation diffusion and urban system in Japan during the Meiji and Taisho era, 1868-1926, *Geographical Report of Tokyo Metropolitan University* **13**, 29-48.

(J = in Japanese)

CHAPTER 3

DEVELOPMENT PROCESS OF THE EAST ASIAN URBAN SYSTEM

3.1. Introduction

Since the 1980s, economic development in the Asia-Pacific region has been dazzling, with all countries of the region maintaining a high rate of growth in the 1990s. Tremendous amounts of capital have been invested here by Japan, Anglo-America and the European Union such that the area is now recognized as the premier center of economic growth in the world. This growth is evident from a variety of statistics for trade, commerce, industry, finance and services. Movement of goods and passengers is one of the best indicators to show this dramatic growth. For example, as we can see in Table 3.1, air passenger traffic in the Asia-Pacific region in 1983 accounted for a mere 16.0% of all such traffic in the world; by 1994, however, this share had increased to 30.5%. Air cargo grew from 22.6% of the world total in 1983 to 33.0%, or one-third of the world total, in 1994. At the same time, the share of air passenger and cargo transport in North America and Europe has been declining.

Table 3.1. Air transportation in the world (scheduled)

(unit: 1 million passenger-kilometers)

	North America	Europe	South America	Africa	Middle East	Asia-Pacific
year	Passengers · kilometers (international + domestic)					
1983	475,689	388,326	60,136	34,359	38,057	190,031 (16.0%)
1988	720,888	508,575	78,669	36,908	45,601	308,091 (18.1%)
1994	867,207	522,524	105,580	47,439	61,079	488,371 (30.5%)
year	commodity ton · kilometers (international + domestic)					
1983	9,943.4	12,445.4	1,719.7	1,079.2	1,871.4	7,897.4 (22.6%)
1988	14,999.0	17,996.8	2,452.6	1,263.0	2,438.9	14,289.3 (26.7%)
1994	20,562.0	22,764.0	3,637.0	1,422.0	3,306.0	25,437.0 (33.0%)

Note: The number in parentheses show the percentage of Asia-Pacific departures + arrivals to the world total.

Source: ICAO (1995).

Within the Asia-Pacific region, East Asia has been experiencing especially remarkable economic growth, with markets being established which are on par with those in North America and Europe. This unprecedented rapid growth has been called "the East Asian Miracle" (World Bank, 1993).

East Asia, which strengthened economic ties with Japan and the United States in the 1980s, has also been intensifying intraregional mutual dependencies in recent years (see Table 3.2). The ASEAN trade structure has gone from exporting raw materials to developed countries and importing finished products, to one of exporting both raw materials and finished products. Semiconductors, parts, and partially assembled electronic goods produced in ASEAN countries are exported within the East Asian sphere, and assembly plants are springing up all over East Asia. Trade within the ASEAN region increased 7-fold from Stage 1 (1984-88) to Stage 2 (1988-92). Trade from ASEAN to NIES countries increased 4-fold. In contrast, trade between Japan and the US on the one hand, and NIES on the other, has been decreasing. The East Asian economy is well on its way to independence.

Table 3.2. The increase of the trade in Asia

(unit: 1 million dollars)

Origin \ Destination		Japan	NIES	ASEAN	China	USA	Total
Japan	stage 1	—	25,484	3,592	2,287	29,846	61,179
	stage 2		22,893	14,594	2,481	6,471	46,439
NIES	stage 1	21,184	19,884	7,315	16,086	58,234	122,703
	stage 2	2,290	19,101	11,574	50,968	9,297	63,230
ASEAN	stage 1	-496	3,378	409	1,033	2,472	6,796
	stage 2	9,088	13,164	2,838	1,468	11,133	37,691
China	stage 1	2,891	11,938	575	—	1,086	16,490
	stage 2	3,645	22,195	893		5,191	31,924
USA	stage 1	14,045	16,939	1,004	2,013	—	34,001
	stage 2	10,144	13,862	6,953	2,453		33,412
Total	stage 1	37,624	77,623	12,895	21,419	91,608	241,169
	stage 2	25,167	91,215	36,852	27,370	32,092	212,696

Note: stage 1: 1984-88, stage 2: 1988-92.

Source: Watanabe (1995). IMF, Direction of Trade Statistics, various issues.

The intensification of mutual economic dependencies in the East Asian region has led from emphasis on only industrial production to the creation of new services related to distribution, wholesaling, and sales, which is in turn increasing the relative role of cities where these functions are concentrating. New jobs are being created in the metropolitan areas which are attracting an influx of job-seekers from the hinterlands, while at the same time, numerous types of "support industries" such as transport companies, distribution centers, banks, hotels, offices, etc., are locating in cities to support industrial development. Major metropolitan areas, which are at the apex of national urban systems, control the economies of agricultural areas and minor cities while strengthening economic ties with major cities in other countries through the activities of multi-national corporations, the export of industrial goods, financial operations, and so on. In other words, the handful of mega-cities where finance and communications are concentrated have become the driving force of the East Asian economy. The great number of minor cities, and farmlands (which comprise 90% of a nation's land area) have been incorporated into the hinterlands which support the intensive economic activities of the mega-cities. As the role of mega-cities in economic development increases, not only will the national unit be used for describing economic growth and international relations — it has now become essential to include major cities as analytical units.

Given this background, this chapter will use urban system theory developed in geography to provide an understanding of the processes of this dynamic economic development in East Asia through the dynamics of interurban linkages. Taking the growth of East Asian mega-cities, the changes in urban functions, and the strengthening of interurban linkages into account, this study considers how each of these cities is being incorporated into the East Asian and upper-level world urban systems.

3.2. Changes in Nodes

3.2.1 GROWTH OF THE EAST ASIAN MEGA-CITIES

In the latter half of the 20th century, there has been a conspicuous trend worldwide toward concentration of population in cities. In 1950, the Asian cities among the 15 most populous cities worldwide were only Tokyo, Shanghai, Calcutta, Osaka, and Beijing. However, by 1994, nine Asian cities were in the top 15, and predictions are that there will be 11 by the year 2015 (Table 3.3) (United Nations, 1995). There were no Asian cities with at least 10 million population in 1950, but by 1990 there were 7 such cities with a total population of 93 million. By 2015, this figure is expected to rise to 18 cities with a total population of 340 million (Table 3.4). The proportion of the Asian urban population living in mega-cities of 1 million or more was 24.6% in 1950 but by 1990, it had grown to 35.4%. It is expected to reach 41.9% by 2015.

In 1950, there were only 24 cities in East Asia with a population of at least 750,000, but by 1990, this number had grown to 84. China accounted for the lion's share, with 11 such cities in 1950 and 49 in 1990. During this 40-year period, 38 more cities in China reached the ranks of mega-cities with 750,000+ population. This was followed by Indonesia with 8 new mega-cities, Korea with 4, and Japan with 3. According to UN estimates, there will be 9 cities in Asia with a population of at least 10 million by the year 2015 (Figure 3.1).

Population is not increasing uniformly in each city (Figure 3.2). Looking at the annual rate of change from 1990-95, we can see that the greatest increase among all these cities was in Tanjung Karang, Indonesia, with 8.43%, followed by Bogor, Indonesia, at 5.50%. All of the Indonesian cities are growing by at least 3% per year (the highest in the region), followed by Philippines, Taiwan, and China. Countries with low annual rates of change include Japan, Hong Kong, and Singapore. Kyoto, Japan is the only mega-city in East Asia which has registered a population loss.

Table 3.3. The 15 largest urban agglomerations, ranked by population size, 1950-2015

(Millions)

Rank	*Agglomeration and country*	*Population*
	A. *1950*	
1.	New York, United States of America	12.3
2.	London, United Kingdom	8.7
3.	Tokyo, Japan	6.9
4.	Paris, France	5.4
5.	Moscow, Russian Federation	5.4
6.	Shanghai, China	5.3
7.	Essen, Germany	5.3
8.	Buenos Aires, Argentina	5.0
9.	Chicago, United States of America	4.9
10.	Calcutta, India	4.4
11.	Osaka, Japan	4.1
12.	Los Angeles, United States of America	4.0
13.	Beijing, China	3.9
14.	Milan, Italy	3.6
15.	Berlin, Germany	3.3
	B. *1994*	
1.	Tokyo, Japan	26.5
2.	New York, United States of America	16.3
3.	São Paulo, Brazil	16.1
4.	Mexico City, Mexico	15.5
5.	Shanghai, China	14.7
6.	Bombay, India	14.5
7.	Los Angeles, United States of America	12.2
8.	Beijing, China	12.0
9.	Calcutta, India	11.5
10.	Seoul, Republic of Korea	11.5
11.	Jakarta, Indonesia	11.0
12.	Buenos Aires, Argentina	10.9
13.	Osaka, Japan	10.6
14.	Tianjin, China	10.4
15.	Rio de Janeiro, Brazil	9.8
	C. *2015*	
1.	Tokyo, Japan	28.7
2.	Bombay, India	27.4
3.	Lagos, Nigeria	24.4
4.	Shanghai, China	23.4
5.	São Paulo, Brazil	21.2
6.	Karachi, Pakistan	20.8
7.	Jakarta, Indonesia	20.6
8.	Beijing, China	19.4
9.	Dhaka, Bangladesh	19.0
10.	Mexico City, Mexico	18.8
11.	New York, United States of America	17.6
12.	Calcutta, India	17.6
13.	Delhi, India	17.6
14.	Tianjin, China	17.0
15.	Metro Manila, Philippines	14.7

Source: United Nations (1995).

Table 3.4. Urban population, number of cities and percentage of urban population, by city-size class, Asia

Size class	*1950*	*1970*	*1990*	*2015*
A. *10 million or more*				
Number of agglomerations	0	2	7	18
Population	0	28	93	304
Percentage urban	0.0	5.5	9.2	14.1
B. *From 5 million to 10 million*				
Number of agglomerations	2	6	11	22
Population	12	41	78	138
Percentage urban	5.2	8.1	7.7	6.4
C. *From 1 million to 5 million*				
Number of agglomerations	26	53	100	232
Population	46	99	188	461
Percentage urban	19.4	19.7	18.5	21.4
D. *From 500,000 to 999,999*				
Number of agglomerations	39	64	121	183
Population	27	43	84	129
Percentage urban	11.3	8.6	8.3	6.0
E. *Fewer than 500,000*				
Population	151	292	571	1125
Percentage urban	64.1	58.1	56.3	52.1

Source: United Nations (1995).

The annual rate of change is strongly connected to the stage of industrial development. From the early to middle stages of this development, population grows rapidly. During the stage of maturation, that is, when the work force shifts from labor-intensive industries to a service-oriented economy, population increase slackens. Therefore, if we take this point into consideration, we can see that the rapidly growing cities of Metro Manila and Davao in Philippines, and the cities of Indonesia, are presently undergoing industrial development. Cities which are just now entering the initial stage of industrial development, such as Ho Chi Minh City and Hanoi in Vietnam and Yangon in Myanmar, are expected to see population increases in the future. According to the United Nations (1995), Ho Chi Minh City experienced an annual rate of 1.88% population growth from 1991-95. However, this rate is expected to accelerate to 2.32% from 1995-2000 and to 2.87% from 2000-05. Yangon is expected to go from its present rate of 3.08% to 3.34% for the period 2000-05. (For its part, Japan experienced its period of high population growth from 1965-70). Furthermore, a trend has been recognized in all these countries in which rapid population begins in the capital cities, eventually reaching the second- and third-largest cities after a time lag of about 4 or 5 years.

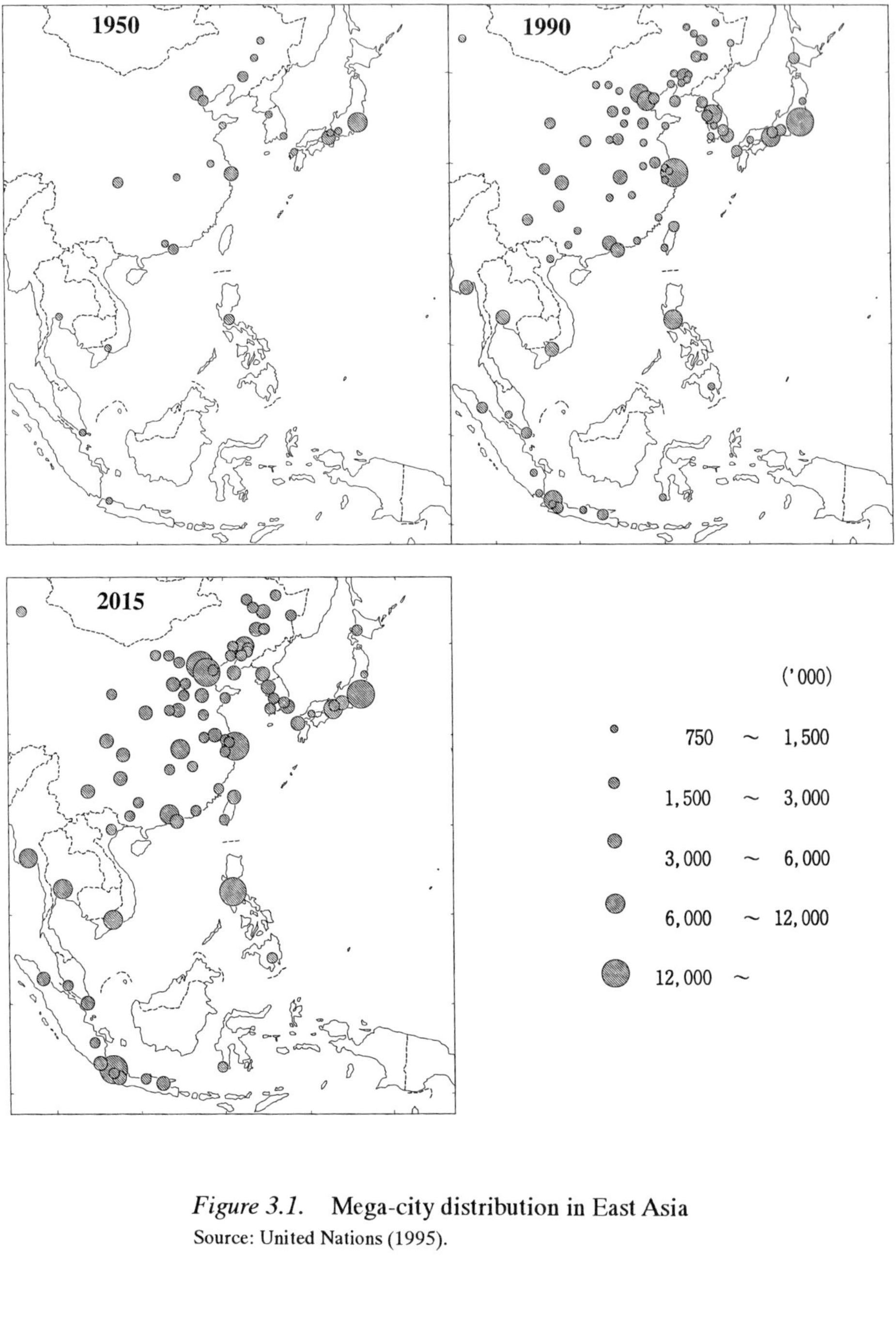

Figure 3.1. Mega-city distribution in East Asia
Source: United Nations (1995).

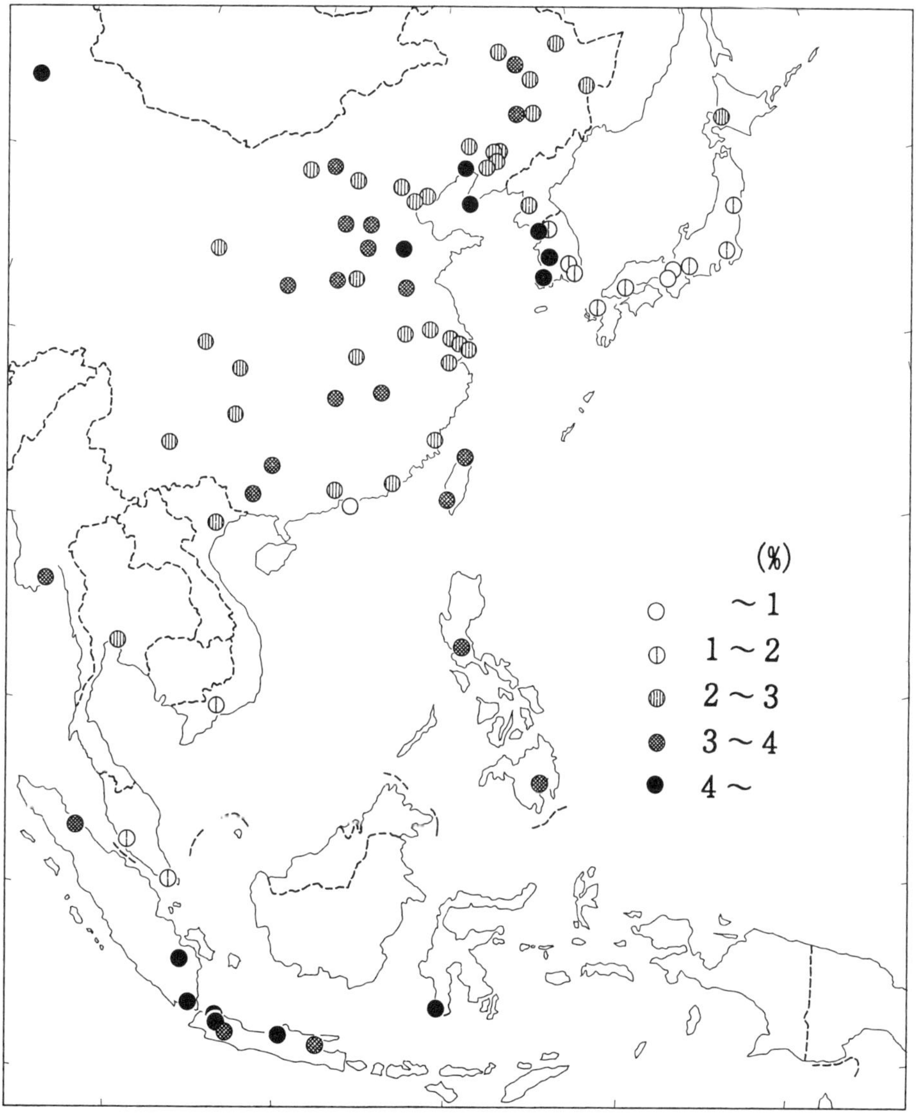

Figure 3.2. Rate of annual population increase in East Asia, 1990-95
Source: United Nations (1995).

3.2.2 INTENSIFICATION OF PRIMACY

One of the distinguishing features of the urban systems of all Asian countries is primacy. For example, Bangkok, the capital of Thailand, has about 30 times as much population as the next largest Thai city, Chiang Mai. In Philippines in 1990, Metro Manila had 9.4 times as many people as the next largest city, Davao. The author would like to point out here that primacy is accelerating at a rapid pace in many countries. For example, the Indonesian capital had 3.4 times as many people as that country's second largest city in 1980, a gap which grew to 3.7 times in 1990 and is expected to reach 4 times by the year 2000. This indicates that industrial development is not progressing uniformly in Indonesian cities but is concentrating in the area around Jakarta, the national capital. If nothing is done and the economy is left to market forces, there is no doubt that this condition will promote unbalanced development within the country. To rectify this situation, the central government will have to invest vast sums of money which will work to the short-term detriment of the economy.

Primacy is also expected to intensify in developed countries such as Taiwan, Japan, and Korea. For example, the United Nations (1995) has predicted that primacy in Korea will go from 2.65 in 1980 to 2.79 in 1990 to 2.89 in the year 2000; in Taiwan, it is expected to go from 1.86 (1980) to 1.96 (1990) to 1.98 (2000). However, the intensification of primacy in these countries is probably resulting from the shift toward the service-oriented economy, clearly unlike the trend in the aforementioned developing countries.

The capital cities of Taiwan, Japan, and Korea are strengthening their economic muscle vis-à-vis the other cities in their respective countries and are improving their ability to control their respective national urban systems. At the same time, they are strengthening their economic ties to world cities such as London and New York and have become nodes in the global urban system. Multinational corporations and their branches and foreign-owned financial institutions are springing up in the capital cities. The major domestic companies have many ties to such economically influential corporations. They value face-to-face contact and thus establish their offices in the center of the capital. The multiplicative increase in such domestic and international functions is promoting the growth of the capital city as the control center of economic activities. A large number of employment opportunities are thus generated in capital cities.

If we measure primacy (capital city/ second largest city) from the concentration of core economic functions, we can see that in Japan and Korea, the gap in economic primacy is even greater than the gap in population primacy. Given this situation, we can see that the enormous influence of a city as a node in the urban system is not just a matter of the scale of population, but also has to do with the concentration of core management functions. One condition for a city to become an important node in the global urban system is to attract numerous financial institutions and headquarters and regional coordinating branches of multinational corporations.

In terms of concentration of core management functions, the acknowledged East Asian nodes in the international urban system include Tokyo, Osaka, Nagoya, and Fukuoka in Japan; Seoul and Pusan in South Korea; Beijing, Shanghai and Hong Kong in China; Manila in the Philippines; Taipei and Kaohsiung in Taiwan; Singapore; Jakarta, Indonesia; Kuala Lumpur, Malaysia; and Bangkok, Thailand. Although there are 47 mega-cities in China, there are only two on this list because the other 45 lack the core management functions and level of internationalization that are required for East Asian nodes in the global urban system. Except for China, nodes in developing nations are limited to the respective capital cities.

3.3. Changes in Linkages

Cities which are tightening their economic control over other domestic cities are, as would be expected, strengthening their mutual dependencies with foreign cities. Here, the author would like to explain what kinds of changes have occurred in the patterns of interurban linkages in the early 1980s (before the rapid economic development began in East Asia) and in the 1990s. Linkages have been measured from population movements and from the flow of goods, people, information and capital. We use two types of indices to explain interurban linkages: flow of information (telephone calls, fax messages, mail volume), which is important in transmitting decisions and an important aspect of daily business operations; and traffic volume (goods and passengers).

3.3.1 INTER-NATION LEVEL

First, let us examine telecommunication flow as an index by general spatial flow patterns at the national level in East Asia. The transmission of information is made possible by undersea cables. Figure 3.3 shows the Trans-Pacific and Eurasian cable systems which have been established in the East Asian region. The "Main Street" of this undersea system is the first route that was established, that is, the one linking Japan, Taiwan, Hong Kong and Singapore. As the costs for laying undersea cable decline each year, voice path capacity, thanks to technological innovations, has been increasing. With the increasing economic vitality and accompanying increases in demand, along with lowering costs, the volume of communications has skyrocketed.

In 1994, the largest area of international communication transmissions was Hong Kong. In 1991, the total transmission time was 913 million minutes (hereafter abbreviated as M.m.); by 1994, this figure had jumped to 1578.4 M.m. Next came Japan, with 1160.5 M.m. in 1991 and which, until FY 1993, had been Number 1 in this area. The rapid growth in communications transmissions in Hong Kong can be mainly attributed to increased communications with Mainland China. In the 1990s, economic growth in the coastal cities of China has been occurring at a dazzling pace, with Hong Kong acting as the gateway. In 1991, 41.6% of all communications originating out of Hong Kong were directed at Mainland China, but by 1994, this figure had increased to 52%. Coming in at a distant second was the United States of America, which in 1994 received a mere 6% of transmissions originating from Hong Kong. Next came Japan at 4%, with Taiwan and Canada tied for the 5th place. The share of communications from Hong Kong to Japan in 1994 (4%) fell from the 4.9% of 1991.

Trans-Pacific and Eurasian Cable Systems

Year in Service	Cable System	Cost (US$) per voice path	Capacity (voice paths)
1957	Hawaii 1*	378,000	91
1964	TPC-1*	406,000	167
1974	Hawaii 2*	41,000	1,690
1975	TPC-2*	73,000	1,690
1988	TPC-3	16,000	37,800
1991	North Pacific Cable	5,000	85,000
1992	TPC-4	5,500	75,600
1996	TPC-5/6	2,000	605,000
1997	FLAG	1,500	605,000

* No longer in service.

Notes: Costs are capital and construction costs only, stated in US$ to the nearest $500, unadjusted for inflation. Current technology permits approximately 5 virtual voice paths to be derived from a digital channel operating at 64,000 bits per second (64 kbit/s). Fiber optic cables are expected to have a useful life of at least 25 years. Table reports average cost per voice path for cables with multiple landing points. For example, the TAT-9 system connects the U.S. and Canada with the U.K., France and Spain. The average U.S.-U.K. cost per voice path is approximately $4000. Reserve capacity of cables is generally excluded.

Source: FCC and carriers.

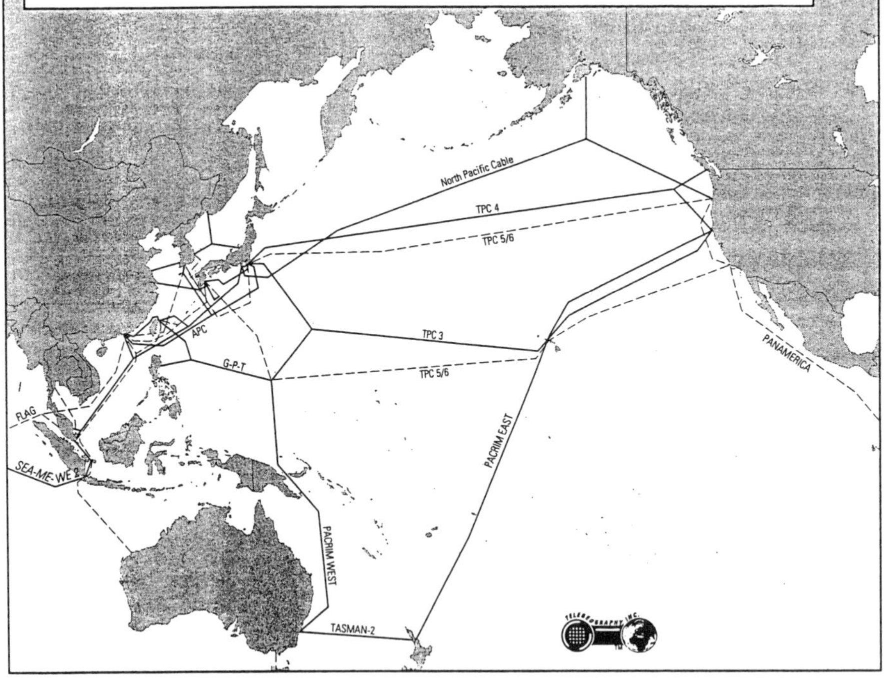

Figure 3.3. Transpacific and Eurasian cable systems
Source: Telegeography (1995).

In 1994, the second greatest source of communications was Japan, which directed 19.4% of its transmissions to its largest communications partner, the United States. The next largest destinations were Korea (9.9%) and China (9.8%). It should be noted that communications volume with China was increasing; for example, in 1991, China-destined communications were only 5.5%, a distant fifth. Since 1992, however, large numbers of Japanese-owned companies have been moving to China, accounting for the dramatic increase in communications with that country. Compared with 1990, the share of communications going to the US fell by 3.1%, going instead to East Asia. This is one bit of evidence to show that the overseas activities of Japanese companies are shifting away from the United States and toward East Asia.

Even within East Asia, however, there has been a marked increase in communications going from Japan to developing countries such as Philippine, Thailand, Malaysia, and Indonesia, and a relative decline in transmissions directed toward developed countries such as Korea, Hong Kong, Taiwan and Singapore. This can be explained in part because in recent years, Japanese companies have been shifting their activities from NIES countries to countries where labor and production costs are lower.

International communications flow in East Asia is greatest between Hong Kong and China, followed by Singapore-Malaysia communications. Communications from Hong Kong to China account for the fourth-largest volume of international communications in the world, behind US-to-Canada, US-to-Mexico, and US-to-Great Britain.

Per capita international communications volume in Hong Kong and Singapore is far greater than in other countries in region (see Table 3.5). Each person makes an average of 272 international communications per year in Hong Kong, followed by 229 in Singapore. In these two countries, there is a large number of multinational corporations and financial institutions which have established offices there. In contrast, Japan's per capita international communications transmissions are only 12 per year, a mere 5% of the rate in Hong Kong and Singapore and quite low if viewed on a national level. (However, if we limit the discussion to Tokyo, then such communications would probably jump to 200 per person per year, although unfortunately there are no city data to confirm this). At the same time, the per capita rates in Thailand, Philippines, Indonesia, and China are low, especially in the last two, where it is believed that the per capita rate of international telephone transmissions is less than one per year.

Given this situation, what are the spatial features of origin-destination patterns in East Asia? First of all, we can say that compared with the early 1990s, the share has been declining but in actual terms, the communications volume with the United States has reached unprecedented highs (Table 3.6). As of 1994, the United States was the Number 1 destination for communications originating from Japan, Korea, and Philippine, and the second-largest destination for transmissions from Hong Kong, Indonesia, Taiwan, and Thailand.

Secondly, though Tokyo is a tame second to Hong Kong in volume of transmissions, it is increasingly directing these communications to a more diverse lot of destinations, including the entire East Asian region. In contrast, Hong Kong directs the lion's share of its communications to neighboring Mainland of China.

Table 3.5. International telecommunication traffic in East Asia (1994)

Origin	Outgoing volume (Millions)	Outgoing volume Per capita
1. Hong Kong	1578.4	272.13
2. Singapore	643.0	229.64
3. Taiwan	498.5	23.40
4. Malaysia	342.3	17.55
5. Japan	1502.0	12.03
6. Korea	440.4	9.89
7. Thailand	173.2	2.95
8. Philippines	160.0	2.41
9. Indonesia	182.5	0.96
10. China	1135.0	0.95

Note: Data are in millions of minutes for public voice circuits.
Source: Telegeography (1995).

Table 3.6. Largest telecommunication routes in East Asia (1994)

Origin	The largest destination (% of outgoing traffic)	The second largest destination (% of outgoing traffic)
China	Hong Kong (59.0%)	Japan (11.5%)
Hong Kong	China (52.0%)	USA (6.0%)
Indonesia	Singapore (23.0%)	USA (11.0%)
Japan	USA (19.4%)	Korea (9.9%)
Korea	USA (27.8%)	Japan (24.2%)
Malaysia	Singapore (46.7%)	Japan (6.7%)
Philippines	USA (34.4%)	Japan (15.6%)
Singapore	Malaysia (29.5%)	Indonesia (8.1%)
Taiwan	China (21.9%)	USA (19.0%)
Thailand	Japan (16.1%)	USA (12.6%)

Source: Telegeography (1995).

Thirdly, the more developed a country is, the more transmissions than receptions of communications there will be. For example, in 1993, there were 1.44 times more communications originating from Japan than destined to that country, an increase over the 1.07 ratio of 1989 and the 1.39 ratio of 1991. In the economic superpower United States, twice as many transmissions as receptions are made. In specific terms, the ratio of transmissions to receptions was 2.12 in 1991, 2.28 in 1993, and 2.41 in 1994, a trend which is increasing year by year and which is thought to be the result of the intensive efforts the US has made in global economic development.

In FY 1986, there were more communications than receptions than transmissions in Hong Kong, but by FY 1994, the ratio had become reversed (1.09 times more transmissions).

Regarding communications between economically advanced and developing countries, far more transmissions are made from the former to the latter (see Figure 3.4). For example, Japan transmits far more information to Thailand, Malaysia, Philippines, etc., than it receives from those countries. This ratio is especially high vis-à-vis the Philippines, where 81% of the 126.5 M.m. (i.e., 103.6 M.m.) of communications between those two countries originated from Japan. This means that a mere 19%, or 22.9 M.m. of communications, originated from the Philippines (see Figure 3.4). Important information in the decision-making process is sent by the controller of this information, showing that in the case of Japan and the Philippines, there is a very strong dominant-subordinate economic relationship which has developed.

3.3.2 INTERURBAN LEVEL

Next, let's consider the flow of information at the interurban level in East Asia. Unfortunately, origin-destination data for telecommunication flow between cities has not been collected as of yet. Therefore, let's consider another index for interurban linkages: data for interurban air mail flow, which the ICAO yearly compiles and publishes in an annual.

Figure 3.5 shows mail flow patterns between the major cities of East Asia in 1982 and 1994. The nodes in 1982 were Tokyo, Hong Kong, Singapore and Bangkok, but by 1994, this list had grown to include Seoul, Taipei, and Manila. In both years, the flow of air mail in Tokyo, Hong Kong, and Singapore was prominent (Figure 3.5). Moreover, there has been a high rate of increase in the flow of mail moving among these three cities.

Figure 3.6 is a graphic representation of the mail flow between Tokyo, Hong Kong, and Singapore. Especially noteworthy is that between Tokyo and Hong Kong in 1982, the former accounted for most of the transmissions, but by 1994, this had made a large-scale turn-around. The same thing has also been happening with Hong Kong's communications with Singapore. If we also consider the telecommunication flow mentioned previously, we can see that Hong Kong has overcome Tokyo as the information transmission base of East Asia. However, we must understand that this development has been supported by economic exchanges with China and cities within a relatively close proximity such as Tokyo and Seoul, while Hong Kong has not attained the power to control the international urban system in East Asia as a whole. Although we can say that Hong Kong is a center of the Far Eastern economic sphere, the best situated city in the Southeast Asian sphere is Singapore. Tokyo has equal linkages with and plays roles in both economic spheres.

Figure 3.7 shows interurban linkages in terms of movement of air freight. The dominant flow in 1982 was along a linear route linking Tokyo, Taipei, Hong Kong, and Bangkok. This flow was one-directional, from Bangkok to Hong Kong, Hong Kong to Tokyo, and from Taipei to Tokyo, indicating that Tokyo was the final destination for freight in this region. By 1994, Seoul and Singapore had been added to this linear route. Seoul has strong linkages with Tokyo and Hong Kong and is firming ties with Bangkok

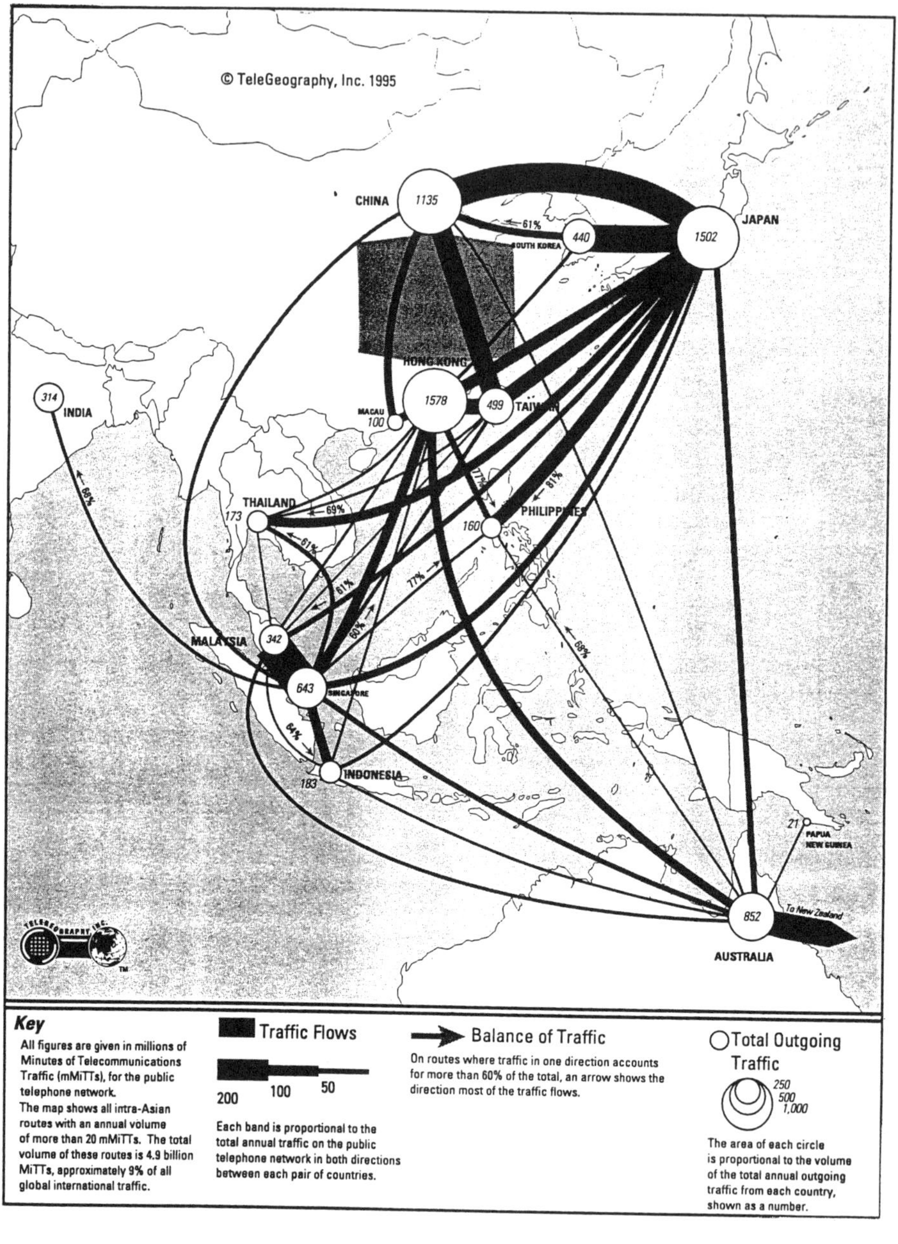

Figure 3.4. East Asian telecommunication flows (1994)
Source: Telegeography (1995).

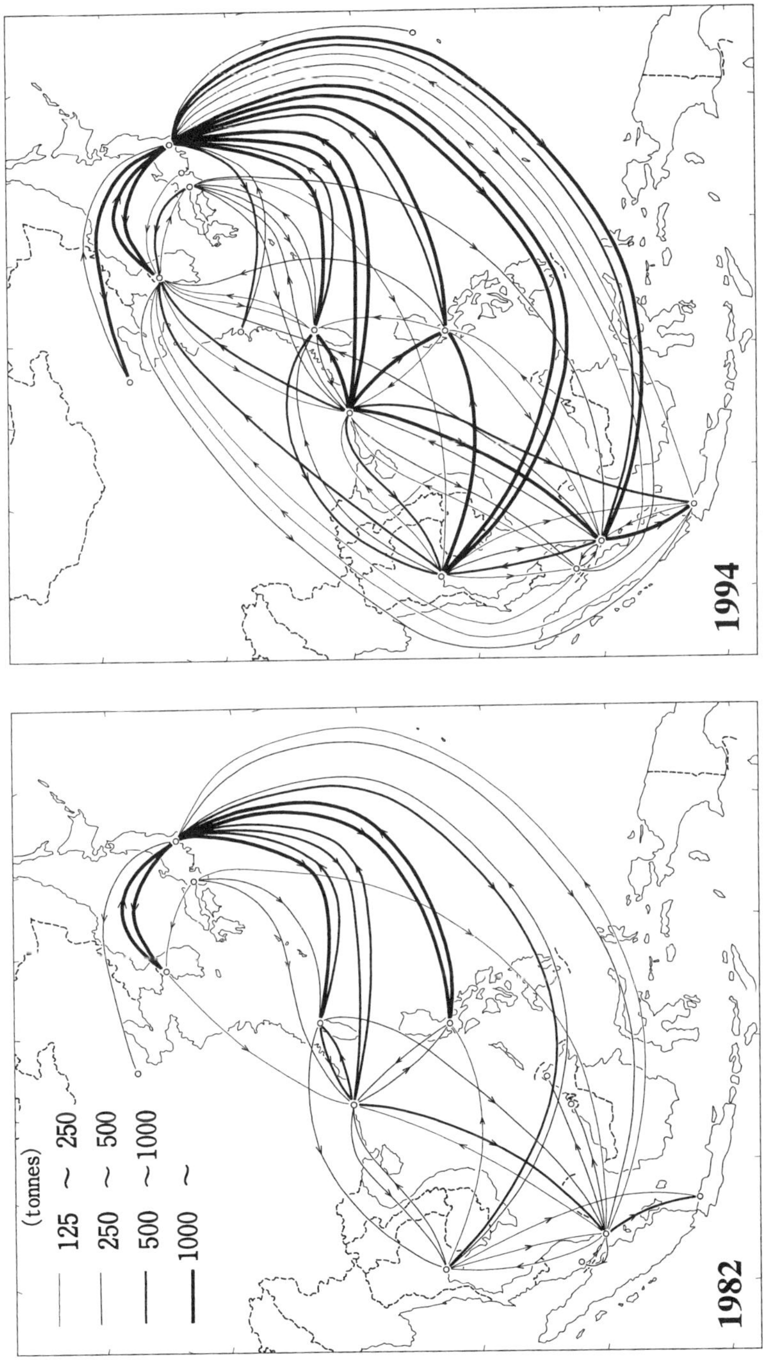

Figure 3.5. Interurban linkages in terms of air mails
Source: ICAO (1995).

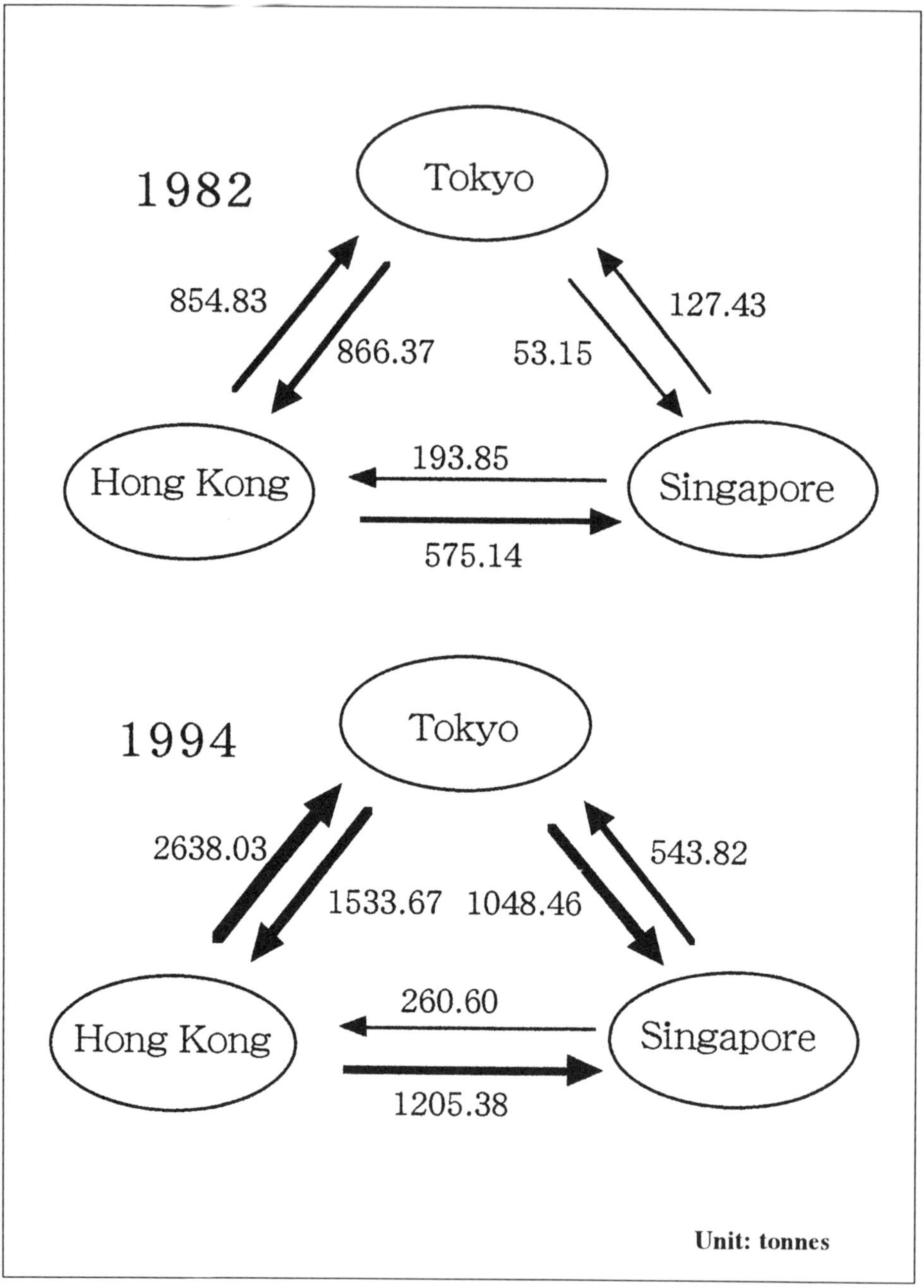

Figure 3.6. Air mail flow between Hong Kong, Tokyo and Singapore
Source: ICAO (1983, 1995).

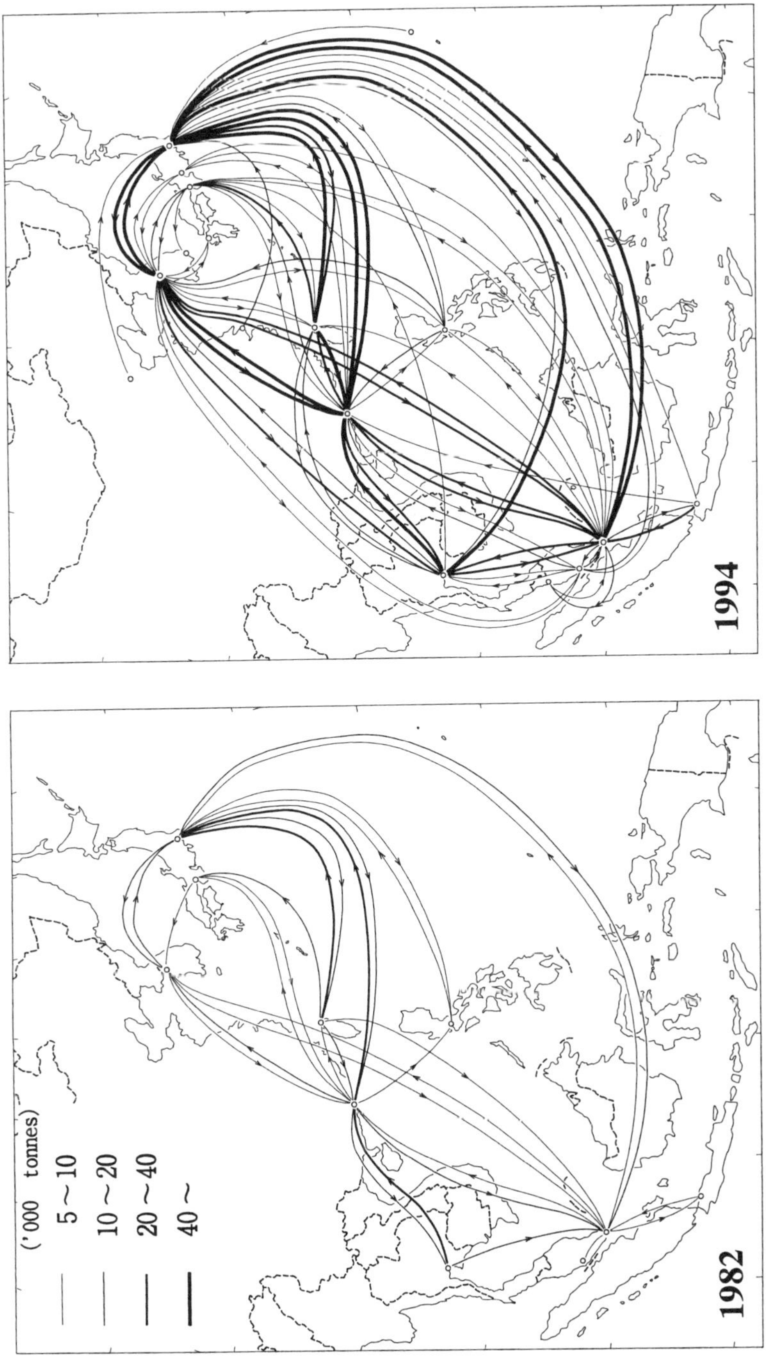

Figure 3.7. **Interurban linkages in terms of air freight**
Source: ICAO (1995).

and Singapore as well as with the Japanese cities of Osaka, Nagoya and Kitakyushu. It is developing as an international hub with functions rivaling those of Tokyo.

Singapore has dual-directional linkages with Tokyo, Seoul and Hong Kong, and has other linkages with the nearby cities of Jakarta, Kuala Lumpur and Penang. Beijing, on the other hand, only has linkages with Tokyo, and its economic exchanges with other East Asian cities are very weak.

Figure 3.8 is a graphic representation of air passenger flow. Passenger flow in the O-D air passenger chart is roughly equal in the two directions (origin and destination) and is depicted by a triangular matrix. Therefore, there are no vectors as there are in Figures 3.8 and 3.9. The number of passengers embarking and disembarking at Hong Kong is the third highest in the world, behind only London and Frankfurt. Among the top 15 airports in the world in this category, 6 (Hong Kong, Tokyo, Singapore, Bangkok, Taipei, and Seoul) are in East Asia.

In terms of air passenger flow, we can see in Table 3.7 that East Asian cities have become important nodes in the global urban system. The highest intraregion passenger flow is between Hong Kong and Taipei, which is second in the world only behind the London-Paris route.

Of the top 25 air routes in either destinations or arrivals in the world, 15 are in East Asia. If both destinations and origins are considered, there are 12 (see Table 3.8). Using the graph theory method developed by Nystuen and Dacey (1961) to determine the world nodal system, we can see that in both 1983 and 1991, the world system was comprised of the three subsystems of Europe, North America, and East Asia (see Figure 3.9). However, the following must be pointed out in regard to this last subsystem. That is, in both years, there was no change in the nodes of the European (London) and North American (New York) subsystems, but with the emergence of Hong Kong as an independent node, there has arisen a three-pronged nodal system in East Asia consisting of Tokyo, Hong Kong, and Singapore.

Looking collectively at the flow systems of air mail, air freight, and air passengers, we can see a hierarchical urban system developing within the East Asian international urban system (see Figure 3.10). The difference between 1982 and 1994 is that Hong Kong and Singapore had been subordinate to Tokyo in the earlier year, but by 1994, they had reached the same level as Tokyo at the top (fourth stage) of the hierarchy. Other than that, however, there were no other changes in the hierarchical relationship.

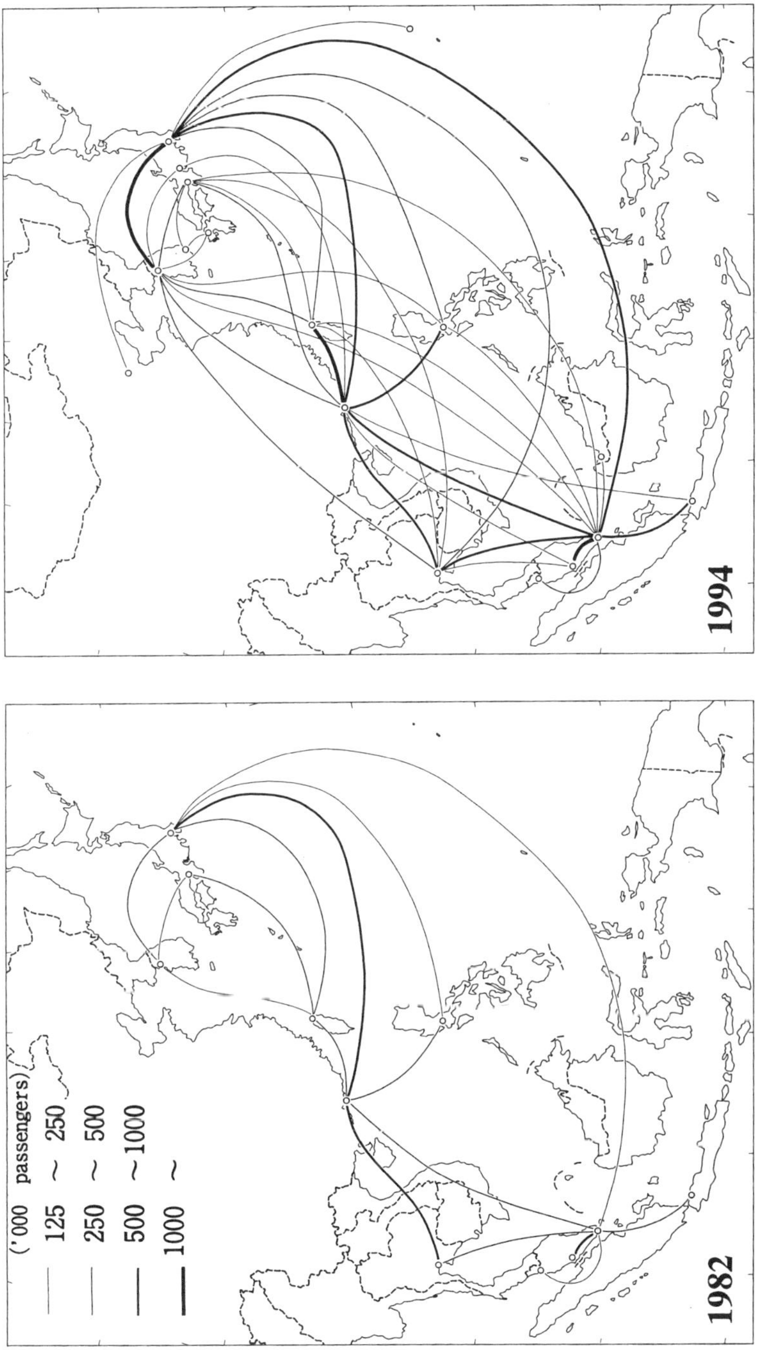

Figure 3.8. Interurban linkages in terms of air passengers
Source: ICAO (1995).

Table 3.7. Cities with highest international passenger traffic (1993)

Rank	City	Number in 000's	Number / Population
1	London	59,504	8.43
2	Frankfurt	25,119	7.27
3	*Hong Kong	24,421	4.55
4	Paris	33,473	3.59
5	Amsterdam	20,658	19.62
6	*Tokyo	18,947	0.76
7	*Singapore	18,796	6.95
8	New York	14,821	0.92
9	*Bangkok	12,789	2.19
10	Miami	12,373	6.43
11	Zurich	12,255	14.84
12	Los Angeles	11,945	1.04
13	*Taipei	11,154	3.85
14	Manchester	10,791	4.74
15	*Seoul	10,599	1.00

Note: * cities in Asia.

Source: ICAO (1994), United Nations (1995).

Table 3.8. City-pair with highest traffic in international passenger traffic (year-ending 30/06/1994)

	City-Pair			Passengers – On-Flight Origin and Destination		
	A		B	A→B	A←B	A⇔B
1	London	—	Paris	1,719,441	1,821,978	3,541,419
2	*Hong Kong	—	Taipei	1,266,937	1,213,213	2,480,150
3	London	—	New York	1,224,842	1,201,078	2,425,920
4	*Kuala Lumpur	—	Singapore	1,174,531	1,112,337	2,286,868
5	*Seoul	—	Tokyo	1,086,964	1,076,360	2,163,324
6	Honolulu	—	Tokyo	1,009,508	999,961	2,009,469
7	*Hong Kong	—	Tokyo	938,172	922,637	1,860,809
8	*Bangkok	—	Hong Kong	885,947	857,741	1,743,688
9	Amsterdam	—	London	865,086	860,968	1,726,054
10	*Jakarta	—	Singapore	777,619	809,710	1,587,329
11	Dublin	—	London	773,620	787,006	1,560,626
12	*Bangkok	—	Singapore	703,798	741,421	1,445,219
13	Frankfurt	—	London	671,325	679,578	1,350,903
14	*Hong Kong	—	Singapore	658,834	667,185	1,326,019
15	*Hong Kong	—	Manila	579,072	581,947	1,161,019
16	Brussels	—	London	581,141	572,370	1,153,511
17	*Singapore	—	Tokyo	555,619	534,602	1,090,221
18	New York	—	Paris	536,830	548,683	1,085,513
19	Los Angeles	—	Tokyo	505,948	512,212	1,018,160
20	London	—	Los Angeles	489,980	496,753	986,733
21	*Taipei	—	Tokyo	495,517	484,155	979,672
22	Chicago	—	Toronto	487,527	484,525	972,052
23	*Hong Kong	—	Seoul	481,798	487,578	969,376
24	London	—	Tokyo	466,415	482,550	948,965
25	New York	—	Toronto	458,259	455,721	913,980

Note: * indicates city-pairs within East Asia.

Source: ICAO (1995).

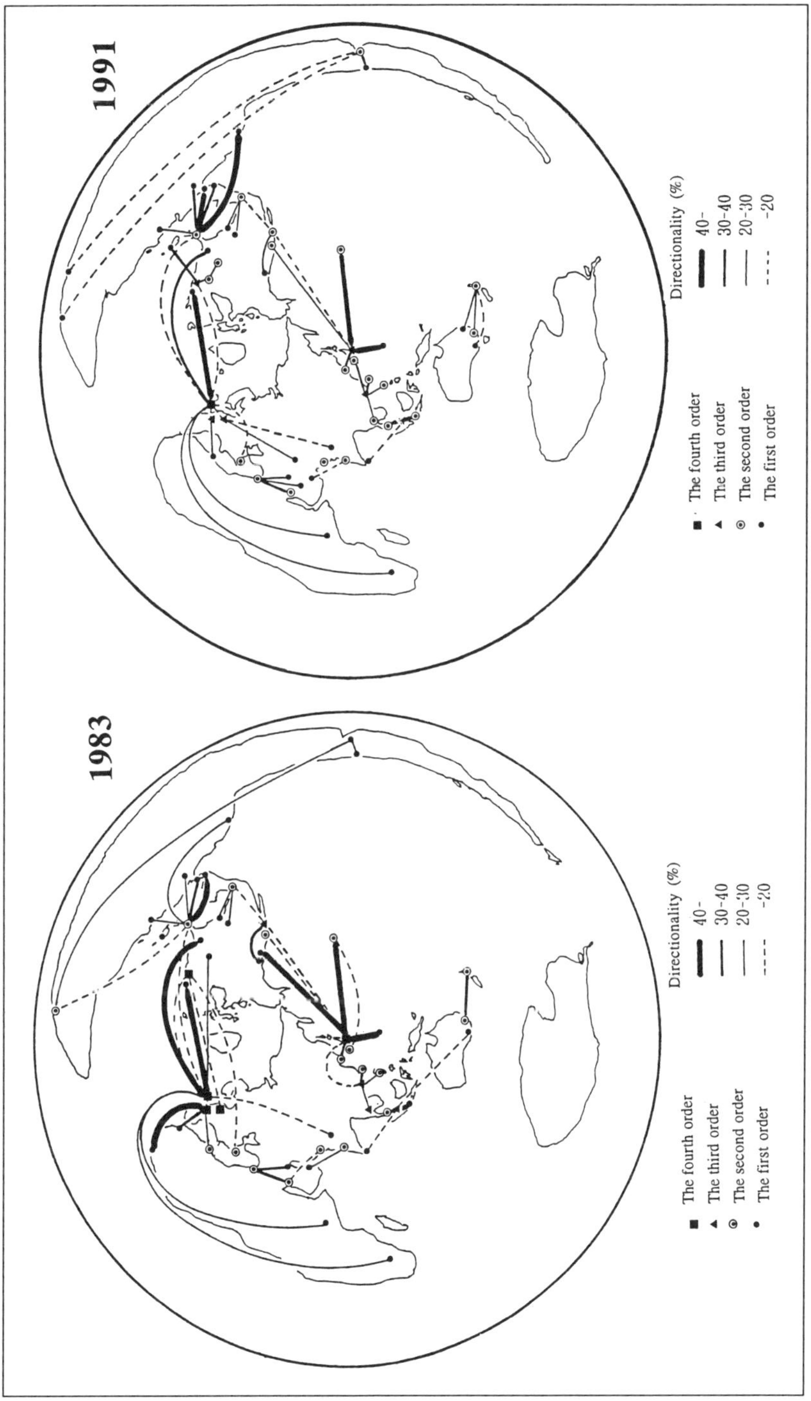

Figure 3.9. Interurban linkages in terms of air passengers flows

Note: In Europe, only London, Frankfurt, Paris and Athens are shown.

Source: Jochi (1995).

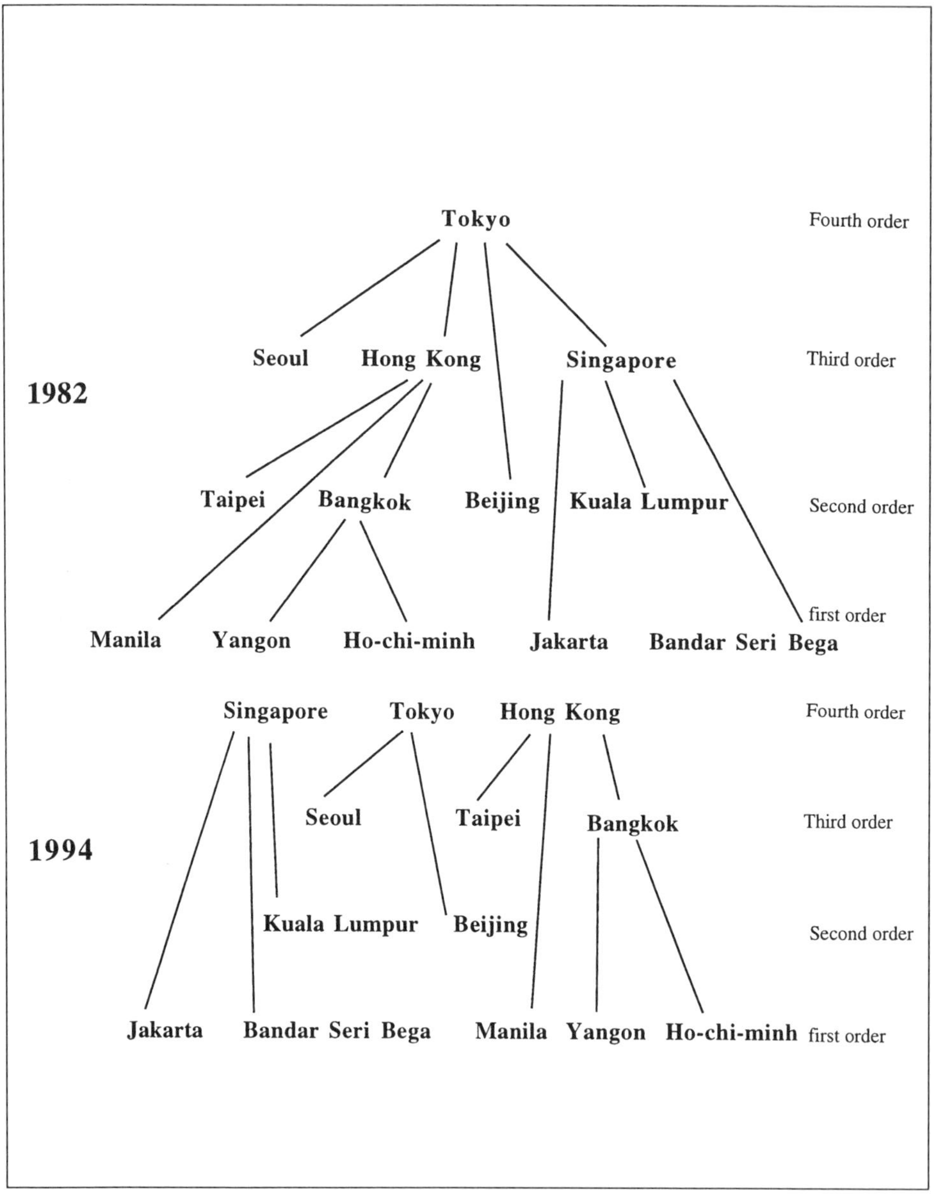

Figure 3.10. The East Asian international urban system

3.4. Driving Forces of Urban System Change

3.4.1 THE ACTIVITIES OF MULTINATIONAL CORPORATIONS

The driving force supplying the power to East Asia's mega-cities is, without a doubt, activities of foreign-owned businesses, especially multinational corporations. Until the early 1980s, these corporations had been building assembly plants in large Asian cities or surrounding areas to take advantage of the relatively cheap labor. This system was nothing more than assembling parts and sending them back to the consumer country in Europe, North America, etc., for final assembly. However, with the economic development of the latter half of the 1980s, local (intraregional) demand arose, and companies began building factories in many different locations. Companies started developing comprehensive strategies in which intra-company divisions of labor arose which were enhanced by the addition of service sector activities such as distribution and sales. As markets matured, regional coordinating headquarters were established in East Asia; with these as a focal point, companies increasingly began to perform "total" operations. For example, until the 1980s, Japanese-owned companies had been establishing regional coordination branches in Europe and North America, with Tokyo controlling all activities within the East Asian region. However, in recent years regional coordination branches have been established in Singapore and Hong Kong, with an increasing number of companies establishing a 3-polar system consisting of European, North American, and East Asian branches. With regional operational coordinating headquarters (hereafter called "operational headquarters") as their focus, companies began to construct an inter-city network of subordinate organizations such as branches and regional and local offices.

The author would like to emphasize here that the high-level management functions are in special primate cities; therefore, these mega-cities, with the concentration of these functions in their centers, have begun to acquire to ability to control economic activity in lower-ranking cities in other countries. In this way, urban functions have been transcending national boundaries and urban systems are becoming more hierarchical. The spatial organization of these cities is also occurring within their respective national urban systems. The rush to establish foreign-owned or multinational corporations in capital cities has led to regional economic imbalances. Gaps have arisen in wages and operational costs, encouraging large numbers of people to migrate to the capital cities. National governments have tried to rectify this economic polarization by investing in smaller cities. Since wages and industrial land are cheaper in these smaller cities, the development of transportation access networks would spur industrial development there. Numerous regional cities have grown in this fashion, and economic networks have formed between the capital and other cities through various media such as information transmissions, shipment of goods, and the comings and goings of workers. The interurban division of labor has been proceeding, with core managerial functions increasingly concentrating in the mega-cities which are at the apex of the world urban system.

Let's consider the example of Japanese-owned companies "setting up shop" in

Malaysia (Table 3.9). According to JETRO Kuala Lumpur (1995), there were 477 such companies in 1986, 846 in 1992, 960 in 1993, and 1014 in 1994; by March 1995, this figure had grown to 1070. However, the number of Japanese companies moving to Malaysia may have reached its peak in the last 2 or 3 years. This is due to the declining attractiveness of Malaysia, such as frequent labor shortages and rising wages, which are turning Japanese capital toward Vietnam, China and so forth.

Table 3.9. Japanese related companies in Malaysia

	Kuala Lumpur		Selangor		Johor		Penang		Other areas		Total	
	1991	1995	1991	1995	1991	1995	1991	1995	1991	1995	1991	1995
Total	251	310	226	365	61	102	77	127	114	166	729	1070
Manufacturers	19	21	156	245	57	97	63	102	72	131	367	596
Non-manufacturers	232	289	70	120	4	5	14	25	42	35	362	474
Construction & Engineering	55	65	12	25	1	2	0	4	7	7	75	103
Foreign trade & Commerce	60	58	5	8	0	0	5	7	30	25	100	98
Banking / Finance & Insurance	37	45	1	2	0	0	0	0	0	0	38	47
Services	28	51	3	14	1	1	0	4	1	1	33	71
Others	52	70	49	71	2	2	9	10	4	2	116	155

Source: JETRO Kuala Lumpur (1995).

The establishment of Japanese-owned businesses in Malaysia is becoming more spatially diverse. For example, in the 1980s, most such companies tended to concentrate in either Kuala Lumpur or Selangor, but in the 1990s they started moving to other places such as Johor and Penang. Thus, regional cities have been gaining economic power and have been strengthening their linkages with Kuala Lumpur. Malaysia is presently seeking spatially-balanced economic development in an effort to relieve the over-concentration of activities in the capital. Under this policy, the national urban system, which is currently overemphasizing Kuala Lumpur, should gradually become more balanced. The completion of a north-south expressway should ameliorate access discrepancies, and industrial development will probably become more spatially-dispersed in the future.

Almost no manufacturers are now setting up in Kuala Lumpur, but are turning toward the nearby cities of Johor and Penang, or even toward cities outside this area. In contrast, non-manufacturers overwhelmingly choose the capital Kuala Lumpur, which, together with the surrounding Selangor area, accounts for more than 80% of all such operations. Of the 47 Japanese banking, finance, and insurance companies in Malaysia as of 1995, 45 were in the capital (the other two companies were both involved with leasing). Banking, finance, and insurance companies are all in Kuala Lumpur, and most of these have offices in the central business district (CBD).

Since 1991, Japanese-owned service industries in Kuala Lumpur have been increasing dramatically as more and more Japanese are being sent there for long- or short-term business or are going there on long vacations. This has led to growth in businesses supporting the daily lives of these people, such as food stores, supermarkets, department stores, hotels, consultants, bookstores, travel agencies, and so on. Non-Malaysian families, including Japanese, have been moving to the area around central Kuala Lumpur to be close to educational facilities for their children, access to transportation, and other conveniences of daily life.

3.4.2 DOMESTIC AND INTERNATIONAL MIGRATION

Until the 1970s, there was a conspicuous out-migration of labor from East Asia to Europe, North America, Australia, and the Middle East, but in the 1980s, this movement began to turn inward, that is, within the East Asian region. The movement of people from Taiwan, Singapore, Hong Kong, etc., to Europe and North America slowed, and these Asian countries themselves became the destination for migrant laborers coming from Indonesia, the Philippines, Bangladesh, and other developing Asian countries. Malaysia became both supplier and recipient of these population migrations. During the five years from 1986-91, 299,490 people entered Malaysia legally. Among these immigrants were 162,554 Indonesians, 44,874 Filipinos, 22,759 Thais, and 14,868 Singaporeans.

International migration patterns are closely related to economic development. The greater a country's economic development is, the greater its population influx will be, and the less developed or more stagnant a country's economy is, the less dramatic this influx will be (see Table 3.10).

Table 3.10. International migration in East Asia (1990)

	Number of migrants [1]	% of total pop.	% of females [2]
China	345,518	0.0	46.4
Dem. People's Rep. of Korea	36,861	0.2	33.0
Japan	867,667	0.7	49.4
Republic of Korea	900,229	2.1	41.5
Bangladesh	800,093	0.7	46.4
Brunei	76,832	30.2	42.1
Cambodia	22,064	0.3	40.3
Indonesia	95,837	0.1	49.1
Malaysia	745,401	4.2	50.8
Myanmar	100,000	0.2	46.1
Philippines	37,860	0.1	45.7
Singapore	417,808	15.5	52.6
Thailand	313,750	0.6	49.4
Vietnam	21,000	0.0	49.0

Note: 1) The estimated total number of international migrants who lived in the country on 1 January 1990.
2) Percentage of females in migration stock.

Source: United Nations (1995).

Migrants tend to concentrate in large cities, especially capital cities. For example, the average migration rate (migrants/total population) in Malaysia is 4.2%, but the rate for

Kuala Lumpur is nearly 50% higher at 6.2%.

One feature of recent international population movements is that the trend to move from minor cities and farming communities to large cities in other countries has been decreasing, being replaced by movement from a large city in one country to a large city in another. Furthermore, this movement has become two-directional. Typical of this pattern is movement between Singapore and Kuala Lumpur. In the 1980s, there was an overwhelming movement out of Kuala Lumpur to Singapore, but now, this movement is becoming more two-directional.

The same trend can be seen in movement between Singapore and Jakarta. In this case, Singapore was experiencing a chronic shortage of cheap, unskilled labor, while Kuala Lumpur and Jakarta were experiencing shortages in skilled workers, technicians, and managers who were needed to staff the foreign-owned subsidiary companies that were locating in those cities. Therefore, there was a migration of unskilled labor from Kuala Lumpur and Jakarta to Singapore, and a migration of white collar technicians and managers from Singapore to those cities.

The fundamental causes of international population movement is discrepancies in wages between both countries and cities, as shown in Table 3.11. What should be especially noted here is the great differences between wages for different occupations. In Thailand and Malaysia, which suffer chronic shortages of managers and technicians, there is at least a 1,000% differential between the salaries of managers and the initial wages of junior high school graduates, and this gap is increasing every year. Recent trend within the inter-city network of increasingly strong linkages in the East Asian region, is that white collar workers with desirable technical and managerial skills are exhibiting an increasing ability to move around the area. Even in industrial development within a country, economic gaps are created between cities, and this trend becomes conspicuous, especially among the management classes. For example, monthly salaries of managers in the Chinese city of Guangzhou average US$437, but in Beijing, they are a mere US$243 (see Table 3.11). However, there are almost no differences between the initial wages of junior high graduates in these two cities.

In the 1980s, most domestic population migration involved movement from rural areas to big cities, but by the 1990s, most of this movement was between big cities (see Figure 3.11). In addition to this, there has also been, in recent years, a growing trend of migration to border areas. This is because in border areas between two or even three countries of great economic disparity, multinational corporations have brought about an international division of labor (see Figure 3.12). Typical of this pattern is the Singapore-centered "growth triangle," where in-migrants have been moving in record numbers to P. Batam in Indonesia and Johor in Malaysia.

The reason why major cities, rather than minor cities, have become the prime source of international population migration is that most of these migrants originally moved from farming villages and smaller cities to the mega-cities where, after saving some money and getting job information, moved on to the international cities of Tokyo, Taipei, Hong

Kong, Singapore, etc., in a sort of "stepwise" movement. This trend is particularly evident among overseas Chinese, who are especially prone to move between cities with an already substantial Chinese population. This is particularly true of Hong Kong and Taipei, which have seen a large influx of blue and white collar workers of Chinese descent. High mobility is a distinguishing feature of the population movements of these overseas Chinese.

Table 3.11. Monthly salaries in different countries and cities in 1995 (Japanese manufactures)

(unit: dollars)

	Initial salary (A)	Initial salary (B)	Factory workers	Engineers	Managers
Thailand	166	197	215	456	1, 756
Malaysia	157	218	221	553	1, 562
Philippines	136	147	191	212	910
Indonesia	71	93	115	250	1, 132
Singapore	548	714	1, 247	1, 204	3, 157
Ho-Chi-Minh	N.A.	N.A.	150-250	230-320	300-450
Dalian	80	91	N.A.	N.A.	256
Beijing	84	97	N.A.	N.A.	243
Shanghai	90	113	N.A.	N.A.	326
Guangzhou	82	103	N.A.	N.A.	437

Note: (A) after junior high school. (B) after high school.
Source: JETRO (1995)

Massive population movements have caused immense changes in population structure. Figure 3.13 shows changes in the population pyramids of the major mega-cities of East Asia, where the ratio of people in their 20s and 30s is increasing. This is the result of young, unmarried laborers coming to cities in search of work. This trend is especially conspicuous in Bangkok, Jakarta, and Ho Chi Minh City. In Manila between 1978 and 1992, there was no major change, and the population pyramid in both years resembled a classic volcano shape. This suggests that compared to other large cities in the region, Manila is still in the initial stages of industrial development. Since the ratio of young children is high in Manila, its population is expected to increase rapidly in the future when there will also likely be an influx of in-migrants.

Rapid population increases have also been affecting the male-to-female population structure (Figure 3.14). For example, in cities specializing in assembly industries, there has been a notable influx of females in their teens and twenties who make up the bulk of workers at assembly plants. It goes without saying that changes in the male-to-female population structure in large cities have also affected this structure in the small cities and rural areas from which most of these workers come. Overall, there is a balance between the male and female population on a national scale, but there are noticeable differences from one city to another. In Singapore, however, there is a balance between the male and female populations because of its distinguishing geographical feature as a city-state.

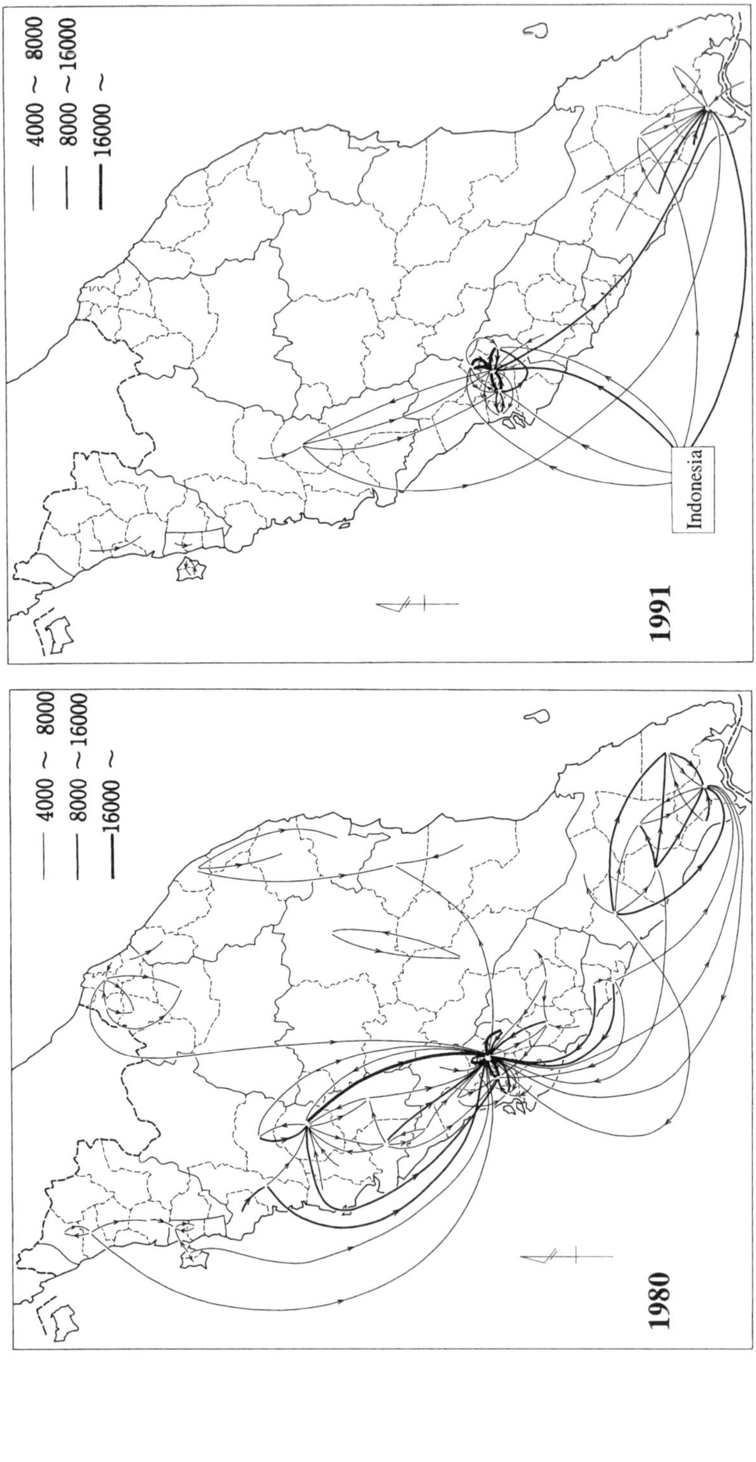

Figure 3.11. Migration flow in Malaysia
Source: Department of Statistics Malavsia (1995).

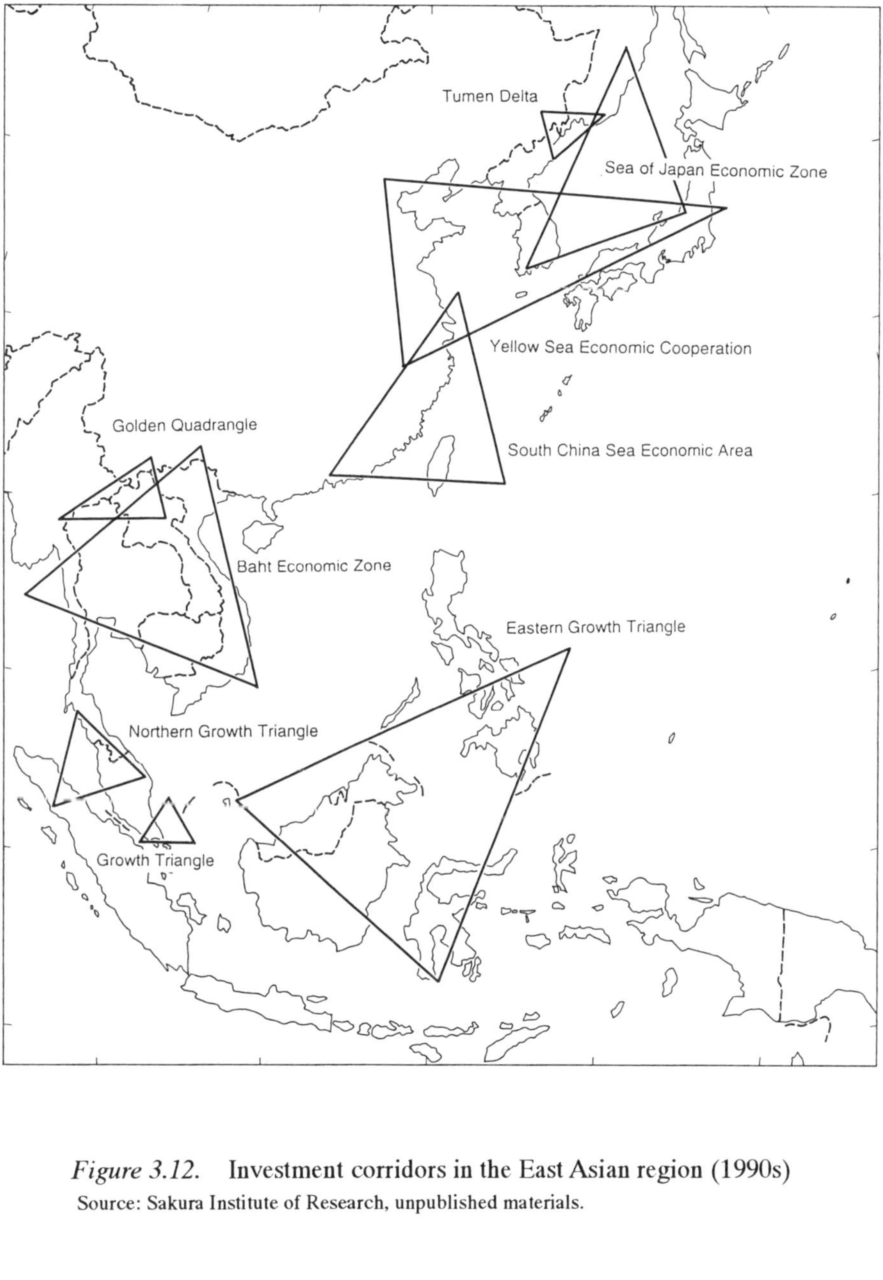

Figure 3.12. Investment corridors in the East Asian region (1990s)
Source: Sakura Institute of Research, unpublished materials.

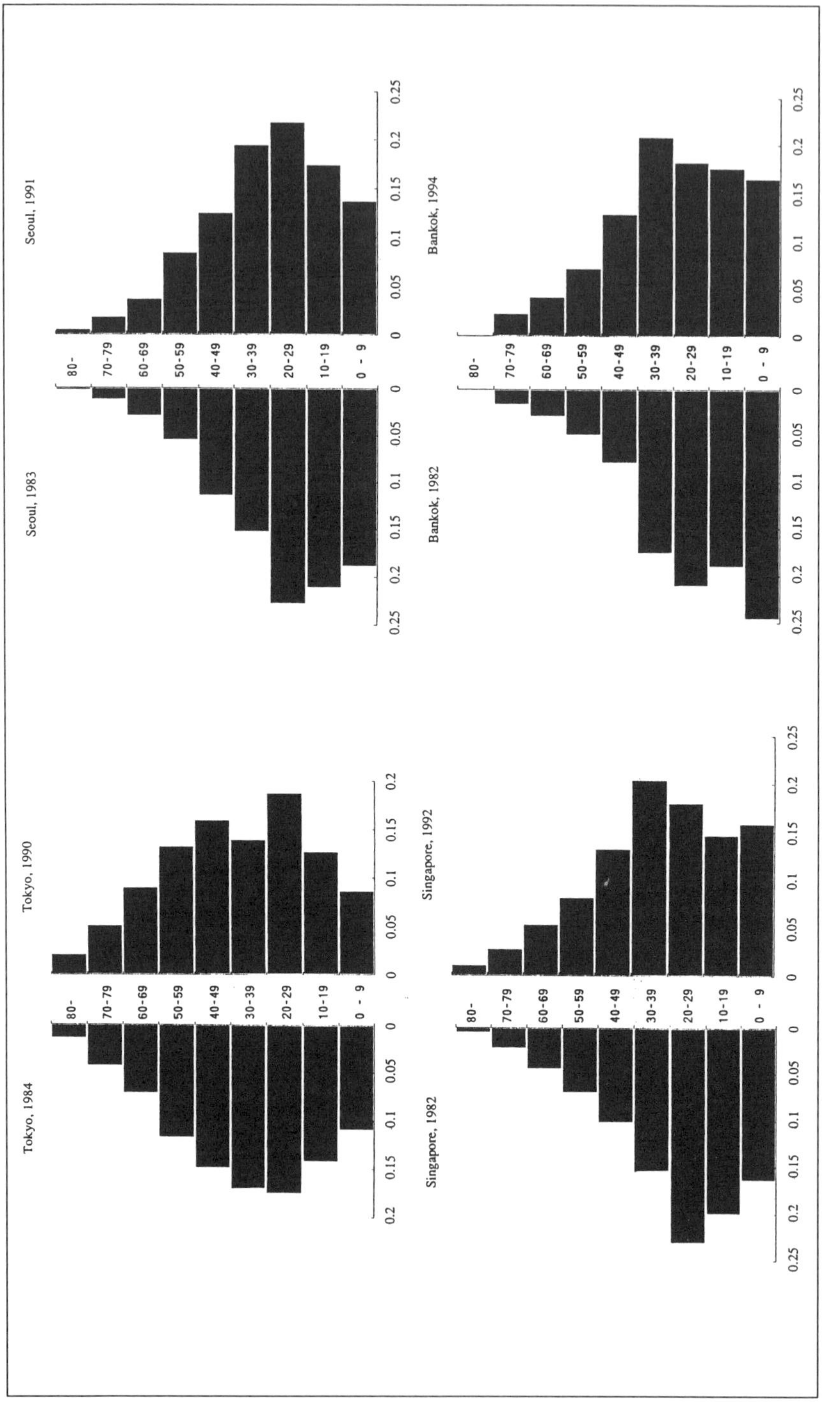

Figure 3.13. Change in the age composition of major mega-cities in East Asia
Source: TMG, 1984, 1993, and 1994.

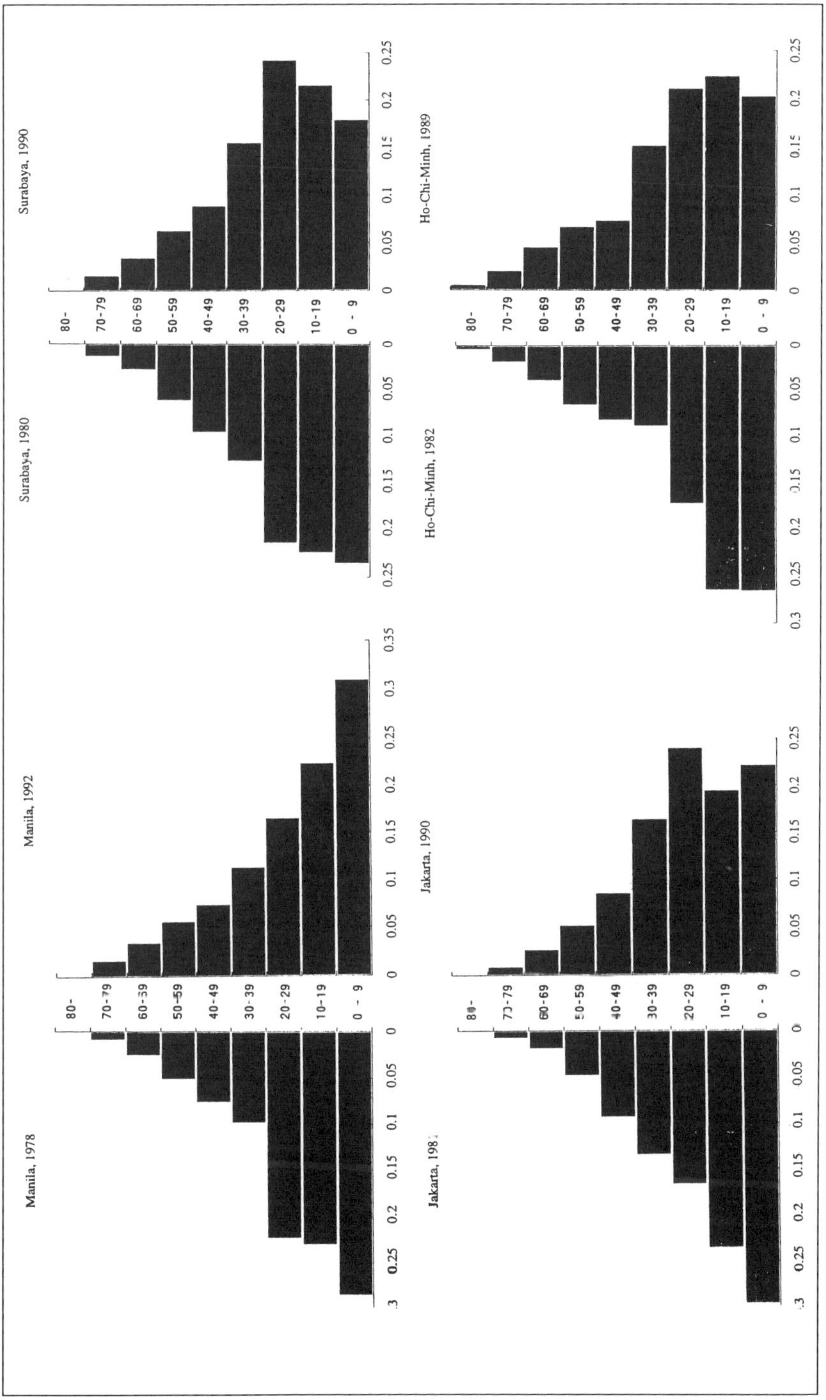

Figure 3.13. Change in the age composition of major mega-cities in East Asia *(continued)*.
Source: TMG, 1984, 1993, and 1994.

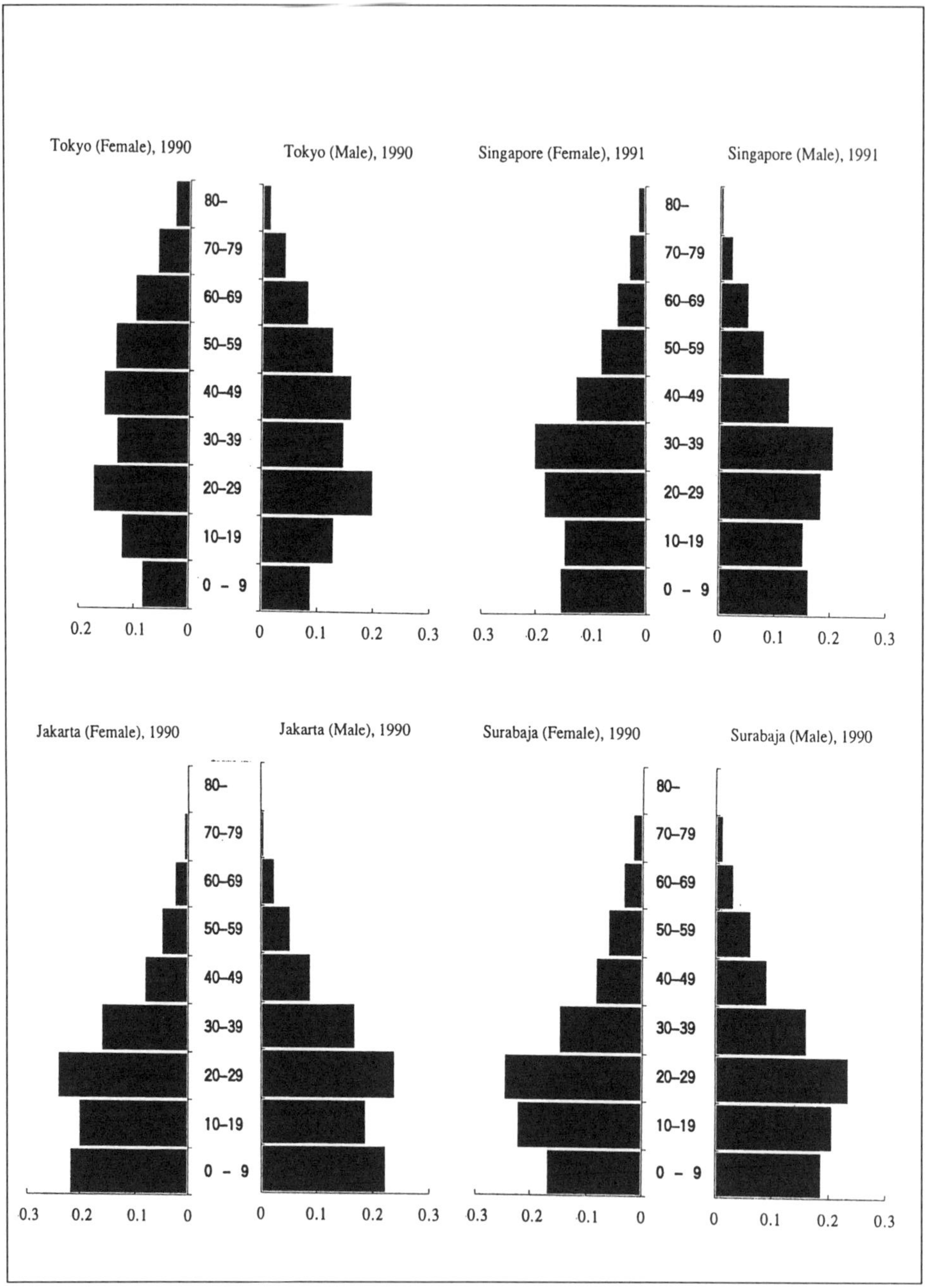

Figure 3.14. Age composition by sex in four mega-cities in East Asia

Source: TMG, 1984, 1993, and 1994.

3.5. Development Process of the East Asian Urban System

Since industrialization and the development of infrastructure such as road and information networks has increased inter-city exchanges among the cities of East Asia, the national urban system hierarchy, focused on the capital, has been completed. Furthermore, the strengthening of economic linkages between these and foreign mega-cities, which transcend national boundaries, is helping to form a global urban system which encompasses all of East Asia. In the 1980s, this world system contained only two subsystems: Europe and North America. However, by the early 1990s, East Asia had been upgraded to the rank of subsystem.

This section will use simple diagrams to explain the formation processes of world and national urban systems in East Asia. The discussion will classify Asian countries according to their level of economic development: Japan; NIES (Korea, Taiwan, Singapore, Hong Kong); ASEAN (the Philippines, Malaysia, Indonesia, Thailand); and other Asian countries (China, Vietnam, Myanmar, etc.). NIES countries will be subcategorized as either city-states (Singapore and Hong Kong) or as nations having national urban systems (Korea and Taiwan). City scale will be divided into 4 levels based on the respective concentration of international economic functions. The thickness of linkages will illustrate the flow of capital and the locational development of multinational corporations (main office-, regional office-, and local office relationship). Depending on their power, they will be divided into 3 levels: high, medium, and low. Openness will be used to indicate the degree of economic linkages with cities overseas.

Figure 3.15 shows the formation process of national urban systems, while Figure 3.16 shows the same for the international urban system.

Stage I: the 1970s

In the 1970s, East Asia had not yet developed an international urban system. Japanese companies began light industrial production, such as textiles and clothing, in areas around the capitals of NIES countries, but Tokyo had weak linkages to these capitals.

At the same time, there was very little openness in ASEAN and other Asian countries. In these countries, it was not the inland mega-cities but the coastal trading cities that had most of the contacts with foreign cities. Local offices and warehouses were established in major port cities to help trading companies in their purchase of primary industrial products. The Japanese ports of Yokohama and Kobe, the destination of much of the material, acted as bases for information exchange. There were also many foreign businessmen in these two cities.

At this time, the national urban systems of East Asia, with the exception of Japan, were not fully developed. In Japan, Tokyo had strong links with the central and northeast regions of the country, while Osaka had strong economic links with the west. There was a two-apex national urban system with Tokyo and Osaka at the top.

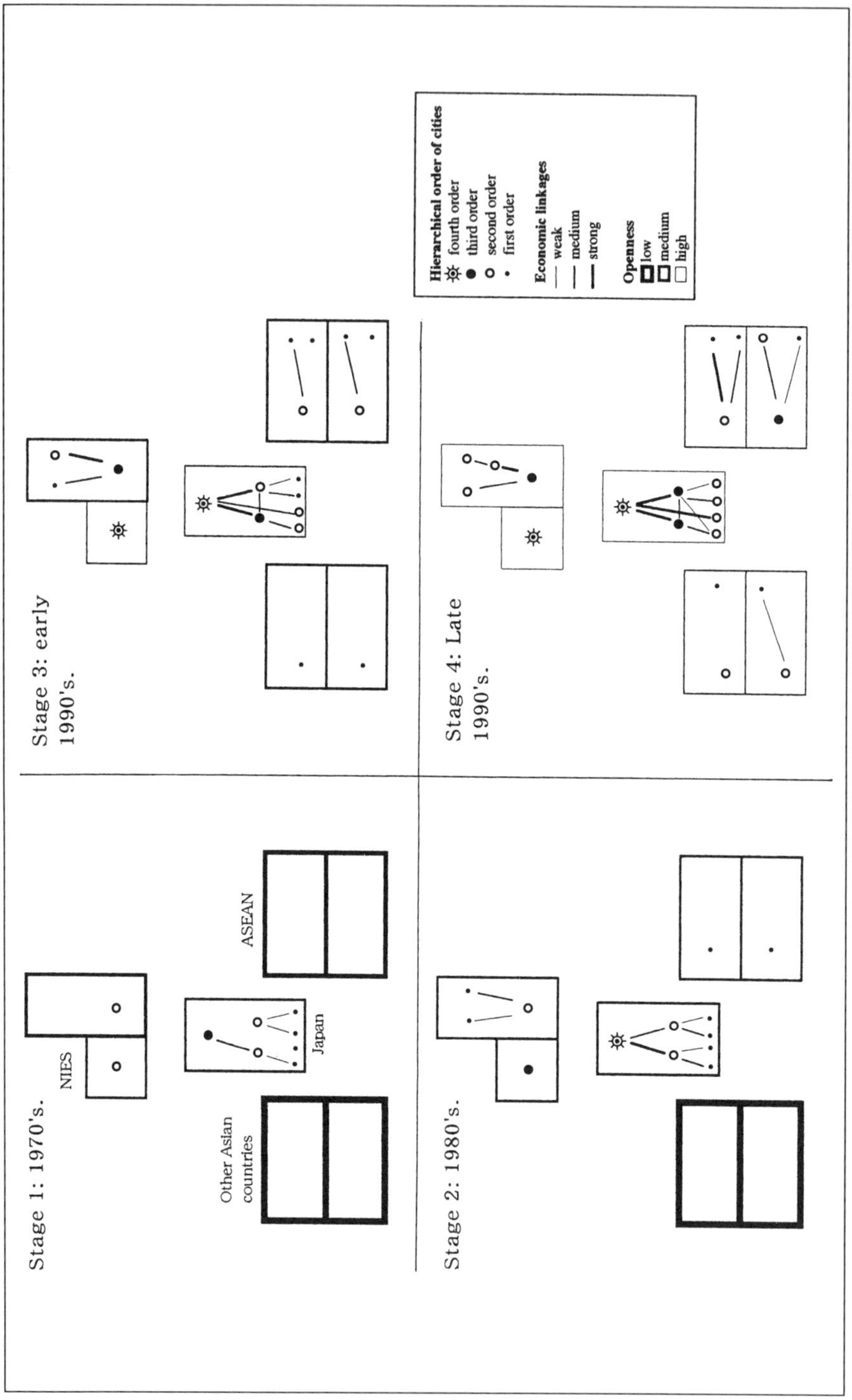

Figure 3.15. Development process of the national urban system in East Asia

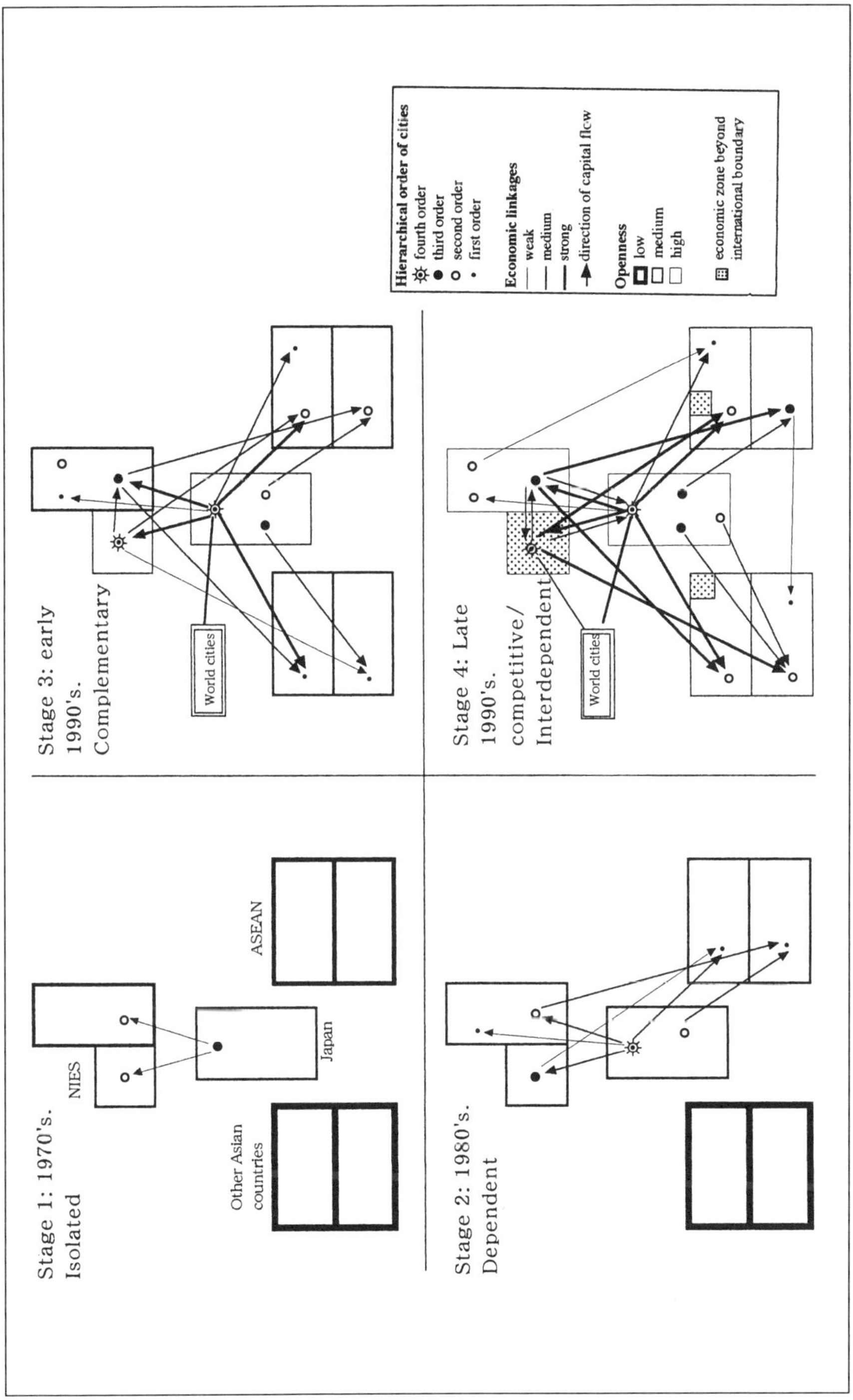

Figure 3.16. Development process of the international urban system

Stage II: the 1980s

During the 1980s, multinational corporations began to intensify their activities, which promoted capital and information flow, investment, and dispatches of businessmen on a global scale. Economic linkages were formed between large cities in the industrialized countries such as London, New York, Los Angeles, and Tokyo on the one hand, and cities, especially capitals, in NIES and ASEAN countries. By the latter half of the decade, linkages were forming between Tokyo and the second-largest cities in Taiwan and Korea. In these two countries, a hierarchical system of cities was established, and the capacity of the respective capitals to control domestic cities was enhanced. At the same time, city-states with no hinterlands (Singapore and Hong Kong) began intensive investments in ASEAN cities. Domestic population began to in-migrate to ASEAN capitals and primacy took firm root. This intensification of primacy led to increased economic disparity between the capitals and other cities. At the same time, the openness of ASEAN countries increased from Stage I due to expanded economic relations with Japan and NIES countries.

In the Japanese urban system, economic and managerial functions began to concentrate in Tokyo; as Osaka fell into relative decline, Tokyo assumed the sole position at the top of the Japanese urban hierarchy. By the latter half of the decade, Tokyo had strengthened mutual links with European, North American, and East Asian cities, and branches of multinationals, foreign-owned banks, and international hotels began to locate in Tokyo in increasing numbers. International businessmen also began to congregate there in large numbers, making Tokyo a world city in the same class with New York and London.

Stage III: the early 1990s

In the early 1990s, both national and international urban systems in East Asia were developing hierarchies. The international system in East Asia was comprised of Tokyo at the top (fourth level), followed by Singapore, Hong Kong, Taipei and Seoul at the third level. Below these were the cities of the ASEAN countries. World city Tokyo began to form financial networks with the world cities of Europe and North America such as London, New York and Frankfurt. The city-states of Singapore and Hong Kong, which did not have hinterlands, increased their investment not only in ASEAN countries, but also in other Asian countries.

In Japan, international functions were concentrating not just in Tokyo and Osaka, but also in Nagoya and Fukuoka, which rose in status as nodes in the East Asian international urban system. More frequent air transport was established between these cities and other cities of East Asia, stimulating the movement of people and goods.

During this time, the so-called "bubble economy" in Tokyo burst, dampening the city's high rate of growth and allowing for Singapore and Hong Kong to catch up. At the same time, Tokyo solidified its control of the Japanese national urban system; that is, the former pyramid, in which low-ranking cities were subordinate to mid-ranking cities, which were in turn subordinate to Tokyo, had now changed so that Tokyo was now directly controlling all subordinate cities in Japan.

As various types of industries, such as trading, real estate, accounting, insurance, leasing, banking, and securities were being established to support the economic activities of foreign-owned companies in the ASEAN countries, businesses to support the lives of foreigners and their families, such as restaurants, food stores, travel companies, international home delivery services, and educational organizations for children were also established. This phenomenon was quite conspicuous in Kuala Lumpur, Bangkok, Jakarta, and Manila. During this time as well, investment from Japan, Singapore, and Hong Kong was beginning to show its effects in the economic growth of cities in other Asian countries.

Stage IV: the late 1990s (present and future)
In Japan and all NIES and ASEAN countries, openness is on the rise, with not only capitals, but also the second- and third-largest cities of these countries establishing linkages with major cities overseas. With more and more East Asian cities becoming nodes in the world urban system, restrictions imposed on economic activity by national boundaries have weakened. Singapore and Hong Kong, which have emerged as world business centers but have no hinterlands, are relying on neighboring countries for labor-intensive production functions, and have been creating economic spheres of influence in their respective national boundary regions.

Tokyo, Singapore, Hong Kong, Taipei and Seoul, all of which have a high concentration of high-level core managerial functions, now have two-way economic interrelationships. For example, while Hong Kong has established branches of financial institutions and large corporations in Singapore, Singapore has done the same in Hong Kong, resulting in a two-way flow of capital. Tokyo, Singapore and Hong Kong also are competing with one another both to attract international functions and to bring as many East Asian cities as possible under their respective economic control. In all three cities, movement away from secondary (manufacturing) to tertiary (service) industries is rapidly taking place.

As the international urban system of East Asia reaches maturity, not only economic but also cultural and social linkages will become stronger. More sister city relationships will be initiated and the exchanges of goodwill ambassadors and students will increase.

3.6. Conclusion

The features and formation processes of the international urban system in East Asia can be summarized as follows:

Industrialization has led to the growth of cities, and numerous mega-cities have appeared in East Asia. Manufacturing and population, as well as a variety of service industries, are concentrating in these mega-cities, which have incorporated smaller domestic cities into their economic spheres. These mega-cities, especially the large metropolitan areas which have seen a concentration of economic and core managerial functions, are strengthening their economic ties with large cities in other countries, contributing to the spatial organization of the world urban system in East Asia. As interurban linkages become stronger, the economic growth of these cities is being determined by the strength of their economic ties with cities overseas.

In the 1980s, a hierarchical urban system was forming in East Asia in which the capitals of all East Asian countries were coordinated with their respective national urban systems but had weak linkages with world cities. Presently, however, even the second- and third-largest cities of these countries are establishing ties with foreign cities, resulting in a horizontal network-type urban system. The flow of information and capital and the dissemination of technology is going from world cities to the capitals of all countries, and the second- and third-largest cities of countries are moving away from one-directional hierarchical flow to two-way flow. Furthermore, the "friction" imposed on economic activities by national boundaries is decreasing as cities in different countries are increasing their mutual dependencies.

In the future, increasing interurban dependencies and competition will help East Asian mega-cities, where core finance and managerial functions are concentrating, to solidify their control over the international urban system of East Asia. In the 1980s, only Tokyo exercised dominance in this area, but now it is sharing those duties with Singapore and Hong Kong. As we approach the 21st century, the author believes that Singapore will surpass the other two due to its excellent geographical situation. Singapore is surrounded by countries such as Indonesia, Vietnam and the Philippines which have abundant labor forces, developing markets and high growth potential. Furthermore, accessibility, which is an important factor to facilitate the activities of international businessmen, will favor Singapore's situation in the middle of Asia even more should India begin its expected rapid growth, since the city-state will become the most convenient place for business activities. Singapore even has convenient access to Europe and North America. Compared to this, in terms of accessibility, Tokyo, which is situated at the far end of the Far East, will become much less attractive.

References

Department of Statistics Malaysia (1995): *Population and Housing Census of Malaysia 1991, General Report of the Population Census Vol. 2*, Department of Statistics Malaysia, Kuala Lumpur.

ICAO (1983): *On-Flight Origin and Destination*, International Civil Aviation Organization, Montreal.

ICAO (1994): *Civil Aviation Statistics of the World 1994*, International Civil Aviation Organization, Montreal.

ICAO (1995): *On-Flight Origin and Destination*, International Civil Aviation Organization, Montreal.

JETRO (1995): *Survey on the Japanese Related Manufactures Activity in ASEAN*, Japan External Trade Organization, Tokyo.

JETRO Kuala Lumpur (1995): *Japanese Related Companies in Malaysia*, JETRO Kuala Lumpur Center, Kuala Lumpur.

Jochi, T. (1995): Global urban system and its evolution process — an approach from air flows, unpublished graduation thesis, University of Tsukuba. (J)

Nystuen, J.D. and Dacey, M.F. (1961): A graph theory interpretation of nodal regions, *Papers Proc. Reg. Sci. Ass.* **7**, 29-42.

TeleGeography Inc. (1995): *TeleGeography — Global Telecommunications Traffic Statistics and*

Commentary.

TMG (1984): *Statistics of World Large Cities*, Tokyo Metropolitan Government, Tokyo.

TMG (1993): *Statistics of World Large Cities*, Tokyo Metropolitan Government, Tokyo.

TMG (1994): *Statistics of World Large Cities*, Tokyo Metropolitan Government, Tokyo.

United Nations (1995): *World Urbanization Prospects: the 1994 Revision*, United Nations, New York.

Watanabe, T. (1995): *Scope for the New Century Asia*, Chikuma Shobo, Tokyo. (J)

World Bank (1993): *The East Asian Miracle, Economic Growth and Public Policy*, Oxford University Press, New York.

(J = in Japanese)

PART III

NATIONAL URBAN SYSTEM

CHAPTER 4

REGIONAL STRUCTURE OF THE INFORMATION FLOW

4.1. Introduction

Through the exchange of information, cities are forming interconnections with other cities. Cities which transmit a large amount of information to one another also have strong social and economic ties. In this day and age, when spatial interdependencies are intensifying, it would not be an exaggeration to say that information and communications are the keys to urban and regional development. Information is truly the driving force of a society and its economy.

How then does information flow in Japan? Which cities and regions are transmitting and receiving large amounts of information? What types of spatial regimentation and order exist between transmitting areas and receiving areas? Using the flow of telephone communications[1] that individuals, local governments and companies widely use during the course of their daily lives and business as an index, this study will use the flow of information to clarify the regional (i.e., nodal and hierarchical) structure of Japan.

4.2. Data and Methods

Since the deregulation of electronic communications which began on April 1, 1985, three new long-distance domestic carriers (the "Shin-Denden) have appeared in the Japanese telecommunications industry — Daini Denden (DDI), Japan Telecom (JT) and Nihon Kosoku Tsushin (TWJ). Including NTT (Nihon Telephone and Telegraph), there are 4 long-distance telecommunications carriers operating nationwide. Transmission flow data collected from each of these companies is released on an annual basis by the Association of Electronic Communications Companies. The data collection unit is the "message area" (MA) established by NTT, of which 567 exist. Using these data, this study will perform analyses using a 567 row and 567 column origin-destination (O-D) chart for telephone calls made in FY 1992. Each of the cities or regions to be introduced later constitutes one MA in this chapter.

The methodology for the analysis was as follows: First, the intraregional directionality (intraregional call volume / total calls received) was derived in order to understand whether short-distance or long-distance communications were dominant in each city or region. In the O-D chart (matrix), the column totals represent the volume of in-coming calls in core cities or regions, while the diagonal element is the volume of telephone calls within a region. Therefore, the intraregional directionality can be derived by dividing the diagonal element by the column totals. The intraregional directionality can be considered to define the scale, location, and socio-economic characteristics of a city or region.

Next, the most dominant telecommunications flows were investigated to find spatial linkage relationships. In other words, in each city or region, the origin of the maximum volume of in-coming calls (primary telephone communications flow) was sought, and links (vectors having direction and thickness) were drawn from points of origin to points of destination (core regions). At that time, links were divided into 4 groups (1-2%, 2-5%, 5-10%, >10%) to gain an understanding of the strength of these linkages. For example, a link of 10% indicates that the destination receives 10% of all its in-coming calls (including intra-regional calls) from the originating point with which it has a linkage.[2] Then, secondary flow of in-coming calls was subjected to the same procedure to derive spatial flow patterns. It should be noted that in this study, the spatial realm of a "city" is based on the MAs of NTT mentioned earlier, and does not always coincide exactly with administrative boundaries.

4.3. Spatial Flow Patterns

4.3.1. RATIO OF CALLS WITHIN A REGION

There are 58.77 million telephones (18.47 million for commercial use, 40.3 million for personal / home use) in Japan. Since there is an average of 77,235,480,000 telephone calls made each year, each telephone was considered to have made 1,314 calls during the course of a year (FY 1992), and the ratio of intraregional calls reached 65.23%. This means that two-thirds of all telephone calls were actually completed within the same region. Of the 567 MAs, only one (Ishikari in Hokkaido) showed more calls destined for outside its region than within its region (there were 9.61 million calls made within the Ishikari region, but there were 16.22 million calls which originated from Sapporo). Major metropolitan areas showed a higher proportion of long-distance calls than smaller cities, and also tended to direct these calls to more dispersed destinations.

Figure 4.1 shows the regional distribution of the ratio of intraregional calls. In the Tokaido Megalopolis, which has a high density of urban systems, interregional calls were dominant because social and economic interaction among these cities is quite robust. At the same time, outside of this megalopolis, especially in the fringes of the country where the main industries are primary and secondary, the ratio of intraregional calls was also generally high. This is probably because smaller cities and rural areas make relatively few long-distance calls that are related to business operations. In nearly all areas of Okinawa Prefecture, the ratio of intraregional calls exceeded 80%. Okinawa is a group of islands located on the fringe of Japan and has an economic structure which is heavy with primary industries, a situation which is thought to be contributing greatly to the social and economic isolation of the prefecture.

The peripheral areas of cities are characterized by a low ratio of intraregional calls. In metropolitan suburbs there are strong social and economic linkages with the major

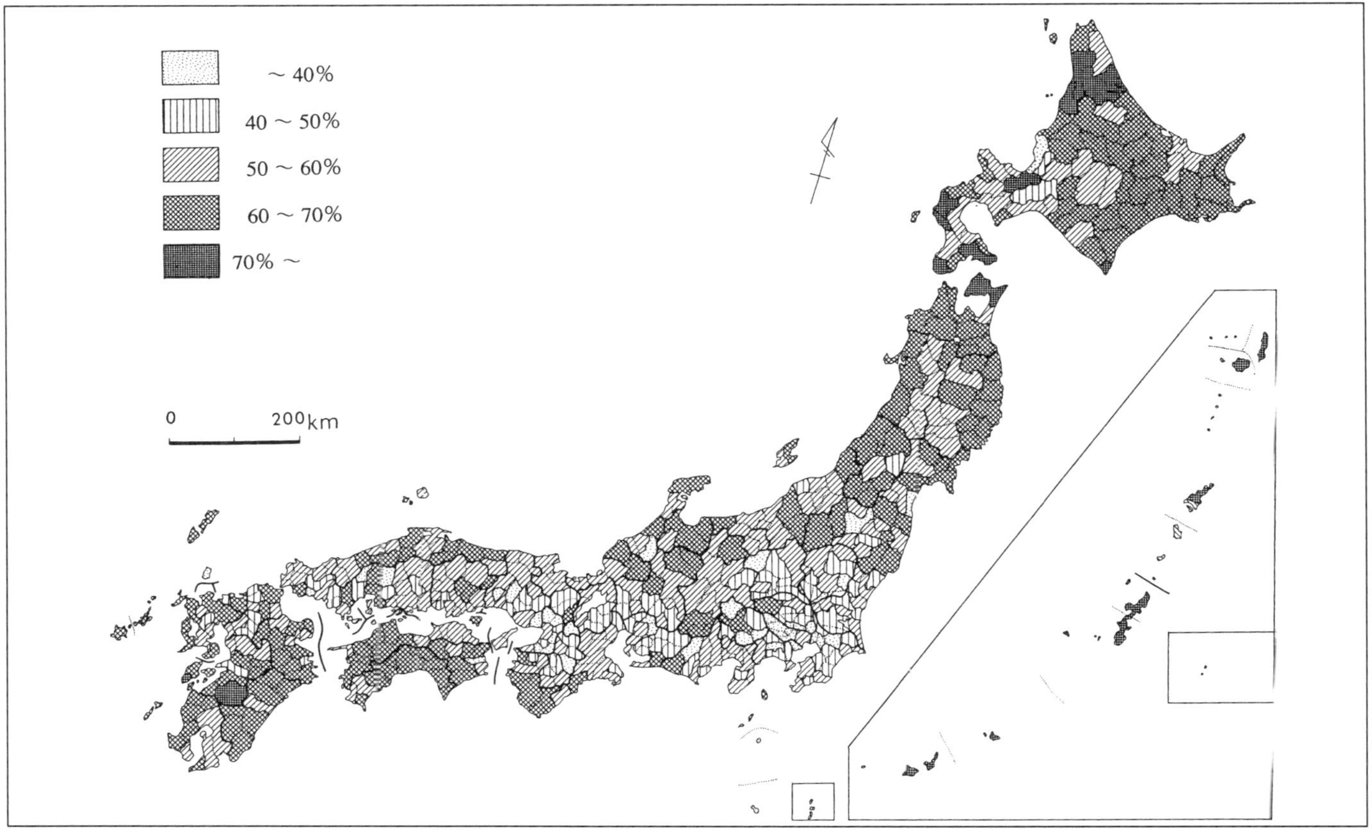

Figure 4.1. Intraregional distribution of directionality (1992)

Note: Intraregional directionality refers to the volume of telephone messages within a region divided by the total volume of telephone messages.

cities, so the ratio of intraregional calls is especially low, but this has an inverse relationship[3] with the ratio of calls directed at major cities. In the areas adjoining Tokyo, the ratio of intra-regional calls is 50% or less in Ichikawa (44.1%), Soka (48.7%), Kawaguchi (45.0%), Musashino-Mitaka (37.1%), Kawasaki (45.0%). In the areas neighboring Osaka, MAs with ratios at the 40%-50% range included Hachio (45.1%), Ibaraki (43.3%), Ikeda (43.9%), and Nishinomiya (47.5%). In addition, in both Tokyo and Osaka there is 50-60% directionality in the areas surrounding the adjacent suburbs. Thus in both metropolitan areas the ratio of intraregional calls extending the central city to suburban areas is gradually increasing, forming a concentric structure. This concentric structure is much more defined in Tokyo than in Osaka, and the radii in the former are also larger.

The ratio of intra-regional calls within Tokyo and Osaka proper stood at 69.3% and 65.3%, respectively, much higher than in the neighboring or suburban areas. The reason for this might lie in the fact that in Tokyo and Osaka, where office and service functions are concentrated, short-distance business calls within the respective urban regions are dominant. However, even if these ratios of calls is high, they are only slightly higher than the national average of 65.23%, but they are much higher than the same ratios in neighboring or suburban areas. This reason why the 70% level was not attained is that both Tokyo and Osaka make frequent calls to areas throughout Japan and function as the country's socio-economic core regions, meaning that an equivalent value is reached not only in intraregional calls, but also in interregional calls.

In FY 1992, the annual number of calls made per telephone in Tokyo and Osaka stood at 1,630 and 1,702, respectively, much higher than the national average of 1,314. Both cities truly function as bases for the transmission and reception of information.

The concentric patterns based on the ratio of intraregional calls are small in scale, but they can also be seen smaller metropolitan areas such as Sapporo, Hakodate, Morioka, Sendai, Niigata, Fukuoka, and Kagoshima. The ratios of intraregional calls for Sapporo (80.3%), Hakodate (81.6%), Morioka (70.8%), Sendai(72.0%), Niigata (69.8%), Fukuoka (73.2%), and Kagoshima (75.8%) are much higher than those for Tokyo and Osaka. The reason for this is that, unlike Tokyo and Osaka, destinations for interregional calls are not so dispersed, that is, the majority of calls are limited to the local area.

4.3.2. GREATEST FLOW OF TELEPHONE COMMUNICATIONS

Figure 4.2 shows the spatial linkage relationships for the greatest flow of telephone communications. The arrows in the Figure indicate links to core cities (destinations) from the origin of greatest volume of in-coming calls in each city. Most of the call ratios exceed 5%, and links are generally short. There are also faint long-distance links which exist with major cities as origins, but the ratio of calls is low. As we can clearly see in this Figure, origins are concentrated in a small number of urban systems, but the destinations surround the origins and they have a dispersed distribution. In this way, the flow of telephone communications shows a distinct nodal structure, forming "message

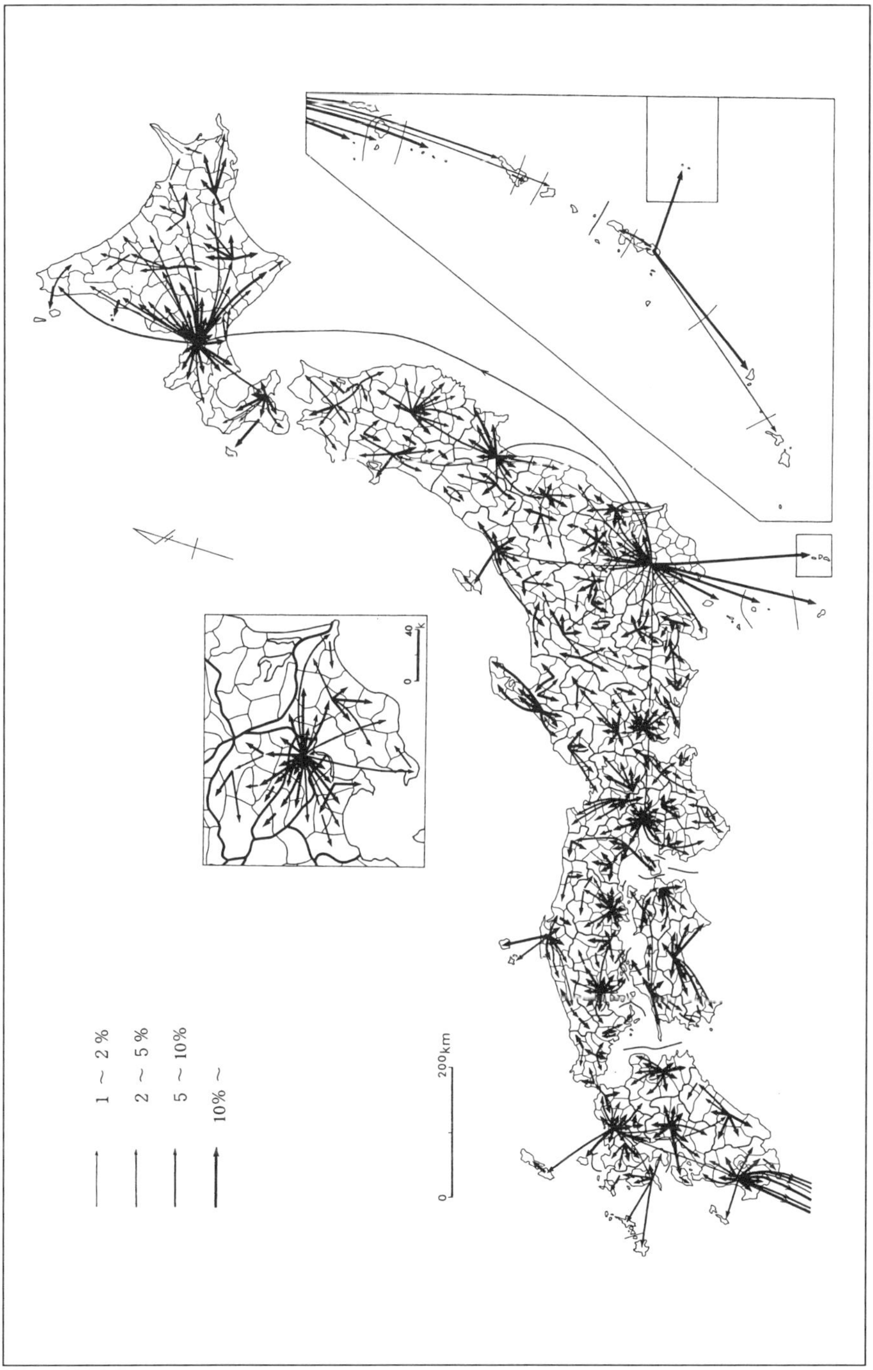

Figure 4.2. Spatial flow patterns in terms of primary (maximum) telephone message volume (1992)

areas," or aggregates of destinations of links. In this chapter, these message areas are defined as the aggregates of destinations receiving primary message flow, while the origins are referred to as the cores of these message areas. We can find 6 cores that have extensive message areas that transcend prefectural boundaries: Tokyo, Osaka, Fukuoka, Nagoya, Sapporo, and Sendai.

Tokyo forms the most extensive message area. With a calling ratio of at least 5%, it forms linkages with the other cities of the Kanto region, and extends its linkages with regional core cities such as Osaka, Sapporo, Sendai, Niigata, Nagano, and Kofu.

The city which is the core of the second largest message area is Osaka, which is only about half the scale as that of Tokyo. The nearby cities of Kobe and Kyoto form their own message areas, sandwiching in the Osaka region. Unlike Tokyo, Osaka does not have the kind of centripetal force to bring in regional core cities in "leapfrog" fashion.

Fukuoka serves as the core of Japan's third largest calling region. Fukuoka is extending its linkages to 5 of the 6 other prefectural capitals of Kyushu (the exception is Oita City). Since there is no other city in Kyushu that can stand head and shoulders with Fukuoka, Fukuoka is playing the leading role as the base for information transmission and reception in the entire Kyushu region.

In terms of scale, the calling region of Nagoya, one of the so-called "Big Three" metropolitan areas of Japan, has to settle for fourth behind to Fukuoka. The volume of telephone communications originating from Nagoya in only about 2/5 that of Osaka, and a mere 1/7 that of Tokyo (see Table 4.1). Seen only in terms of message flow, Nagoya's position is much lower than what its population would indicate. Outside of its own Aichi Prefecture, Nagoya has extended links only with the cities of Gifu, Tajimi, and Nakatsugawa, all in Gifu Prefecture. In terms of areal extent, the Nagoya calling area is much smaller than the message areas of the core cities of Sapporo and Sendai, not to mention Morioka, Niigata, and Hiroshima.

Table 4.1. The six major calling regions in Japan (1992)

Core city	Transmissions within region (100's)	Number of cities	Number of telephones within calling region
1. Tokyo	193,312,116	45	22,224,026
2. Osaka	69,860,994	16	8,055,662
3. Fukuoka	37,593,682	16	3,500,837
4. Nagoya	28,754,443	13	3,439,119
5. Sapporo	24,240,864	32	2,390,498
6. Sendai	16,781,940	13	1,726,413

Note: 1) "Calling region" refers to the aggregates of destinations (cities and regions) receiving primary message flow originating from core cities.
2) The volume of transmissions within a calling region refers to the total number of telephone communications being sent from the core city to other places within the region.
3) The "number of cities (regions)" corresponds to the number of links.

Even though Sapporo is in only the fifth largest intraregional transmission zone, it has the second largest number of linkages after Tokyo, and serves a larger area than Tokyo. The following two factors may be considered reasons why Sapporo has formed such a large message areas. First of all, there are relatively few cities in Hokkaido, and they are not spaced close together, so there is no city that can compete with Sapporo. Secondly, Hokkaido has formed an independent socio-economic sphere, increasing its isolation, while the prefectural capital Sapporo administers an area that would cover several prefectures on the main island of Honshu.

Number 6 Sendai not only incorporates all of Miyagi Prefecture, it is also extending linkages to other core cities of the Tohoku region, namely Morioka, Akita, Yamagata, and Koriyama.

In addition to the 6 extensive message areas of these core cities, there are numerous smaller message areas whose boundaries, as we can clearly see in Figure 4.2, correspond nearly exactly to prefectural boundaries. The existence of prefectural message areas suggests that prefectural regions are spatial units for daily social and business activities. Nearly all the core cities of prefectural message areas are prefectural capitals.

In terms of O-D linkage relationships, we can classify prefectural message areas into the following 4 types:

1) Type in which prefectural capitals have extended linkages throughout nearly the entire prefecture. The dominance of the prefectural capital gives rise to a rigid nodal structure. Prefectures that fit this type are Iwate, Yamagata, Ishikawa, Fukui, Yamanashi, Shiga, Kyoto, Okayama, Kagawa, Tokushima, Kochi, Ehime, Miyazaki, Oita, Kumamoto, Kagoshima, Nagasaki, Saga, and Okinawa.

2) Type in which several sub-message areas link together to form a prefectural calling region. Prefectures that fit this type are Aomori, Akita, Fukushima, Niigata, Shizuoka, Nagano, Gifu, Wakayama, Hiroshima, and Shimane.

3) Type which is comprised of several independent message areas. This is caused by weak nodality arising from the lack of a dominant city in the prefecture. Prefectures that fit this type are Gunma, Toyama, Mie, Tottori, and Yamaguchi. Gunma contains two small, independent message areas of Maebashi; Toyama is comprised of the Toyama and Takaoka regions; and Mie is made up of the Tsu and Yokkaichi message areas. Since the nodality of Tottori City, the capital of Tottori Prefecture, is weak, some areas of the prefecture have been incorporated into the message areas of neighboring prefectures. Yamaguchi, on the other hand, is comprised of the limited message areas of Yamaguchi City, Shimonoseki, Tokuyama, and Hagi.

4) The fourth type includes prefectures which are influenced by metropolitan areas (i.e., Tokyo and Osaka). As mentioned in a previous report, the western part of Chiba Prefecture, the southern parts of Ibaraki, Tochigi, and Saitama prefectures, and nearly

all of Kanagawa Prefecture are contained within Tokyo's information sphere, while western Nara Prefecture and eastern Hyogo Prefecture fall within Osaka's information sphere. The core cities in these prefectures form only partial message areas, which include areas which are facing away from either Tokyo or Osaka.

4.3.3. SECOND GREATEST FLOW OF TELEPHONE COMMUNICATIONS

Figure 4.3 shows the spatial linkage relationship of the second greatest flow of telephone communications. Nodality is not as great as in the primary message flow, and linkages are generally long. Cities with concentrations of multiple linkages include Tokyo at the top, followed by Osaka, Nagoya, Sapporo, Fukuoka, Kobe, Asahikawa, and Hiroshima. The prefectural calling region that manifested itself in the primary message flow is nowhere to be seen. There are few prefectural core cities that are extending linkages to cities (regions) outside their respective prefectures.

There are numerous linkages that are concentrated in Tokyo. In Tokyo Metropolitan Prefecture and Saitama and Kanagawa prefectures, all of the cities which do not show directionality with Tokyo in primary message flow are linked in secondary message flow. Excluding Yokkaichiba (which show directionality with Tokyo as third largest message flow), all cities in Chiba Prefecture have either primary message flow or secondary message flow with Tokyo. In Ibaraki, Tochigi and Gunma prefectures, there are only 13 cities (regions) that do not have either primary message flow or secondary message flow linkages with Tokyo. Thus, by considering the combined destinations of primary and secondary message flow, we can clearly see how enormous the influence of Tokyo is in the Kanto region.

In the Tokyo calling region, cities which have amassed at least 3 linkages in primary message flow are found only in Chiba Prefecture. When secondary message flow is included, the list includes Tsuchiura, Urawa-Omiya, and Yokohama. Looking at the Tokyo MA while ignoring the influence of Tokyo, it becomes clear that these cities, which have been hidden in the shadow of Tokyo's overwhelming nodality, are functioning as secondary information centers.

Tokyo is extending its linkages not only within the Kanto region, but also to the core cities of the Tohoku, Hokkaido, and Chubu regions. This strongly indicates that regional core cities have solid social and economic linkages with Tokyo. Passing over the regions with strong linkages to Osaka, that is, the Kinki, Chugoku, and Shikoku regions, Tokyo is also extending its linkages with Fukuoka in the Kyushu region, as well as with Naha, Okinawa Miyako, Yaeyama, and Nago in Okinawa. It is especially interesting to note that the four districts of Okinawa are establishing linkages not with Fukuoka (the core city of Kyushu) or Osaka (the core city of Western Japan), but with Tokyo.

Osaka is extending linkages to most of the cities in Wakayama and Hyogo prefectures, as well as some of the cities in Nara, Shiga, Kyoto, Fukui, Mie, and Okayama prefectures.[4] It should also be noted that Osaka is also the origin of linkages for the

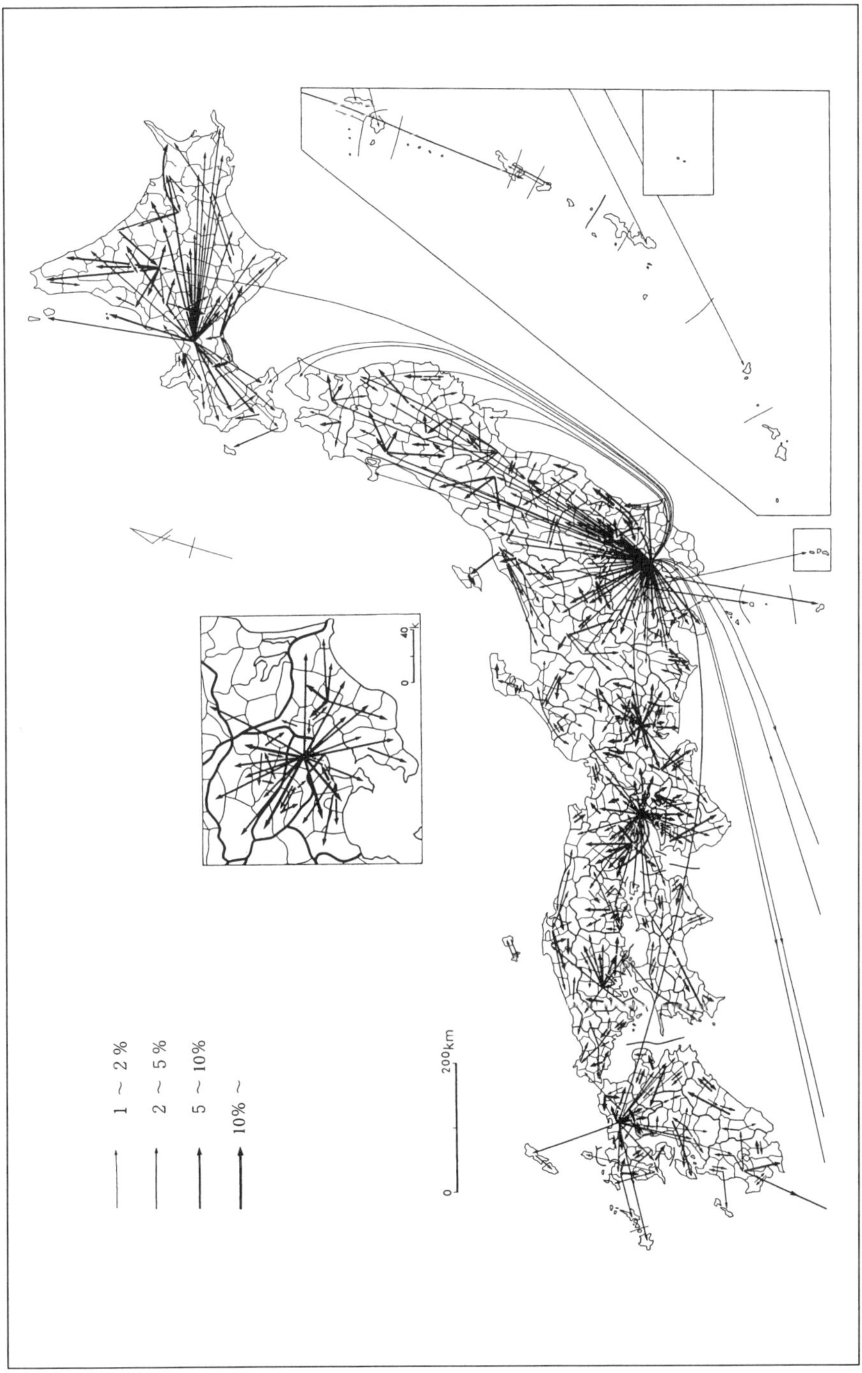

Figure 4.3. Spatial flow patterns in terms of secondary telephone message volume (1992)

major cities of the Chugoku and Shikoku regions, namely, prefectural capitals such as Okayama, Tokushima, and Takamatsu. Nagoya is extending linkages within Aichi Prefecture, as well as to southern Gifu and northern Mie prefectures. In eastern Aichi Prefecture, Okazaki has amassed 3 linkages and is building itself as a local core there.

Sapporo has amassed 26 linkages. Primary message flow from the cities of eastern Hokkaido is directed toward Obihiro, Kushiro, and Kitami, while secondary message flow is directed toward Sapporo. The primary message flow in cities around Hakodate is directed toward Hakodate, while secondary flow is in all cases directed toward Sapporo. Asahikawa has accumulated 7 linkages from northern Hokkaido, making it the city with the greatest centripetal force behind Sapporo.

In Figure 4.3, there are 9 regions, including mountainous areas in Shikoku and Kinki and isolated islands in Kyushu, which do not attain 1% directionality.[5] All of these areas are far removed from large cities; they are non-urbanized, with comparatively little active of the telephone. These areas are characterized by a high intraregional directionality and the directionality of primary message flow is mainly confined to neighboring regions.

4.4. The Hierarchical Structure of Telephone Message Areas

Telephone message areas in Japan show a hierarchical structure consisting of three levels: prefectural, regional, and national. At the prefectural level, cities conduct most of their communications with the core city(s) of their respective prefecture. These prefectural core cities have high directionality and are linked to regional core cities, which themselves are at the top or near the top in communication flow and are linked to Tokyo. This is the general spatial chain-type relationship of telephone message flow that we can read from Figures 4.2 and 4.3.

The core cities at the regional level are Osaka, Fukuoka, Nagoya, Sapporo, and Sendai. As would be expected, Tokyo, which is the core city at the national level, also functions as a regional core,[6] covering not only the 7 prefectures in the Kanto region but also Shizuoka, Yamanashi, Nagano, Niigata, Toyama, Ishikawa, and Okinawa for a total of 14 prefectures. The Osaka message area includes the 16 prefectures of the Kinki, Chugoku and Shikoku regions, while the Fukuoka message area covers the 7 prefectures of the island of Kyushu. It should be noted that there are two sub-cores in the Osaka MA: Hiroshima and Takamatsu. Takamatsu includes message areas at the prefectural level in Shikoku, while Hiroshima acts in the same capacity for the prefectures of Hiroshima, Yamaguchi, and Shimane. The Nagoya regional MA is comprised of Aichi, Gifu, and Mie prefectures, Sapporo oversees the entire island of Hokkaido, and Sendai covers the 6 prefectures of the Tohoku region.

Tokyo has enormous influence; of the 567 message areas nationwide, over 70% have at least 1% directionality with the city.[7] The highest directionality to Tokyo is from the other areas of the Kanto region, most of which are linked to the city by primary message flow, followed by Okinawa and most of the Tohoku MAs having secondary message flow, and most of Hokkaido and Hokuriku having tertiary message flow, each with directionality toward Tokyo.

Chubu and most of Kinki show the 4th largest message flow, while most of Shikoku and Chugoku are linked to Tokyo by the 5th largest flow. In Kyushu, there are very few regions that show even a 6th largest flow linkage with Tokyo. Therefore, we can rank regional linkages with Tokyo in the following order, from greatest to least: 1) Kanto, 2) Okinawa and Tohoku, 3) Hokkaido and Hokuriku, 4) Chubu and Kinki, 5) Shikoku and Chugoku, and 6) Kyushu.

Directionality toward Tokyo also depends on the size and attributes of a region. Cities which transmit and receive large volumes of information, as well as prefectural and regional administrative and economic centers, have high directionality toward Tokyo. This directionality is especially strong in prefectural capitals and core regional cities.

Compared with Tokyo, the information absorption capacity of Osaka is very weak. Regions with at least 1% directionality toward Osaka are limited to Kinki, Chugoku, Shikoku, and part of Chubu,[8] there is no directionality at all from Kanto, Tohoku, and Hokkaido, and there are only 4 districts in Kyushu with directionality toward Osaka.

The core city at the national level (i.e., Tokyo) tends to transmit an excess of messages to core cities at the regional and prefectural level than it receives from them, while cities at the regional level transmit more messages to prefectural level cities than vice versa. Let us take the example of Tokyo-Osaka. In 1992, there were 155.18 million messages transmitted from Tokyo to Osaka, but only 149 million transmissions originating from Osaka; thus, in electronic communications between the two cities, Tokyo made 6.16 million more transmissions to Osaka than vice versa.

Another example is Sendai, a regional core city, and Morioka, a prefectural core city. There are 12.11 million transmissions annually from Sendai to Morioka, but only 10.21 million originating from Morioka. We can also see this in the communications between Fukuoka (regional) and Kagoshima (prefectural), in which 14.61 million messages originate each year from the former, but only 12.3 million from the latter.

The same kind of pattern is apparent at the prefectural level. Let's take the case of Iwate Prefecture. Setting the index of transmissions from Morioka to other areas within Iwate at 100, the index for transmission from selected cities to Morioka can be expressed as follows: Hanamaki, 90.1; Iwate City, 86.6; Mizusawa, 68.7; Kitakami, 71.0; Miyako, 83.6; Ichinoseki, 79.4; and Ninohe, 92.5.[9] Such cities at the top of the urban hierarchy transmit an excess of messages to lower-ranking cities, suggesting that in Japan, a pyramidal structure exists for the transmission of information from the top cities of the hierarchy to those at the bottom.

4.5. Conclusion

This chapter has attempted to explain the hierarchical and nodal structures of telephone message areas (MAs) in Japan using the flow of telephone communications as an index. The analysis revealed the following:

Telephone communications are concentrated in major cities. The volume of message transmission and reception greatly exceeds what one would expect from the size of the population. Furthermore, large cities tend to have a greater share of long distance communications and more dispersed destinations than smaller cities. Extraregional communications dominate in the so-called *Tokaido Megalopolis*, likely because of vibrant social and economic activities interaction between cities in that region.

Telephone message flow shows a clear nodality which forms message areas. There are 6 core cities having regional message areas that transcend prefectural boundaries: Tokyo, Osaka, Nagoya, Fukuoka, Sapporo and Sendai. Besides these major MAs, there exist smaller MAs which roughly correspond to prefectural boundaries, suggesting that prefectural regions are becoming spatial units of daily business and social activities.

Message areas show a three-level hierarchy of national, regional, and prefectural levels. Cities within a prefecture are linked to the core city for their prefecture by primary message flow. In addition, prefectural core cities show high directionality in their linkages with regional core cities, which in turn are linked to Tokyo via either primary or secondary message flow.

The core city at the national level, Tokyo, tends to transmit far more messages to the regional and prefectural levels than it receives from them, while the regional level transmits far more messages to the prefectural level than vice versa. We can assume there are spatial links in the transmission of official government and corporate information at three levels: from Tokyo to regional cores; from regional cores to prefectural cores; and from Tokyo directly to prefectural cores. Although this may illustrate one aspect of the Japanese national urban system of cities, more in-depth analysis must be undertaken to prove that such spatial links exist. That is a topic for future research.

Notes

1. Although there are many media for conveying information besides telephones, such as postal services, facsimile communications, and e-mail, telephone message flow, being used at all levels of the national system, is considered to be the most appropriate as an index for understanding information flow at the national level.

2. In the present paper, this proportion is referred to as "directionality."

3. Instead of being low, intraregional directionality in the cities adjoining Tokyo (and Osaka) is markedly high. For example, in all the cities surrounding Tokyo (Ichikawa, Soka, Kawaguchi, Musashino-Mitaka, and Kawasaki), it exceeds 20%, ranging from 21.5 at Kawasaki to 30.7% at Musashino-Mitaka. In the case of Osaka, which is adjoined by the cities of Sakai, Hachio, Nemurogawa, Ibaraki, Ikeda, and Nishinomiya,

intra-regional directionality ranges from 19.5% at Nishinomiya to 30.5% at Hachio.
4. However, it should be noted that there are many areas in Nara which do not show directionality toward Osaka; there are also independent message areas within the prefecture, with neither primary message flow (with Yamato-Takada as the core) nor primary message flow (with Nara City as the core) directed toward Osaka.

5. Therefore, there are 558 links, as shown in Figure 4.3.

6. Sapporo, Sendai and Osaka are all linked to Tokyo by primary message flow; Nagoya and Fukuoka, by secondary flow.

7. Areas with less than 1% directionality include the smaller cities of Hokkaido, Chugoku, Shikoku and Kyushu, and most of the regions where primary industries predominate.

8. In the Kinki region, nearly all areas are linked by at least tertiary message flow to Osaka, while most areas in the Shikoku and Chugoku regions are linked to Osaka by at least 5th largest message flow.

9. However, this does not mean that core cities in major metropolitan areas always account for the lion's share of transmissions to neighboring cities. In the top 8 cities in the Tokyo area, for example, there are only two which receive more transmissions to Tokyo than they send: Chiba and Urawa. The other 6 cities (Yokohama, Kawasaki, Musashino-Mikawa, Ichikawa, Kawaguchi, and Kokubunji) transmit far more to Tokyo than they receive from there. At the same time in the Osaka area, Osaka accounts for more transmissions to Hachio and Kyoto than vice versa, yet there are far fewer transmissions from Osaka to Kobe, Sakai, Nemurogawa, Ibaraki, Ikeda, and Ninomiya than there are from those other cities to Osaka. Thus we can see that there is a complex flow pattern in major metropolitan areas.

CHAPTER 5

REGIONAL STRUCTURE OF THE COMMODITY FLOW

5.1. Introduction

The flow of goods or commodities is not in a random, irrational pattern but rather a phenomenon which maintains a certain spatial order. In regions where this flow originates, supply exceeds the demand for these goods, so they are treated as excess and sent to other regions. Therefore, regions which are absorbing commodities must obtain some of them from outside, regardless of whether or not they have adequate facilities to process them.

Geographical studies of commodity flows have been extremely rare. The conceptual and theoretical systematization appeared in the 1950s (Ullman, 1954), which led to the introduction of quantitative methods and methodological expansion in the 1960s. With the development of large computers in the 1970s came increasing emphasis on substantiative research using multivariate analyses such as factor analysis (Black, 1973) and multiple-regression analysis (Holsman, 1979), and quantitative methods such as graph theory (Holsman, 1975) and linear programming (Barr, 1970). This has become the area of geographical research where the greatest results are expected.

Of particular interest is the *General Field Theory of Spatial Behavior* (hereafter called *Geographical Field Theory* (Berry, 1966, 1968a, 1968b)) which, advanced by Berry in the mid-1960s, has made an especially significant contribution to this field. Hypothesizing that the socio-economic characteristics of a region and interregional interactions (flow patterns) are mutually regulatory and have an isomorphic relationship, Berry proved his hypothesis using multivariate analysis. Berry was able to derive the spatial flow patterns of various goods, in this case in India, using factor analysis.

He then used canonical analysis to explain the mutual dependency between these patterns and the regional socio-economic characteristics which had also been derived by factor analysis. The validity of this *Geographical Field Theory* was subsequently proven by numerous verification studies, and many attempts have been made to improve the methodology.

Clark (1973) established functional regions based on O-D data on telephone communications in Wales, United Kingdom, and examined the relationship between these functional regions and homogeneous regions (in terms of occupations of telephone subscribers) using *Geographical Field Theory*. On a global scale, Freeman (1973) discussed the relation between the socio-economic characteristics of various nations and the flow of people, goods and capital among these nations. Schwind (1975), modifying Berry's analytical procedure, found that in the United States there is a strong link between population movement patterns and the socio-economic characteristics of a region. Kanno (1976), invoking Schwind's analytical framework, revealed the relationship between commodity flow patterns among major metropolitan regions of the United States and the socio-economic structure of these regions.

Among the studies done in Japan are those of Ichiminami (1978), who analyzed the relation between socio-economic characteristics and commuter and population flow between regions to explain the regional structure of the Nagoya Metropolitan Area, and Higashi and Moridaira (1980), who investigated the mutual dependency between commuter flow on the one hand, and population and employment characteristics on the other in the city of Osaka.

Finally, there are the regional analyses of Holsman (1979), and McConnell (1967) among others which, while not employing Berry's analytical framework, do confirm the validity of *Geographical Field Theory*.

Invoking the analytical framework of *Geographical Field Theory*, this chapter will attempt to explain the regional structure of commodity flow in Japan. This will involve identifying the systems of interregional linkages of commodity flow in Japan using various types of O-D data, then considering the correlation between these systems on the one hand, and the amount of production and absorption of various commodities and the regional characteristics (e.g., socio-economic characteristics, geographical location) which are believed to regulate the destinations of this flow. At the national level, there are just a few studies (most notably that of Mitsuhashi (1978)) which attempt to provide a holistic understanding of regional flow patterns of various types of commodities in Japan.

The analytical materials used here consist of the Ministry of Transport's Survey of Commodity Flow in Japan. Data will be from 1980 and comes from 46 of the 47 prefectures (i.e., excluding Okinawa) in Japan. The 30 types of commodities listed in Table 5.1 are used to analyze commodity flow. Finally, 40 socio-economic variables and proximity are used as indices of regional characteristics.

The analytical procedure is shown in Figure 5.1. The flow chart on the left side of the figure indicates the order in which the commodity flow patterns were identified. First, a 46 (origins) × 46 (destinations) × 30 (commodities) O-D matrix was prepared, then the diagonal element of the matrix indicating flow within a single region (that is, outside the realm of this analysis) was removed to transform the 3-dimensional matrix into one that had only two dimensions. In other words, each interregional pair (link) was shown in the rows, with goods shown in columns, resulting in a matrix consisting of 2,070 (46 × 45) rows and 30 columns. This new matrix was then subjected to factor analysis, and the degree of similarity of flow patterns was used to summarize linkage systems of commodity flow. The analysis of such a data matrix, in which links are shown in the rows and variables are represented in the column direction, is called *Dyadic Factor Analysis*, a method whose validity has been proven by Black (1973), Chojniki and Czyz

(1973), Davies and Thompson (1980), and Morikawa (1977), among others. Then, based on the extracted factor score matrix (2,070 × p), each score of each factor was mapped out to analyze the regional flow patterns of commodities. Next, the factors accounting for these regional flow patterns were identified using the analytical procedure shown in Figure 5.1. The regional characteristic matrix was constructed of 46 rows (prefectures) and 40 columns (variables) and subjected to factor analysis, then the regional social and economic characteristics were derived as a small number of significant common factors.

Table 5.1. Commodities employed in the analysis

	Commodities	Examples
1	Grains	rice, wheat, cereals, beans
2	Fruits and vegetables	potatoes, vegetables, fruits
3	Livestock products	livestock, meat of birds and beasts, raw milk, eggs, pelt
4	Marine products	fish and shellfishes
5	Miscellaneous agricultural products	industrial crops, processed agricultural products, agricultural products unclassified
6	Timber	material wood, lumber, other forest products
7	Charcoal products	wood for fuel, charcoal
8	Coals	coal, lignite
9	Metallic ores	iron ore, nonferrous metal ore
10	Non-metallic minerals for construction	gravel, sand, building stone
11	Non-metallic ores (industrial use)	limestone, rock phosphate stone, crude oil, salt, other non-metallic minerals
12	Metals	nonferrous metal, steel
13	Metal products	metal products for construction, metal products for building, wire rod, cutlery, other metal products
14	Machineries	industrial machines, electric machines, transport machines
15	Ceramic products	cement, cement products, lime, brick
16	Petroleum products	heavy oil, benzine, volatile oil
17	Coal products	cokes, other coal products
18	Pharmaceuticals	sulphate acid, soda
19	Chemical fertilizers	nitrogenous fertilizers, potassic fertilizers, phosphatic fertilizers
20	Miscellaneous chemical products	dyes, paints, cosmetics, synthetic resin
21	Paper and pulp	paper, pulp
22	Textiles	yarn, woven goods
23	Foodstuffs	food processing products, drinks
24	Daily goods	books, clothes, shoes, stationery, furniture, daily necessities unclassified
25	Miscellaneous manufactured products	rubber, leather products, wooden ware
26	Trash	metal trash, other trash
27	Feeds and manures for animals	manures, fodder, feed grains
28	Waste products	industrial waste products
29	Transport containers	transport containers made of metals
30	Miscellaneous goods	mail matters, property to be moved, baggage

After that, the 46-row, q-column factor score matrix was used to create a new 2,070-row, q-column matrix in which values of socio-economic characteristics of region-pairs were represented in the column direction. Values for these region-pair characteristics were the totals of the factor scores of the respective regions. Apparently, the higher scores of both regions represented an accordingly greater volume of interregional flow. Nevertheless, similarities and/or differences in characteristics between these region-

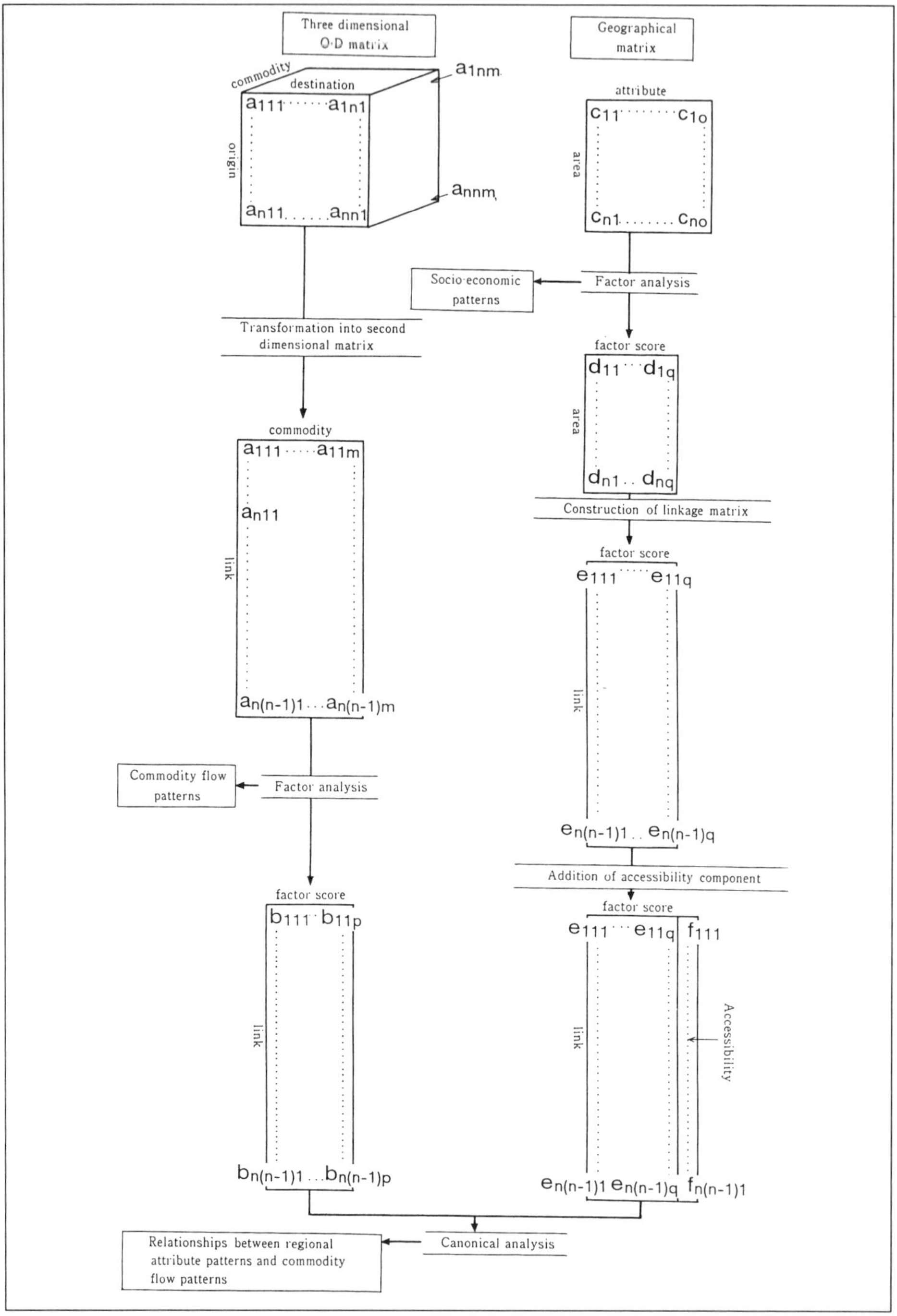

Figure 5.1. Framework of the analysis

pairs are believed to indicate the geographical relation between these regions, that is, their to one another, as a factor regulating extraregional flow. Furthermore, when the concept of proximity was included in the consideration, the value for accessibility of regions was added to the aforementioned 2,070×q matrix to create a final matrix of 2,070 rows and q+1 columns. Then, using this final 2,070×(q+1) matrix and the 2,070×p matrix showing commodity flow patterns, the significance of the interrelationships of these regions was explained using canonical analysis.

5.2. The Structure of Regional Commodity Flow Patterns

Based on the 30 commodities listed in Table 5.1, the region-pairs of the 2,070 sets relating to commodity flow volume were used as sample values, and the correlation coefficients between pairs of commodities were calculated. Table 5.2 shows these values in a triangular matrix. Since no correlation coefficient was less than -0.4, there were no combinations of commodities which had a highly negative correlation.

The highest positive correlation was seen in the combination of metal products and machineries, which had a value of 0.79. In other words, the flow patterns of metal products and machineries showed the greatest degree of similarity. The next highest correlation was seen in the combination of metals and miscellaneous chemical products, as well as in the machineries-daily goods combination, which both showed values of 0.76.

Foodstuffs showed a positive correlation of 0.4 or greater with 21 of the 29 other commodities, and had a flow pattern which was similar to that of most of the others. Overall, it can be said that the highest correlation coefficients were among industrial goods, while neither fruits and vegetables, charcoal products, coals, metallic ores, non-metallic ores (industrial use), coal products, chemical fertilizers, nor waste products had a correlation coefficient of at least 0.4 with any other commodity; each of these had its own unique flow patterns vis-à-vis every other commodity.

Next, in order to describe the flow patterns of the 30 commodities in a significant number of flow patterns, principle factor analysis was conducted on a 2,070-row (links) × 30-column (commodities) matrix in an effort to extract the main factors. As a result, 6 factors with an eigenvalue of 1.0 or higher were found, which reached an accumulated explanatory value of 59.4% up to the 6th factor. Next, to facilitate the interpretation of factors, the varimax method was used to directly substitute a loadings matrix to bring about a simple structure. Table 5.3 shows the factor loading matrix obtained in this fashion.

Factor I had a 38.5% total variance, and was given attributes by the numerous variables which have positive correlations. The variables with the highest loads included machineries (0.72), miscellaneous chemical products (0.71), daily goods (0.69), metal products (0.66), textiles (0.63), metals (0.58), and miscellaneous manufactured products (0.56). Therefore, we can interpret Factor I to refer to the flow of

Table 5.2. Correlation coefficients between the commodities (1980)

Variables	1	2	3	4	5	6	7	8	9	10	11	12	13	14	15	16	17	18	19	20	21	22	23	24	25	26	27	28	29	30
1	1.00				0.57							0.45											0.53							
2		1.00																												
3			1.00	0.42		0.41						0.47	0.55	0.52	0.43					0.48	0.44	0.41	0.58							
4				1.00									0.40								0.51		0.51							
5					1.00							0.49	0.48	0.50						0.47	0.40		0.58							
6						1.00				0.42		0.43	0.50	0.48						0.43	0.48	0.40	0.52							
7							1.00																							
8								1.00																						
9									1.00																					
10										1.00			0.40	0.43	0.43								0.46	0.42					0.51	
11											1.00																			
12												1.00	0.71	0.70	0.57	0.49		0.69		0.76	0.55	0.52	0.70	0.64	0.60	0.42	0.49		0.63	
13													1.00	0.79	0.58			0.56		0.73	0.58	0.57	0.75	0.75	0.65	0.45	0.48		0.68	0.46
14														1.00	0.56			0.53		0.75	0.60	0.61	0.75	0.76	0.69	0.43	0.41		0.68	0.41
15															1.00	0.40		0.54		0.57	0.46	0.42	0.62	0.53	0.57		0.43		0.53	
16																1.00		0.57		0.49			0.43				0.45			
17																	1.00													
18																		1.00		0.69	0.46	0.43	0.57	0.48	0.50		0.46		0.51	
19																			1.00											
20																				1.00	0.59	0.61	0.71	0.70	0.64	0.47	0.46		0.63	
21																					1.00	0.51	0.66	0.64	0.62				0.51	
22																						1.00	0.55	0.61	0.57				0.46	
23																							1.00	0.75	0.66	0.43	0.54		0.70	0.43
24																								1.00	0.66	0.45			0.62	0.44
25																									1.00	0.40	0.40		0.57	
26																										1.00			0.41	
27																											1.00		0.44	
28																												1.00		
29																													1.00	
30																														1.00

1: Grains, 2: Fruits and vegetables, 3: Livestock products, 4: Marine products, 5: Miscellaneous agricultural products, 6: Timber, 7: Charcoal products, 8: Coals, 9: Metallic ores, 10: Non-metallic minerals for construction, 11: Non-metallic ores (industrial use), 12: Metals, 13: Metal products, 14: Machineries, 15: Ceramic products, 16: Petroleum products, 17: Coal products, 18: Pharmaceuticals, 19: Chemical fertilizers, 20: Miscellaneous chemical products, 21: Paper and pulp, 22: Textiles, 23: Foodstuffs, 24: Daily goods, 25: Miscellaneous manufactured products, 26: Trash, 27: Feeds and manures for animals, 28: Waste products, 29: Transport containers, 30:Miscellaneous goods

Note: Correlation coefficients exceeding ±0.4 are indicated.

TABLE 5.3. Factor analysis of commodity flows (1980) *(continued).*

	Variables	Communality	Factor loadings					
			I	II	III	IV	V	VI
1	Grains	0.49			0.72			
2	Fruits and vegetables	0.20						
3	Livestock products	0.45		0.45				
4	Marine products	0.38		0.53				
5	Miscellaneous agricultural products	0.49			0.62			
6	Timber	0.43		0.43				
7	Charcoal products	0.15						0.55
8	Coals	0.11						(0.35)
9	Metallic ores	0.13						
10	Non-metallic minerals for construction	0.37						
11	Non-metallic ores (industrial use)	0.24				0.40		
12	Metals	0.71	0.58			0.41		
13	Metal products	0.75	0.66					
14	Machineries	0.77	0.72					
15	Ceramic products	0.52				0.45		
16	Petroleum products	0.43			0.41	0.51		
17	Coal products	0.17						
18	Pharmaceuticals	0.63	0.40			0.66		
19	Chemical fertilizers	0.21						
20	Miscellaneous chemical products	0.76	0.71			0.44		
21	Paper and pulp	0.57	0.49	0.51				
22	Textiles	0.50	0.63					
23	Foodstuffs	0.79	0.51	0.48	0.40			
24	Daily goods	0.74	0.69	0.45				
25	Miscellaneous manufactured products	0.63	0.56	0.45				
26	Trash	0.32						
27	Feeds and manures for animals	0.51						
28	Waste products	0.19			0.63		0.45	
29	Transport containers	0.64	0.43				0.50	
30	Miscellaneous goods	0.30					0.40	
	Eigenvalue		11.56	1.61	1.26	1.21	1.14	1.04
	Percent of total variance (%)		38.5	5.4	4.2	4.0	3.8	3.5
	Cumulative percent (%)		38.5	43.9	48.1	52.2	55.9	59.4

Note: Only loadings exceeding ±0.4 are indicated.

in the graphic form shown in Figure 5.2-a.[1] In all, there were 85 links with factor scores of 2.0 or higher, of which 7 were 5.0 or higher. These 7 links were Tokyo-Saitama (two-way), Tokyo-to-Kanagawa, Aichi-Osaka (two-way), Osaka-to-Tokyo, and Osaka-to-Hyogo. The greatest flows were the links going from Saitama to Tokyo, with a factor score of 7.4. The region with the greatest origination was Osaka, with 14 originating links, followed by Tokyo, Aichi, Saitama, etc.; together, they formed the main origination regions.

At the same time, Osaka was also the greatest destination region, followed by Tokyo, Saitama, and Hyogo. The flow of Factor I (manufactured products) showed mutual dependency, with a relatively high proportion of both regions in the region-pairs functioning as both point of origination and destination.

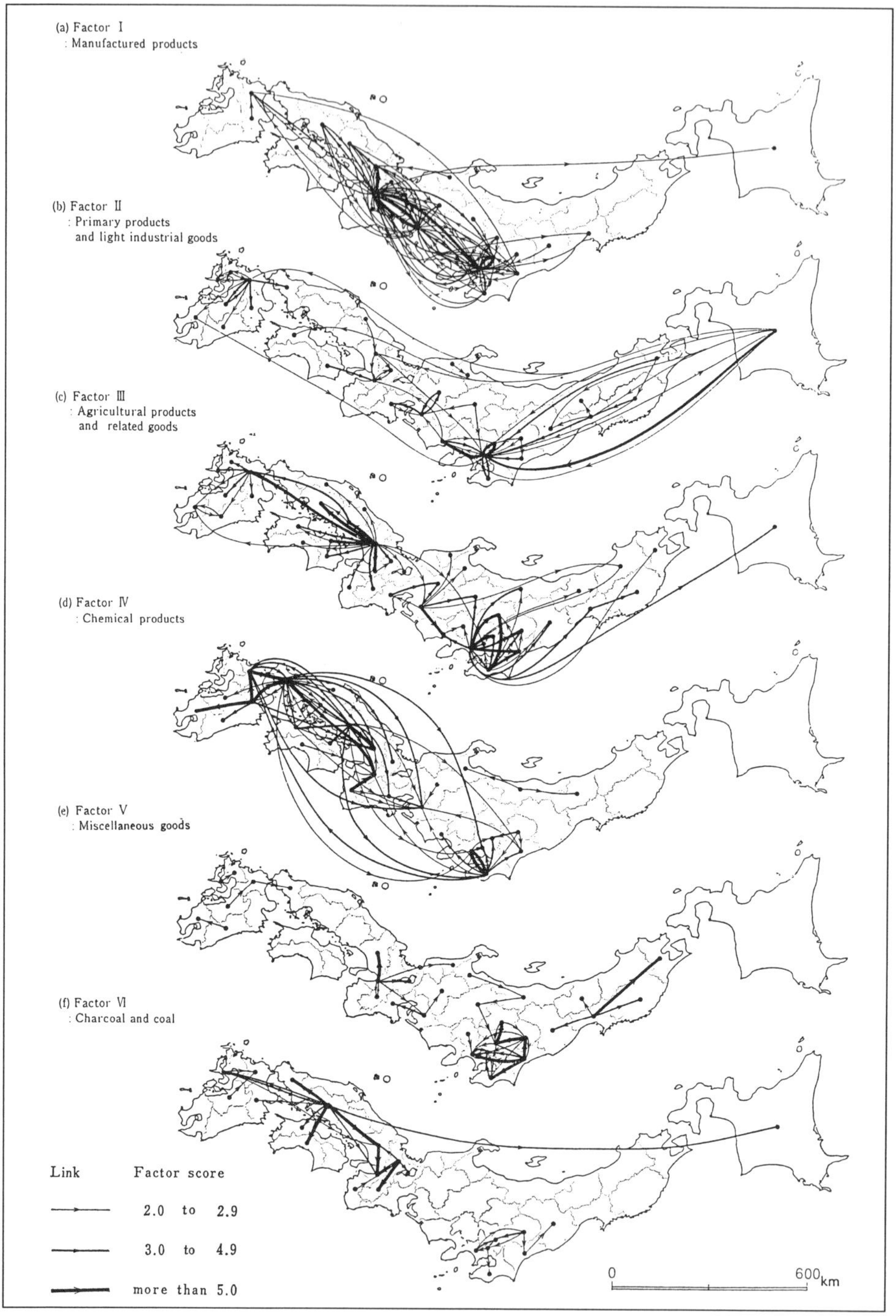

Figure 5.2. Regional commodity flow patterns (1980)

Flow links were predominantly distributed in the Pacific Belt Zone, especially in the three major metropolitan areas (i.e., Tokyo, Nagoya, and Osaka). Another feature is that Tokyo and Osaka each maintained strong linkages with their respective surrounding areas.

Factor II had a total variance of 5.4%. There were 7 variables which had a factor loading of 0.4 or higher, of which marine products was highest at 0.53. In terms of variable structure, Factor II is believed to indicate the flow of primary products and light industrial goods. The factor score was used to graphically illustrate the regional flow patterns of Factor II, which are shown in Figure 5.2-b. These differ from the flow patterns of Factor I, and flow linkages are dispersed over a wide area of the country.

There were 60 linkages in all, with 5 of these having factor scores of 5.0 or more. The highest factor score was for linkages from Tokyo to Saitama (9.7), where the flow volume of primary products and light industrial goods was greatest. Tokyo, which is the largest node for flow, had linkage relationships with all regions of the country, including Hokkaido, though the greatest linkages were with neighboring areas such as Saitama, Chiba, and Kanagawa.

Hokkaido had 8 origination linkages, making it the second-largest origination point, but flow links extended over an extremely long distance. Miyagi is the distribution center of the Tohoku area, receiving commodities from Hokkaido, Iwate, etc., while sending goods to Yamagata, Fukushima, and Tokyo. Aichi is the core flow region of the Chubu area, having especially strong ties to Gifu and Mie. In Kyushu, Fukuoka occupies an important position as the origination point for the rest of regions on the island.

Factor III can explain 4.2% of the total variance. Five variables had a factor loading of at least 0.4, with the maximum being 0.72 for grains. From the structure of variable groups, we can conclude that Factor III represents the flow of agricultural products and related goods. Factor scores were used to identify regional flow patterns for Factor III, which are illustrated in Figure 5.2-c. There were 66 linkages in all, of which 12 had factor scores of 5.0 or more. The greatest region of origination was Hyogo, which provides a wide area, especially in western Japan, with agricultural products and related goods. The second-largest point of origination was Kanagawa, whose hinterlands include the Kanto and Tohoku areas. Chiba and Aichi are also important points of origination.

While Fukuoka is the most important point of origination for the island of Kyushu, it is also a destination point for Hyogo and Okayama, playing the role of flow relay center. Domestically-produced agricultural products and associated goods alone cannot meet the total demand in Japan, so there is great dependency on imports. Therefore, one of the most distinguishing features here is that regions having major port facilities act as the main points of origination. Thus, unlike what is seen in flow patterns for Factors I and II, there is no mutual dependency between the regions in region-pairs, and there is a clear distinction between origins and destinations.

Factor IV accounted for 4% of the total variance. Variables showing high positive correlations with all other variables were pharmaceuticals (0.66), petroleum products (0.51), ceramic products (0.45), miscellaneous chemical products (0.44), metals (0.41), and non-metallic ores (industrial use) (0.40). Therefore, we can conclude that Factor IV illustrates the flow of chemical products. Regional distribution of factor scores are shown in Figure 5.2-d. In all, there were 63 linkages with factor scores of 2.0 or more, of which 9 were confirmed to be 5.0 or higher. The highest factor score was the linkage from Yamaguchi to Fukuoka, which showed a value of 7.3. Yamaguchi, which functions as the distribution center for chemical products in western Japan, was the core flow region. Besides Yamaguchi, major origins included Oita, Okayama and Chiba, while major destinations included Fukuoka, Hyogo, Osaka, and Aichi. Flow linkages were predominant in western Japan; hence, this factor is characterized by flows covering long distances.

Factor V accounted for 3.8% of the total variance. Variables showing high positive correlations with all other variables were transport containers (0.50), waste products (0.45), and miscellaneous goods (0.40). Therefore, we can conclude that Factor V represents the flow of miscellaneous goods. Figure 5.2-e shows factor scores of at least 2.0. There was a total of 39 linkages, the greatest of which was from Ibaraki to Chiba. The largest origin was Ibaraki, which supplies mass quantities of specialty products to the rest of the Kanto region. Other major origins included Miyagi and Osaka, the former of which supplies specialty products to 4 prefectures in Tohoku region: Aomori, Iwate, Fukushima and Yamagata. Osaka, on the other hand, functions as the origin for Hyogo, Shiga, Aichi, Ishikawa, and Nara. Unlike the flow patterns of chemical products (Factor IV), a large proportion of goods is sent to neighboring prefectures, that is, most flow linkages cover only short distances.

Factor VI, which accounted for 3.5% of all variance, indicated the flows of charcoal and coal, variables that had factor loadings of 0.55 and 0.35, respectively. There were 27 linkages that had a factor score of at least 2.0, of which 6 were 5.0 or more. The region-pair with the greatest flow was Hiroshima-to-Hyogo, with a factor score of 11.1. By far the greatest origin was Hiroshima, while the greatest destination was Osaka, which received goods from Kyoto, Hyogo, Hiroshima, Wakayama, and elsewhere. At the national level, we can clearly see that there are two regions: the Western Japan Belt, with Hiroshima as the core, where flow is dominant; and flow linkages emanating from the Kanto area, with nodes (albeit on a much smaller scale) including Kanagawa, Saitama, and Tochigi. Factor VI is characterized by the fact that there are no other regions which possess strong flow linkages.

5.3. Regional Socio-Economic Structure

We can see that the regional flow patterns identified from the preceding analyses originated from regional differences related to geographical characteristics. Therefore in this section, 40 socio-economic variables which were believed to be strongly tied to causative factors of commodity flow patterns, were selected and subjected to factor analysis to derive a small number of significant common factors. As a result of principle factor analysis conducted on a 46 × 40 data matrix, 6 factors were found to have an eigenvalue of 1.0 or greater. Next, varimax rotation was conducted to facilitate the interpretation of these common factors. The accumulated percentage reached 90% up to Factor VI (Table 5.4). Factors were interpreted based on the structure of variable groups having an absolute factor load value of 0.4 or higher, then standardized factor scores were used to identify distribution patterns of socio-economic characteristics.

Factor I had a total variance of 46.3%. This factor indicates variables with an especially high positive correlation, such as the number of wholesale and retail establishments (factor loading 0.99), population (0.98), gross prefectural domestic product (secondary industries) (0.98), establishments of services (0.98), gross prefectural domestic products (tertiary industries) (0.97), number of motor vehicles (0.97), and number of light trucks (0.97). Factor I was characterized by the 23 variables with factor loadings of 0.4 or higher. From the structure of these variable groups we can conclude that Factor I indicates urban activities.

Figure 5.3-a is a cartographic representation of regional distribution based on factor scores. The regions with the highest factor scores were Tokyo (5.0), Osaka (2.7), and Aichi (1.6). These were followed by Kanagawa, Chiba, Saitama and Hyogo, showing that urban activities predominate, especially in the 3 largest metropolitan areas of the country. It should be noted that the region with the lowest factor score was Tokushima, followed by Nara. There were numerous regions in western Japan where factor scores were low, indicating a lack of urban activities.

Factor II had a total variance of 21.4%, and there were 15 variables with an absolute factor loading of at least 0.4. Variables with a high positive factor loading were area of cultivated land (fields) (0.94), establishments of forestry & hunting (0.93), area of forest lands (0.92), gross prefectural domestic product (primary industries) (0.71), and establishments of agriculture (0.71). Hence, we can conclude that Factor II represents rural activities.

Figure 5.3-b illustrates regional distribution patterns based on factor scores. Hokkaido had the highest score (6.1) while Yamanashi had the lowest (-2.2). Factor II shows a pattern symmetrical to that of Factor I; the scores are lowest in the 3 major metropolitan areas, and tend to increase as distance away from these centers increases.

Factor III had a total variance of 9.1%. There were 7 variables with a factor loading over 0.4: tonnage of coastwise vessels (0.88), number of coastwise vessels (0.85), number of fishing managements (0.82), number of fishing ports (0.78), tonnage of ocean-going vessels (0.64), number of ocean-going vessels (0.63), establishments of fishery & aquaculture (0.46). Therefore, we can conclude that Factor III represents port activities.

The region with the highest factor score was Hiroshima (2.8), followed by Yamaguchi (2.4), Hyogo (2.4), Kanagawa (2.0), and Mie (2.0). Figure 5.3-c shows the distribution of regional patterns as they relate to port activities. The factor scores of the hinterlands of major trading ports were high as a matter of course, but the factor scores of bases for marine fisheries were also high. Factor scores of landlocked prefectures were, as would be expected, low.

Table 5.4. Factor analysis of socio-economic attributes (1980)

Variables	Factor loadings					
	I	II	III	IV	V	VI
1 Population	0.98					
2 Population increase (1975-80)						0.82
3 Population density	0.86	-0.42				
4 Gross prefectural domestic product (primary industries)		0.71		0.40		
5 Gross prefectural domestic product (secondary industries)	0.98					
6 Gross prefectural domestic product (tertiary industries)	0.97					
7 Establishments of agriculture	0.59	0.71				
8 Establishments of forestry & hunting		0.93				
9 Establishments of fishery & aquaculture		0.66	0.46			
10 Establishments of mining		0.60			0.42	
11 Establishments of construction	0.96					
12 Establishments of manufacturing	0.91					
13 Establishments of wholesale & retail	0.99					
14 Establishments of finance & insurance	0.58					-0.52
15 Establishments of real estate business	0.96					
16 Establishments of transport & communication	0.96					
17 Establishments of utilities	0.78	0.49				
18 Establishments of services	0.98					
19 Area of mining fields				0.88		
20 Number of mining lots		0.60		0.65		
21 Number of fishing managements		0.44	0.82			
22 Number of fishing ports			0.78			
23 Number of forestry households		0.46			0.48	
24 Area of forest lands		0.92				
25 Full-time farm households		0.65				
26 Part-time farm households (mainly farming)		0.58		0.54		
27 Part-time farm households (mainly other jobs)					0.88	
28 Area of cultivated land (paddy fields)		0.66				
29 Area of cultivated land (fields)		0.94				
30 Floor area of ordinary warehouses	0.93					
31 Volume of refrigeration warehouses	0.88					
32 Number of motor vehicles	0.97					
33 Number of ordinary trucks	0.95					
34 Number of light trucks	0.97					
35 Establishments of truck business	0.95					
36 Number of truck terminals	0.85					
37 Number of ocean-going vessels	0.62		0.63			
38 Number of coastwise vessels			0.85			
39 Tonnage of ocean-going vessels	0.59		0.64			
40 Tonnage of coastwise vessels	0.41		0.88			
Eigenvalues	18.51	8.55	3.64	2.46	1.73	1.12
Percent of total variance (%)	46.3	21.4	9.1	6.1	4.3	2.8
Cumulative percent (%)	46.3	67.7	76.8	82.9	87.2	90.0

Note: Only loadings exceeding ±0.4 are indicated.

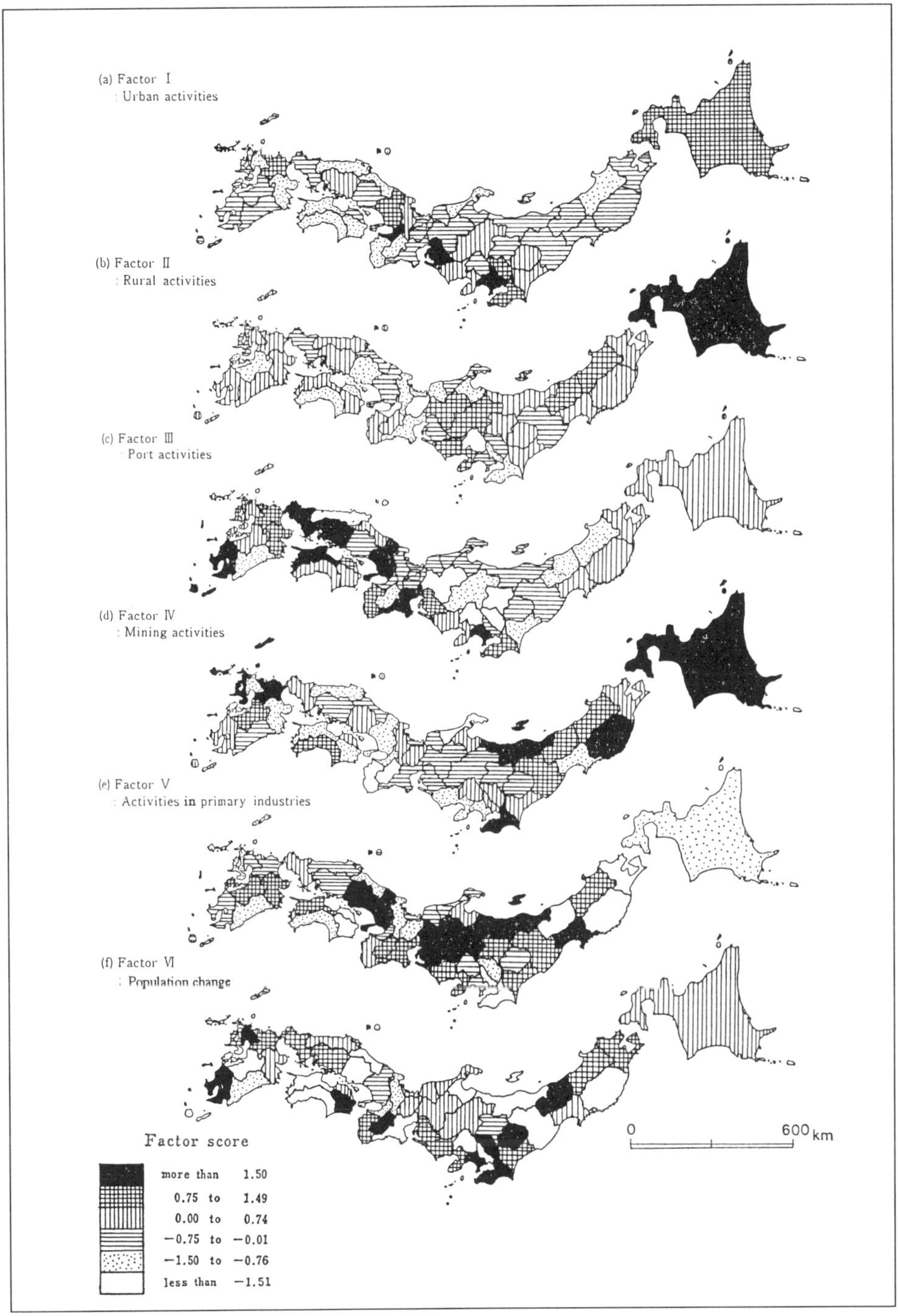

Figure 5.3. Factor score distributions for socio-economic attributes (1980)

Factor IV had a total variance of 6.1%, with area of mining fields (0.88) and number of mining lots (0.65) being the variables with the highest positive correlation. Hence, this factor indicates mining activities. Figure 5.3-d, which is a cartographic representation of regional distribution of mining activities, shows that these activities are noticeably dominant in rural and mountainous areas.

Factor V had a total variance of 4.3%, with the variables of the highest factor loadings being Type 2 part-time farm households (mainly other jobs) (0.88) and number of forestry households (0.48). Therefore, we can conclude that Factor V represents activities in primary industries. Looking at regional distribution patterns derived from factor scores (Figure 5.3-e), we can see that the regions of central Japan all had high factor scores.

Factor VI had a total variance of 2.8% and a high factor loading for population increase (1975 - 80) (0.82); therefore, this factor represents population change. Figure 5.3-f shows regional distribution patterns based on factor scores.

The preceding has been a discussion of the socio-economic attributes of the factors which apparently create commodity flow patterns. The next section examines another probable factor in the analysis, proximity of regions, and discusses the mutual relationships of the commodity flow patterns derived in section 5.2.

5.4. Relationship between Regional Commodity Flow Patterns and Regional Characteristics (Socio-Economic Attributes and Proximity)

Using the 46 × 6 factor score matrix for 2,070 (46 × 45) extraregional pairs, total factor scores for the two regions concerned were derived and used to create a 2,070 row × 6 column matrix. Then, the accessibility matrix (2,070 × 1) was added to the final column to create a new matrix of 2,070 rows × 7 columns. After that, the relation between the variable groups of this new matrix, and those of the 2,070 × 6 factor score matrix showing commodity flow patterns which was derived in section 5.2, was subjected to canonical analysis. The 2,070 × 6 matrix was used to show the criteria variables while the 2,070 × 7 matrix showed the predictor variables.

The results of this analysis (Table 5.5) indicate that there were 5 combinations of canonical variates that had statistically significant canonical correlation coefficients. Table 5.5 also shows the correlation coefficients and coefficient vectors of each canonical variate. Here, Y_1-Y_6 represent commodity flow pattern variables, X_1-X_6 show socio-economic variables, and X_7 is a variable indicating proximity. In more specific terms, Y_1 is the flow of manufactured products, Y_2 is the flow of primary products and light industrial goods, Y_3 is the flow of agricultural products and related goods, Y_4 is the flow of chemical goods, Y_5 is the flow of miscellaneous goods, and Y_6 is the flow of charcoal and coal. On the other side, X_1 is urban activities, X_2 is rural activities, X_3 is port activities, X_4 is mining activities, X_5 is the activities in primary industries, X_6 is population change, and X_7 indicates accessibility, i.e. proximity in terms of commodity flow time. This Table is used in the following to consider the significant canonical variate from the canonical coefficients.

Table 5.5. Canonical analysis of relationships between commodity flows and regional attributes (1980)

Factors		Canonical vectors				
		I	II	III	IV	V
Commodity flows	Y_1	-0.74	0.68	0.11	-0.19	0.02
	Y_2	-0.36	-0.48	-0.75	0.26	0.04
	Y_3	-0.21	-0.01	0.09	0.50	0.20
	Y_4	-0.23	-0.54	0.63	0.30	-0.03
	Y_5	-0.07	-0.40	0.10	-0.78	0.42
	Y_6	-0.09	-0.20	0.09	-0.17	-0.94
Socio-economic attributes	X_1	-0.71	0.59	-0.24	0.18	0.33
	X_2	0.03	-0.62	-0.76	0.43	0.23
	X_3	-0.10	-0.23	0.37	0.78	0.03
	X_4	0.04	-0.45	0.28	-0.15	0.91
	X_5	-0.14	0.13	0.59	0.31	0.10
	X_6	-0.03	0.15	0.66	-0.31	0.34
Accessibility	X_7	-0.58	-0.87	-0.05	-0.18	-0.25
Canonical correlation		0.842	0.545	0.415	0.223	0.111
Wilks lambda		0.159	0.546	0.777	0.938	0.987
Chi-square		3791.5*	1247.3*	521.5*	131.1*	26.0*
Degree of freedom		42	30	20	12	6

Note: * significant at the 0.01 level of confidence.

At 0.842, the canonical correlation coefficient in the first canonical variate is quite high, and shows a highly negative correlation with Y_1 (canonical coefficient of -0.74) vis-à-vis commodity flow patterns. Regarding the one-way regional flow patterns, there is a highly negative correlation with X_1 (-0.71) and X_7 (-0.58). Therefore, we can interpret the first canonical variate as follows:

The regional flow patterns for industrial goods show high interdependency with urban activities and proximity (accessibility) of regions. In other words, the volume of industrial goods flow increases as the dominance of urban activities in the regions of origin and destination increases. Furthermore, the proximity of the two regions is high, that is, the spatial distance that separates them is relatively short. Hence, as illustrated by the flow of manufactured goods shown in Figure 5.2-a, this phenomenon is limited to areas within the so-called Pacific Belt. We can see that in this region, especially in the three major metropolitan areas, urban functions are becoming specialized, and the flow of industrial goods among the regions in this belt is dominant.

At the same time, the flow between regions located outside this belt is nearly insignificant, meaning that urban activities in these regions are laggard and there are no adequate facilities. This apparently makes it difficult to initiate and absorb flows of industrial goods.

The second canonical variate, which has a canonical correlation coefficient of 0.545, shows high canonical coefficients for commodity flow patterns for manufactured products (Y_1), chemical products (Y_4) and primary products and light industrial goods (Y_2), which are 0.68, -0.54, and -0.48, respectively. On the other hand, the highest canonical coefficients as they relate to regional attributes are -0.87 for accessibility (X_7), -0.62 for rural activities (X_2) and 0.59 for urban activities. Taking the flow of manufactured products (Y_1) employed in the analysis of the first canonical variate into consideration, we can make the following comments about the second canonical variate:

First, in the flow of chemical products, and of primary products and light industrial goods, there is an interactive relation with the degree of proximity (distance between two regions) and the dominance of rural and mining activities. In other words, these two commodity flows are dominant between regions with short commodity flow times, so regions which initiate and absorb these flows are vigorously engaged in rural activities. When there is an inverse correlation with X_1, there is little flow between regions where urban activities predominate.

The third canonical variate has a canonical correlation coefficient of 0.415. The main flow variables contributing to this value in commodity flow patterns are primary products and light industrial goods (Y_2), and chemical products (Y_4). In terms of regional attributes, these variables are rural activities (X_2), population change (X_6), and activities in primary industries (X_5). In addition, we can say that Y_2, Y_4, X_2, X_6, and X_5 are interactive. However, the canonical variate has mutually positive-negative relationships, so taking these attributes into consideration complicates the analysis. Therefore in this study, a comparison was made between the commodity flow patterns shown in Figure 5.2 and the regional attributes shown in Figure 5.3, then an interpretation was made of the interactive relationship in the third canonical variate. Excluding the flow of primary products and light industrial goods from consideration because it was previously mentioned in the discussion, we can say that, regarding this canonical variate, the flow of chemical products is greatest between regions where either activities in primary industries or population change is dominant. It should also be pointed out that, unlike the case with the first and second canonical variates, accessibility in the third does not play a major role as a formative factor.

The fourth canonical variate has a canonical correlation coefficient of 0.223. In terms of commodity flow patterns, the flow of agricultural products and related goods, and of miscellaneous goods, are the main contributors to the canonical vectors, while regional attribute patterns are largely effected by port and rural activities. A comparison of Figures 5.2-c and 5.2-e, and Figures 5.3-b and 5.3-c can allow us to make the following conclusions from this canonical variate: First, the flow of agricultural products and related goods are concentrated in regions with thriving port activities, but the destinations of this flow are regions where rural activities are robust. In other words, we can say that agricultural products and related goods flow from regions where port functions are dominant to areas where rural activities are dominant. Moreover, due to the positive-negative relationship among canonical coefficients, the respective activities in these regions diminish as the flow of miscellaneous goods increases.

The fifth canonical variate has a correlation coefficient of only 0.111, meaning that the degree of correspondence is low. This variate, which is characterized by the flow of charcoal and coal (Y_6), is, in terms of regional attribute patterns, closely related to mining activities. Since the canonical coefficients are mutually positive-negative, there is little flow of charcoal and coal in links between regions which both have considerable of mining activity.

From the above extrapolation of 5 interpretable canonical variates of statistical significance, we have been able to confirm a relationship of intense mutual dependency between the patterns of commodity flow and regional attributes. In brief, the social and economic characteristics of each region, as well as similarities or differences in regional distribution as they concern geographic location, play a role in creating an extraregional linkage relationship of commodity flow that has a distinct order. At the same time, this linkage relationship can be said to regulate the spatial arrangement of the social and economic attributes of a region.

5.5. Conclusion

In this chapter, Barry's *Geographical Field Theory* was used to explain the regional structure of commodity flow in Japan. Here, the major commodity flow patterns were derived using O-D matrix data for the movement of 30 types of commodities among the prefectures of the country. The socio-economic and proximal interdependency of each region was analyzed and examined using canonical correlation analysis. The major findings can be summarized as follows:

1) Dyadic factor analysis related to commodity flow revealed 6 major flow patterns, i.e., flow patterns for a) manufactured goods, b) primary products and light industrial goods, c) agricultural products and related goods, d) chemical products, e) miscellaneous goods, and f) charcoal and coal (see Figure 5.2).

2) The regional flow patterns of these 6 types of goods exhibit the following spatial characteristics. A vast majority of manufactured products move within the Pacific Belt Zone, especially among the three major metropolitan areas. While the movement of primary products and light industrial goods occurs throughout the country, it is centered on Hokkaido, Tokyo, Aichi, Osaka, and Fukuoka. Agricultural products and related goods are mostly imported, so the hinterlands of large trading ports (Kanagawa, Hyogo, etc.) act as distribution points for these goods.

Chemical products, whose main center of production is western Japan, are transported relatively long distances, in contrast to miscellaneous goods, which are usually shipped only short distances and for which the cores of each of these regions form a closed system of linkages.

The movement of charcoal and coal can be divided into two regional systems, one based in Hiroshima with the hinterlands of western Japan, and the other based in Tochigi, Saitama, Tokyo and Kanagawa and their respective hinterlands in the Kanto region of eastern Japan.

3) Factor analysis of 40 variables representing the social and economic characteristics of each region was able to provide 6 common factors: a) urban activities, b) rural activities, c) port activities, d) mining activities, e) activities in primary industries, and f) population change. Figure 5.3 shows the spatial distribution of each of these factors.

4) The relation between the commodity flow patterns derived here and regional characteristic patterns including the concept of accessibility was subjected to canonical correlation analysis, resulting in 5 statistically significant canonical variates which could confirm the following interactive relationships:

a) between the flow of manufactured products, and urban activities (related to regional characteristics) and accessibility (first canonical variate),

b) between the flow of chemical products and primary products and light industrial goods, and rural and mining activities and accessibility (second canonical variate),

c) between the flow of chemical products, and activities in primary industries and population change (third canonical variate),

d) between the flow of miscellaneous goods and agricultural products and related goods, and port and rural activities (fourth canonical variate), and

e) between the flow of charcoal and coal, and mining activities.

This evidence indicates that the system of extraregional linkages of commodity flow is the result of a complex interweaving of spatial differences which exist among the regions. On the other hand, spatial differences were formed by differences in flow patterns in what may be called a mutually regulatory relationship.

Using certain research results such as those mentioned above, this chapter was able to reconfirm the validity of *Geographical Field Theory* for regional analysis. Now, I would like to talk about future research.

Research which has been conducted so far, including the present study, has only looked at the interactive relationship between spatial attributes and commodity flow patterns for one time period, that is, it has merely applied *Geographical Field Theory* to a static situation. However, while the flow patterns in one time period might be regulated by regional attributes, the opposite might hold true for the next period. Therefore, we can consider this relationship to be in a state of flux in which one element can affect the other.

Given the increasing use of temporal and spatial analysis (spatial processes research) the dynamic extension of *Geographical Field Theory* will undoubtedly be able to provide an effective framework for regional analysis. In coming study, the author intends to begin formulating such a dynamic field theory.

Note

1. Linkages with factor scores of 2.0 or higher were depicted in graphic form, and the size of the linkages was divided into three levels based on these scores. The higher the factor score, the greater the flow volume between regions. It should be noted that standardized values were used as factor scores.

References

Barr, B.M. (1970) *The Soviet Wood-Processing Industry—A Linear Programming Analysis of the Role of Transportation Costs in Location and Flow Patterns*, University of Toronto Department of Geography Research Publications **5**, Toronto, 134p.

Berry, B.J.L. (1966) *Essays on Commodity Flows and the Spatial Structure of the Indian Economy*, University of Chicago Department of Geography Research Paper **111**, Chicago, 334p.

Berry, B.J.L. (1968a) Interdependency of spatial structure and spatial behavior: a general field theory formulation, *Papers and Proceedings, Regional Science Association* **21** (2), 205-227.

Berry, B.J.L. (1968b) A synthesis of formal and functional regions using a general field theory of spatial behavior, in Berry, B.J.L. and Marble, D.F. eds., *Spatial Analysis; A Reader in Statistical Geography*, Prentice-Hall, New Jersey, 419-428.

Black, W. (1973) Toward a factorial ecology of flows, *Economic Geography* **49** (1), 59-67.

Chojnicki, Z. and Czyz, T. (1973) Structural changes of the economic regions in Poland: a study by factor analysis of commodity flows, *Geographia Polonica* **25** (1), 31-47.

Clark, D. (1973) The formal and functional structure of Wales, *Annals of the Association of American Geographers* **63** (1), 71-84.

Davies, W.K.D. and Thompson, R.R. (1980) The structure of interurban connectivity: a dyadic factor analysis of Prairie commodity flows, *Regional Studies* **14** (4), 297-312.

Freeman, D.B. (1973) *International Trade, Migration, and Capital Flows*, University of Chicago Department of Geography Research Paper **146**, Chicago, 201p.

Higashi, K. and Moridaira, S. (1980) Multivariate analysis of spatial interaction: case study of commuting flow in Osaka, *Research Paper of Computer Center*, Kokushi Kan University, **1**, 31-48. (J)

Holsman, A.J. (1975) Interstate interaction pattern in Australia, *Australian Geographical States* **17** (2), 131-154.

Holsman, A.J. (1979) Freight flows in Australian economy, *Australian Geographical Studies* **13** (1), 41-61.

Ichiminami (1978) Metropolitan structure based on socio-economic characteristics and intrametropolitan journey-to-work in the Nagoya Metropolitan Area, *Geographical Review of Japan* **51** (7), 545-563. (J with E)

Kanno, M. (1976) Canonical analysis of commodity flows and socio-economic structure in major U.S. metropolitan areas, *Geographical Review of Japan* **49** (4), 197-216.

McConnell, J.E. (1967) The middle east; competitive or complementary?, *Tijdschrift voor Econ. En Soc. Geografie* **58** (1), 82-93.

Mitsuhashi, S. (1978) *Japanese Commodity Flows*, University of Chicago Department of Geography

Research Paper **187**, 172p.

Schwind, P.J. (1975) A general field theory of migration; United States, 1955-1960, *Economic Geography* **51** (1), 1-16.

Ullman, E.L. (1954) Geography as spatial interaction, in Revzan, D. and Englebert, E.S. eds, *Interregional Linkages*, University of California, Berkeley, 1-12.

(J = in Japanese; J with E = in Japanese with English abstract)

CHAPTER 6

ACCESSIBILITY CHANGE IN THE JAPANESE URBAN SYSTEM

6.1. Introduction

The Japanese government implemented a policy of national seclusion in 1639 which was maintained until the mid-19th century. This long isolation significantly delayed the modernization of Japan. However, after the Meiji Revolution of 1867 swept away the traditional feudal system, the new government, aware of the serious lag in modernization, swiftly incorporated advanced western technologies. The key to Japanese industrialization in this earliest period was the introduction of railways, which played an important role in economic development by increasing the size of markets and the potential for economic specialization (Murayama, 1992).

The network expansion was isomorphically correspondent with regional development in Japan. The traditional Japanese urban system, linked by coastal shipping and primitive road-system, was completely reorganized by the emergence of railways which created new connections and greatly increased the speed of existing links. Even today, compared with railways, the role of domestic air travel is relatively small. Clearly the growth of a city is influenced by its degree of railway connectivity: rapidly growing cities are those which have dense connectivity.

This chapter examines the integration process of the Japanese urban system after the Meiji period, paying special attention to changes in accessibility of cities as attributes of urban systems. A city is treated here not as an independent entity, but as an element of a larger integrated system.

6.2. Development of Japanese Rail Transportation

Figure 6.1 shows the changes in number of the passengers carried and the passenger-kilometers traveled by railways between 1891-1990. The construction of railways in Japan began in 1872. The first railway connected Tokyo with Yokohama (29 km) in only 53 minutes, about one-fifth of the time previously required. The rail network expanded rapidly, reaching a total length of 158.3 km in 1880. After that, the network was extended eastward and westward on a national scale. In 1889 the Tokaido line reached Kobe and two years later, the Tohoku line from Tokyo to Aomori was completed.

The industrial revolution and events such as the *Sino-Japanese War* (1894-95) and the *Russo-Japanese War* (1904-05) increased domestic traffic flows, with the network attaining a length of 7,793.7 km by 1905. With the nationwide network being constructed within the next decade, the railways enjoyed the prosperity that came with rapid economic development. The Japanese financial panic of 1927 and the beginning of a global depression two years later, however, disturbed railway development between the late 1920s and the early 1930s. The events leading up to World War II produced a rapid increase in traffic followed by a sudden decline in the 1940s as the war dealt a nearly fatal blow to the Japanese economy. Figure 6.1 shows that, during the high economic growth period between the late 1950s and the early 1970s, intercity passenger flows increased. The network reached a length of 27,507.8 km in 1960, before rail transportation began to be overshadowed by road transportation. After peaking in the 1970s, railway has began to decline to the point where by 1985, several hundred km of inefficient lines had been abandoned.

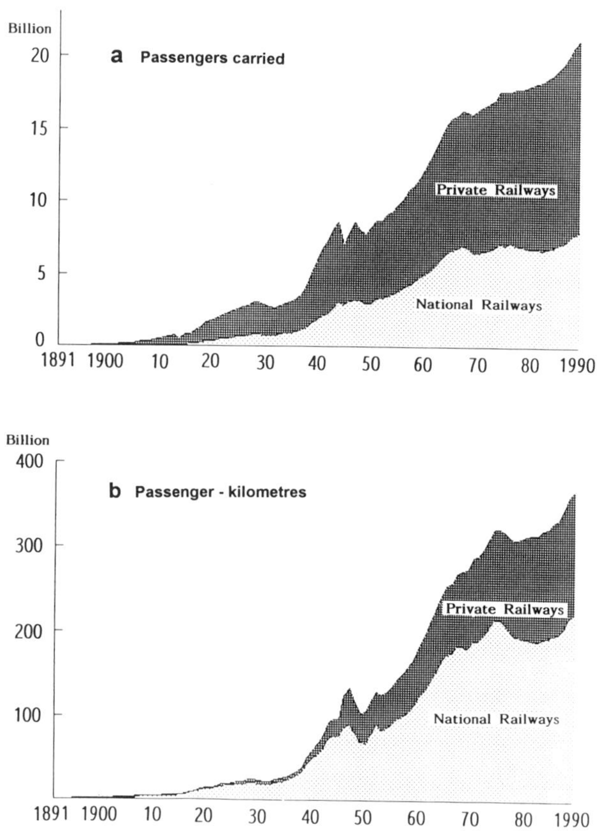

Figure 6.1. Development of Japanese railway transportation (1891-1990)
Source: JNR and others.

Japan has both private and national railways.[1] The latter carry fewer passengers (Figure 6.1-a), but account for a larger proportion of passenger-kilometers (Figure 6.1-b), because the National Railways have been responsible for the longer distance flows, thus making more important contributions to the spatial organization of the national urban system.

6.3. Methodology

The six years, 1898, 1904, 1915, 1934, 1961 and 1990 and the 59 cities with prefectural offices were selected for analysis. The locations of these 59 cities are shown in Figure 6.2. The data employed here were collected from the railway time tables from each study year.

The first stage procedure computed the connectivities and travel time between 59 cities using a time distance matrix of 59 rows and 59 columns to explore the spatial enlargement process of the railway network. The second stage measured the accessibility value for each city employing Shimbel's *Shortest Path Method* (Shimbel, 1953). Using this approach, the changes in accessibility surface from one period to another were calculated to estimate which cities increased or decreased their potential accessibility.

The final stage investigated how the Japanese urban system has been integrated over time, from the perspective of the intercity relationships. ALSCAL, a form of Multi Dimensional Scaling, was employed to delineate the relative locations of cities based on the travel time (Young, *et al.*, 1978).

6.4. Expansion of the Railway Network and the Reduction of Travel Time

Table 6.1 shows the changes in the travel time from Tokyo to five major cities in Japan. Except for the loss of accessibility from Tokyo to Hiroshima and Fukuoka during 1898-1904 when passenger transport was sacrificed in favor of military uses, the increase in accessibility has been continuous and dramatic. For example, the travel time from Tokyo to Osaka, the second largest city, declined drastically from 15 hours and 35 minutes in 1898 to 2 hours and 52 minutes by 1990, and during the three decades between 1961 and 1990, the travel time was cut in half. In 1964 the bullet train, Shinkansen, was opened between Tokyo and Osaka, the opposite edges of the Japanese Megalopolis within which many large cities are densely agglomerated. Soon thereafter the Shinkansen was extended into Fukuoka at the western edge of the Pacific Belt Zone, reducing the travel time between Tokyo and Fukuoka to one third.

Time distances between cities are shown in Figure 6.3.[2] The 1898 map indicates that the network structure was immature. Only 34 cities out of the 59 cities were connected by railways (Figure 6.3-a). Although, access to the other three islands was still inadequate, a spinal trunk line running through Honshu had been completed. Rail travel

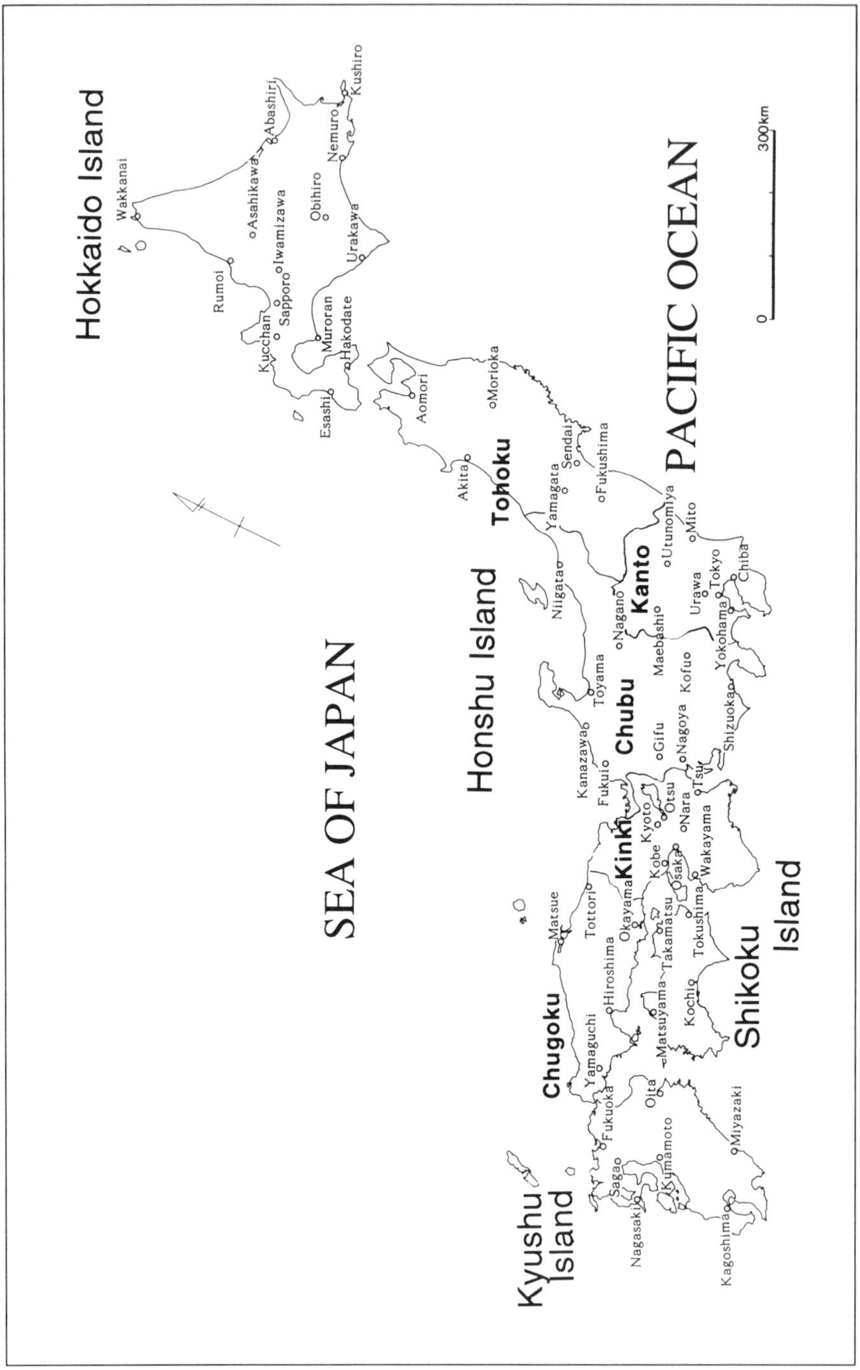

Figure 6.2. Study area

Table 6.1. The travel time from Tokyo to five major cities

Year	Osaka	Hiroshima	Fukuoka	Sendai	Sapporo
1898	15:35	26:57	42:45	12:00	56:40
1904	12:38	28:11	42:59	9:35	50:47
1915	12:33	20:40	30:39	8:08	36:00
1934	8:00	14:39	20:43	5:50	26:14
1961	6:30	11:10	16:55	4:43	19:55
1990	2:52	4:30	5:52	1:50	10:52

Note: hours: minutes.

Source: Railway timetables.

was still slow. For example, it took 16 hours and 27 minutes from Tokyo to Kobe and 25 hours and 10 minutes from Tokyo to Aomori. Rail transportation was still not popular with the average citizen because of its inconvenience and high fares.

Six new cities joined the network by 1904 (Figure 6.3-b), three of which were located in the "Sea of Japan" region. In 1906 the opening of the first super express reduced travel time between Tokyo and Kobe to 13 hours and 40 minutes. Furthermore, it became possible to travel the 1,134 km between Tokyo and Shimonoseki, on the western edge of Honshu in 25 hours. By the end of the Meiji period (1912), the government had nearly completed the main trunk lines in Honshu except for the Sea of Japan region. However, three years later there were still no railways on the island of Shikoku (Figure 6.3-c).

During the Taisho period (1912-1926), the network was expanded into the more isolated regions around the Sea of Japan and the islands of Hokkaido, Kyushu and Shikoku. Many feeder lines were constructed to link small cities with larger ones. The long "Sea of Japan rail-system" connecting Aomori in Tohoku and Maibara in Kinki was completed with the opening of the Hokuriku line in 1923 and the Haetsu line the following year, thus integrating these formerly inaccessible areas into the Japanese urban system. During this period, other new lines were constructed in other parts of the country. In 1923 the Sanin line reached Masuda in Shimane prefecture, while in Hokkaido the Nemuro line went into operation in 1921 and the Soya line in 1922. In Kyushu the Nippou line was opened in 1923 and the Kagoshima line in 1926. The nationwide integration of these local lines promoted smooth exchanges of goods and passengers. The use of railways gradually became part of the daily life of the Japanese people. This period also witnessed a notable increase in train speed; for example, the 12 hours between Tokyo and Osaka in 1912 was reduced to just 11 hours by 1926 (the last year of the Taisho period), while the 20 hours between Tokyo and Aomori in 1912 declined to 17 hours by 1926.

In 1930, a super-express was established, connecting Tokyo with Kobe within 9 hours instead of the 11 hours and 38 minutes previously needed for the journey. By 1934 the network had reached the mature stage (Figure 6.3-d). Only three cities (two in Hokkaido and one in Shikoku) remained without railway connections. Although three cities in Shikoku joined the network for the first time, the linkages between Shikoku and Honshu were still very weak, as it took over two hours by railway ferry to travel

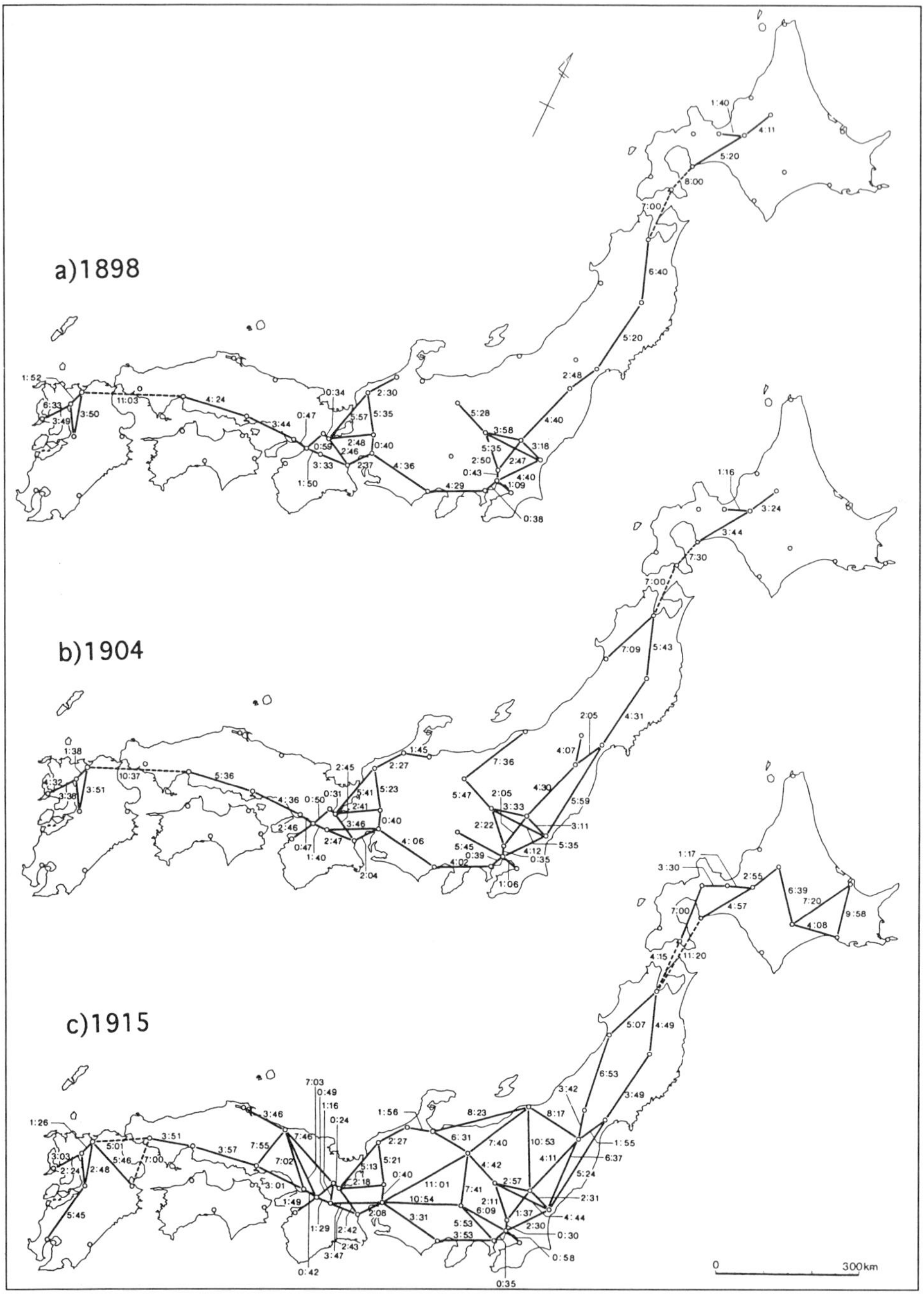

Figure 6.3. Travel time between cities (a, b, c)

Notes: 1) Numbers refer to rail (including ferry) travel time between adjacent cities (hours: minutes).
2) Broken lines indicate indirect connections using ferries.

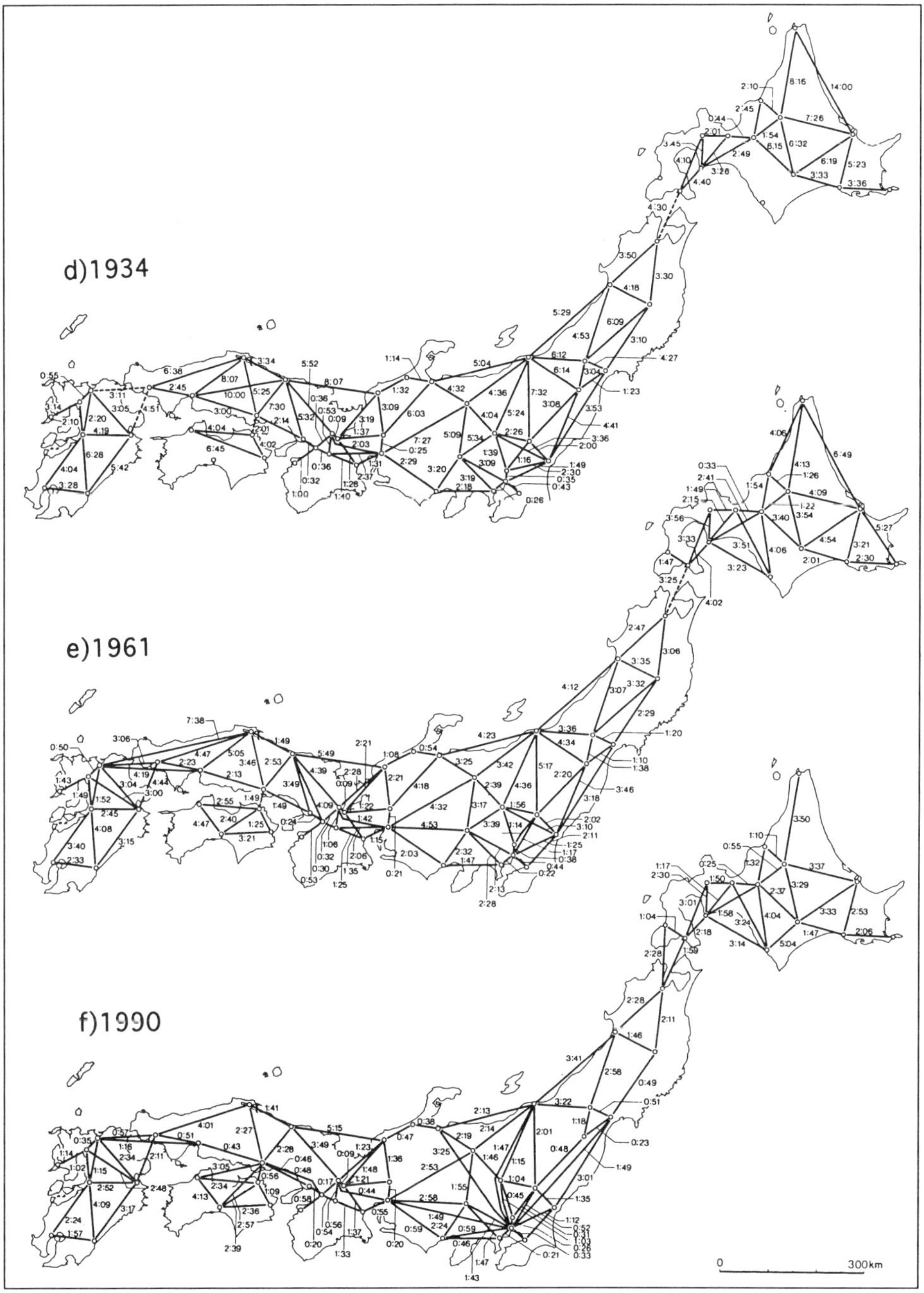

Figure 6.3. Travel time between cities (d, e, f) *(continued).*

Notes: 1) Numbers refer to rail (including ferry) travel time between adjacent cities (hours: minutes).
2) Broken lines indicate indirect connections using ferries.

between the nearest cities in Honshu and Shikoku. It was not until in 1961 that all 59 cities were incorporated into the network system (Figure 6.3-e). The opening of the Kanmon tunnel improved the access between Honshu and Kyushu. The drastic time reduction was evident: only 6 hours and 30 minutes were required from Tokyo to Osaka. This train was called the "business express" because one-day trips became possible for business men between Tokyo and Osaka.

By 1990, the four Japanese islands had been completely connected by railways, eliminating the need for ferries (Figure 6.3-f). Kyushu and Hokkaido were both linked to Honshu by tunnels, while Shikoku was connected to Honshu by a bridge. The rapid extension of the Shinkansen network since 1964 has expanded the people's behavioral space. The previous network structure was swept away by the three Shinkansens, i.e., the Tokaido-Sanyo from Tokyo to Fukuoka, the Tohoku from Tokyo to Morioka and the Joetsu from Tokyo to Niigata. The formation of the so-called Shinkansen Spinal stretching from Morioka to Fukuoka contributed to the increase in accessibility not only to Shinkansen cities, but also to non-Shinkansen cities in remote areas. Accessibility of the latter was improved by using the Shinkansen as a connection instead of relying on old local lines. The dramatic decrease in the use of local lines led to the elimination of many smaller routes (see Figure 6.3-f). For instance, the Yamagata-Morioka route disappeared because the travel time via Sendai using the Tohoku Shinkansen was much faster, saving passengers one hour and 47 minutes. This kind of route deletion can be seen in the 1990 map.[3] Note that the elimination of the previous connections does not always mean the actual disappearance of rail lines.

6.5. Change in the Accessibility Space

Figure 6.4 shows the accessibility space derived from the *Shortest Path Method* which calculates the total travel time from the origin city i to all the other cities, using the shortest path. This method enables us to evaluate the accessibility differences among cities and across time periods. The accessibility value A_i is calculated as follows:

$$A_i = \sum_{\substack{j=1 \\ (i \neq j)}}^{n} t_{ij}$$

where: n is a number of cities, and t_{ij} is the minimum travel time from i to j.

For example, an accessibility value of 335 for Tokyo in 1990 (Figure 6.4-f) means that it would take a total of 335 hours to travel from Tokyo to all the other 58 cities. The smaller the value for a city has, the more advantageous is its position not only for transport but also for efficient economic activity. It is a more favorable place to make business trips, to transmit or receive information and to deliver goods.

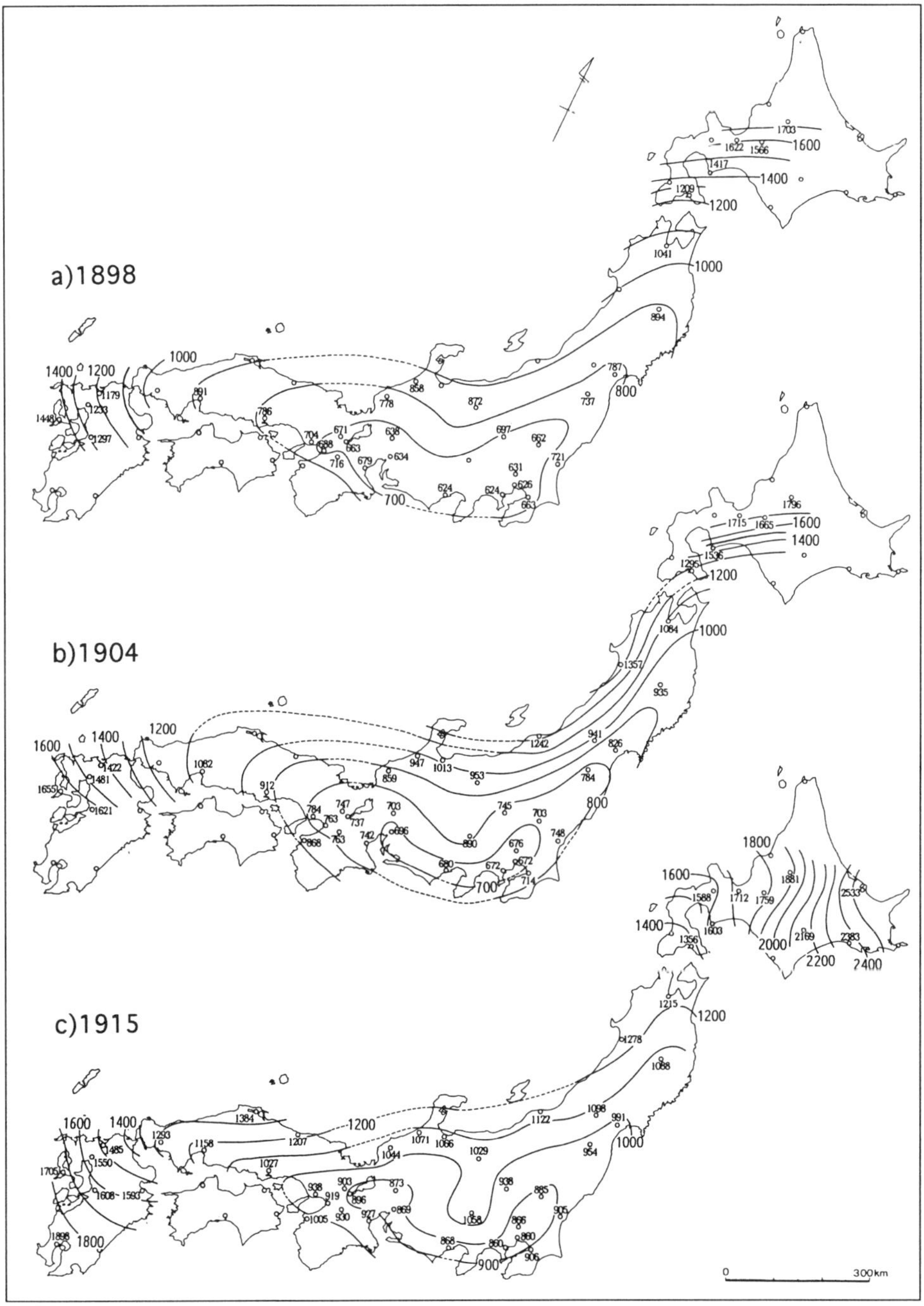

Figure 6.4. Accessibility space (a, b, c)

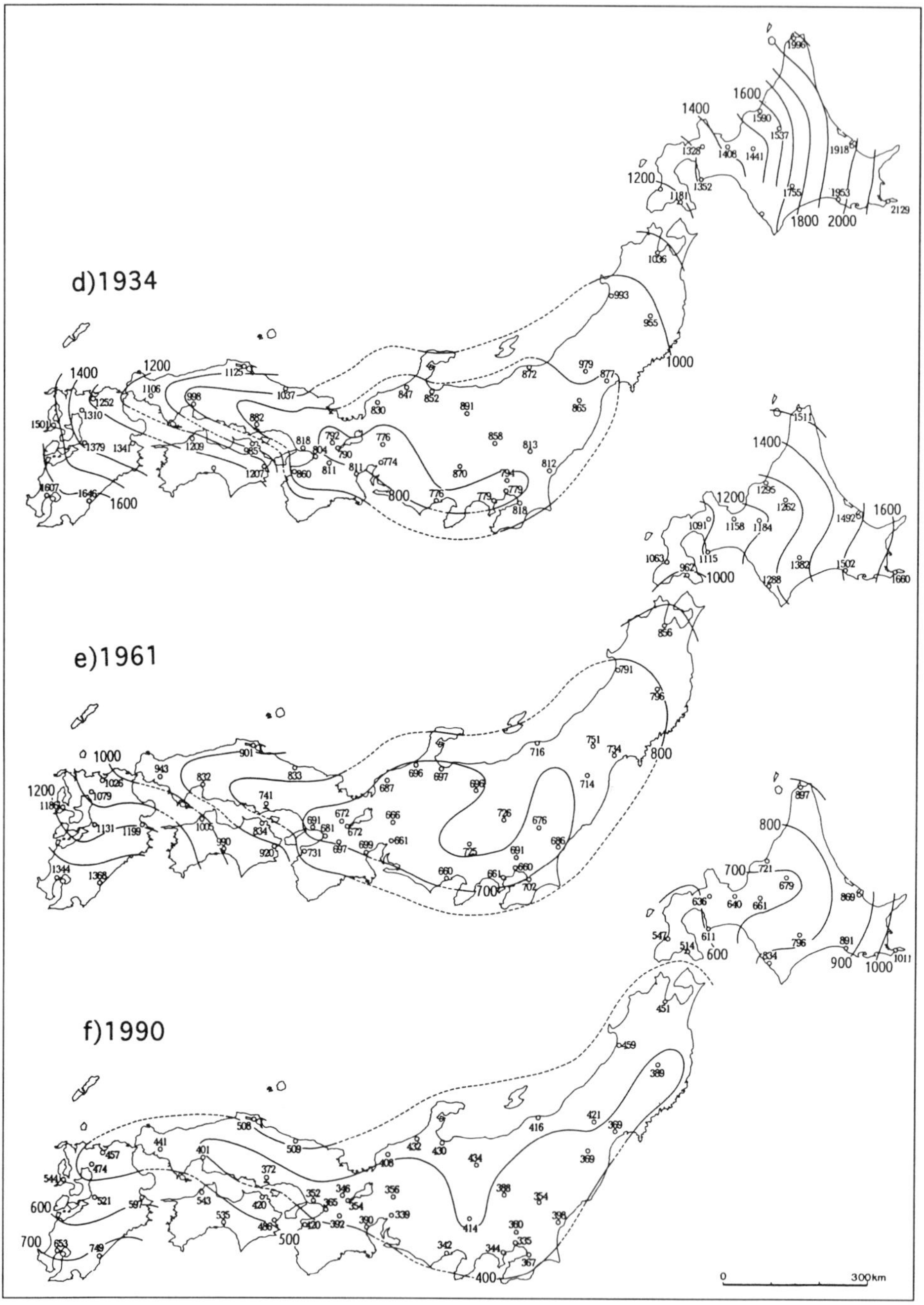

Figure 6.4. Accessibility space (d, e, f) *(continued).*

In 1898, both Yokohama and Shizuoka had the smallest value (624) while Asahikawa in Hokkaido had the largest (1,703). Values were smaller in central Japan which occupied the most advantageous position and larger in the peripheral regions where traveling was less convenient. In particular, the cities along the Tokaido line had high accessibility (Figure 6.4-a).

In 1904 (Figure 6.4-b), the accessibility values of each city increased over those in 1898, as network expansion increased the number of destination, thus the total number of the shortest paths cities by six. Furthermore the decrease in real accessibility was evident: the *Russo-Japanese War* in 1904-5 gave the military transport precedence over the public and the railway transportation in western Japan was particularly distorted by this war priority. A tremendous volume of military goods and armies was transported to the Asian continent from the ports via the Tokaido line.

After the wars, this disadvantage in the western cities was transformed into an advantage (Figure 6.4-c). In particular, the cities in Kinki significantly recovered their accessibility. As was the case in 1904, accessibility values increased with the entry of nine more cities around the country. The low accessibility of Kofu in central Japan is worth noting. Although Kofu was linked with four surrounding cities, many more hours than expected were required to travel to and from these cities, making Kofu an isolated city.

By 1934, cities along the Sea of Japan had increased their accessibility (Figure 6.4-d). Though three cities in Shikoku entered the urban system for the first time, their accessibilities were very low in spite of the relatively short Euclidean distance to the Kinki region. This low accessibility had improved to some extent by 1961 (Figure 6.4-e). The 1961 surface more or less reflects the actual location of cities, revealing the relative equalization of accessibility based upon the time distance from the center. In other words, the elliptical accessibility contours declined moderately in proportion to the distance from the heart of central Japan. This pattern was derived from the slow growth of accessibility in the leading regions and the quick growth in the isolated regions including the Sea of Japan region, Shikoku, Kyushu and Hokkaido. By 1990, however, this well-proportioned pattern had been again distorted by the three Shinkansens (Figure 6.4-f) as Shinkansen cities increased their accessibilities and the cities outside the Shinkansen system decreased their accessibilities.

In comparing the six figures from 1898 to 1990, the following three broad trends can be noted. First, central Japan had high accessibility and the peripheral regions low accessibility over the span of all six periods. Though this fundamental pattern had not changed by 1990, some regional variations were evident. For example, there was a rapid improvement in the accessibility of Tokyo from 1961 to 1990 owing to the concentration of the three Shinkansen terminals. The rapid economic growth of Tokyo since the 1970s is related to this change in transport accessibility. The improvement in accessibility in Kyushu to the west was consistently greater than that in Hokkaido to the north over time, especially from 1961 to 1990 with opening of a Shinkansen terminal in Kyushu.

Second, the number of accessibility contours gradually lessened over time. This reduction in the accessibility gap was a result of the development of a nationwide network and the spread of frequent service in all directions.

Third, the contour configuration evolved from long elliptical contours in the early period before 1915, through the rounder contours in 1934 and 1961, and finally to a linear structure along the Shinkansen route in 1990.

The changes in the network attributes after 1898 are summarized in Table 6.2. The population of Japan has tripled since that time and the operation-kilometers of the JNR (Japanese National Railways) increased from 1,415 km in 1898 to 20,519 km in 1961. The slight decrease from 1961 to 1990 was due to the abandonment of unprofitable local lines. As discussed earlier, the increases in both the maximum and minimum values of accessibility from up to 1915 was attributed to the increase in the number of nodes (cities). The index "coefficient of variation of accessibility" (C.V.) shows the variance of accessibility among cities (nodes), which decreased over time until the construction of the Shinkansens between 1961 and 1990. After 1961, the Shinkansens have expanded the accessibility differences between "Shinkansen" and "non-Shinkansen" cities.

Table 6.2. Changes in the railway network attributes

Factor	1898	1904	1915	1934	1961	1990
Population of Japan (thousands)	42,886	46,135	52,752	68,309	94,287	123,611
JNR network size (km)	1,415	2,449	9,268	16,535	20,519	20,157
Number of nodes	34	40	49	56	59	59
Number of links	40	48	74	104	120	128
Max. accessibility	1,703	1,796	2,533	2,129	1,660	1,011
Min. accessibility	624	672	860	776	660	335
Coefficient of variation of accessibility	0.358	0.342	0.333	0.326	0.295	0.329

Table 6.3. Results of ALSCAL

	1898	1904	1915	1934	1961	1990
Stress	0.021	0.040	0.036	0.042	0.037	0.054
RSQ	0.999	0.994	0.996	0.994	0.996	0.991
Iteration	4	5	5	6	5	4

Notes: 1) Stress values are based on Kruskal's stress formula.
2) RSQ values refer to the proportion of variance of the scaled data in the partition which is accounted for by the corresponding distances of the data.

6.6. Changes in Japanese Urban System

This section examines changes in the shape of the Japanese urban system using the distribution map of cities derived from ALSCAL, a form of MDS. As the stress values

(Kruskal, 1964) indicate, the application of ALSCAL to the data was quite successful (Table 6.3).

The locations of the 59 cities were represented as the three-dimensional pattern shown in Figure 6.5. The first and second dimensions show the locations of cities (latitude and longitude) and the third dimension shows the locality (local variance). In time distance, the size of the Japanese urban system has shrunk to about one fifth during this century. In the ALSCAL analysis, however, the six maps have been standardized to be equal in size for the sake of comparisons over time.

The figure for 1898 indicates that cities ranged east and west like a chain with the exception of three cities in Chubu which were inaccessible from the surrounding cities (Figure 6.5-a). The agglomeration of cities in Kinki suggests that these cities possessed a strong interdependency with each other throughout the region. The actual locations of cities were reproduced well on the MDS space, because the network structure was very simple during this period. Though extending slightly into the north-south direction in 1904, the Japanese urban system did not change its fundamental shape between 1898 and 1904 (Figure 6.5-b). Kofu, Nagano and Niigata in Chubu were plotted far from the actual locations because of low accessibility to adjacent cities.

The 1915 map contains two features (Figure 6.5-c). First, some cities in Hokkaido were poorly integrated with each other on account of weak intercity connections. The population density in Hokkaido was very low and passenger transport was very limited during this period. Second, cities in Chugoku were linearly distributed from north to south. This spatial form is quite different from the actual pattern with which we are familiar. Matsue and Tottori were more accessible to cities in Kinki and Chubu than to cities in Chugoku. This unfamiliar pattern can be attributed to the inaccessibility between Matsue-Tottori and Hiroshima-Yamaguchi. The north-south interaction by rail transportation was very weak in Chugoku. But the opening of three railways connecting these four cities in Chugoku had removed this unfamiliar shape by 1934 (Figure 6.5-d). The increase in accessibility in the Sea of Japan region integrated the Chubu region on an unprecedented scale. But cities in Hokkaido were still mutually inaccessible.

The MDS space for 1961 resembles the actual map except for Shikoku cities which, contrary to our expectations, are plotted in the upper part of Chugoku (Figure 6.5-e). The Shikoku cities plotted by ALSCAL were very unstable in location due to the single route connecting Shikoku with Honshu.

By 1990, the openings of the three Shinkansen lines had changed the shape of the MDS space (Figure 6.5-f). Shinkansen cities were strongly pulled toward the center of the graph due to the time reduction, and are displayed linearly with narrow intervals. This relative convergence pushed non-Shinkansen cities outward especially in a negative direction along the Y axis. This Shinkansen impact also pulled Hokkaido and Kyushu toward the origin of the graph making their shape more linear. But the long linear shapes of these two islands also suggest that inaccessibility was still dominant within their regions.

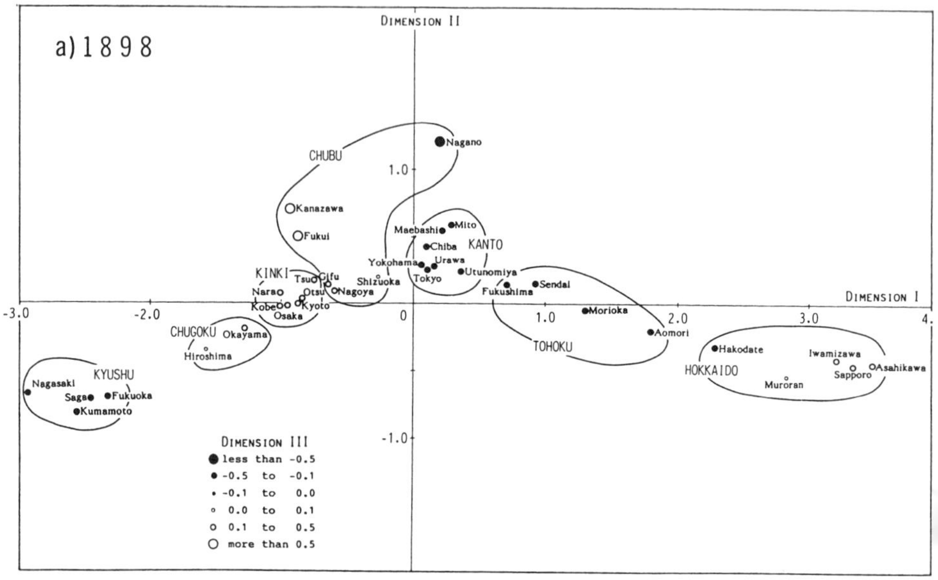

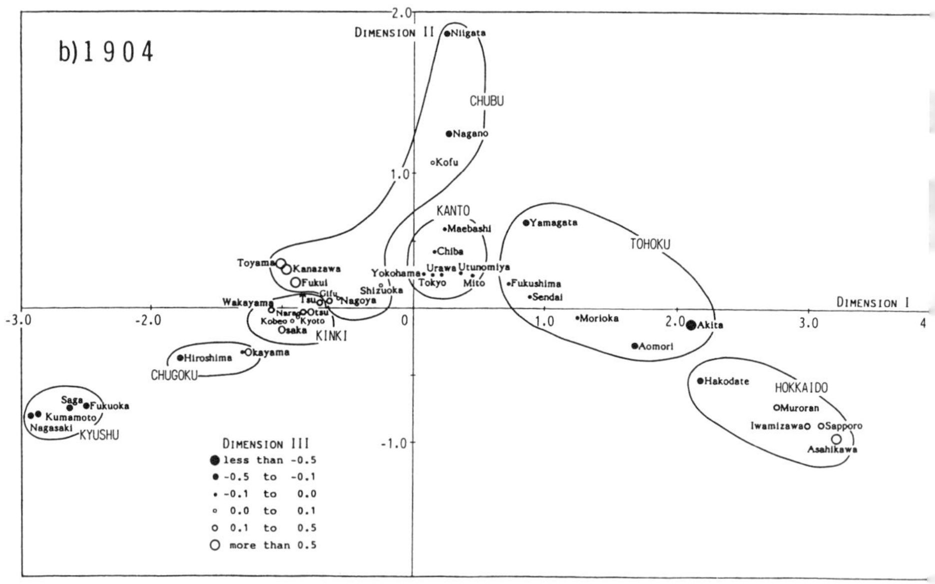

Figure 6.5. Location of cities as derived from ALSCAL (a, b)

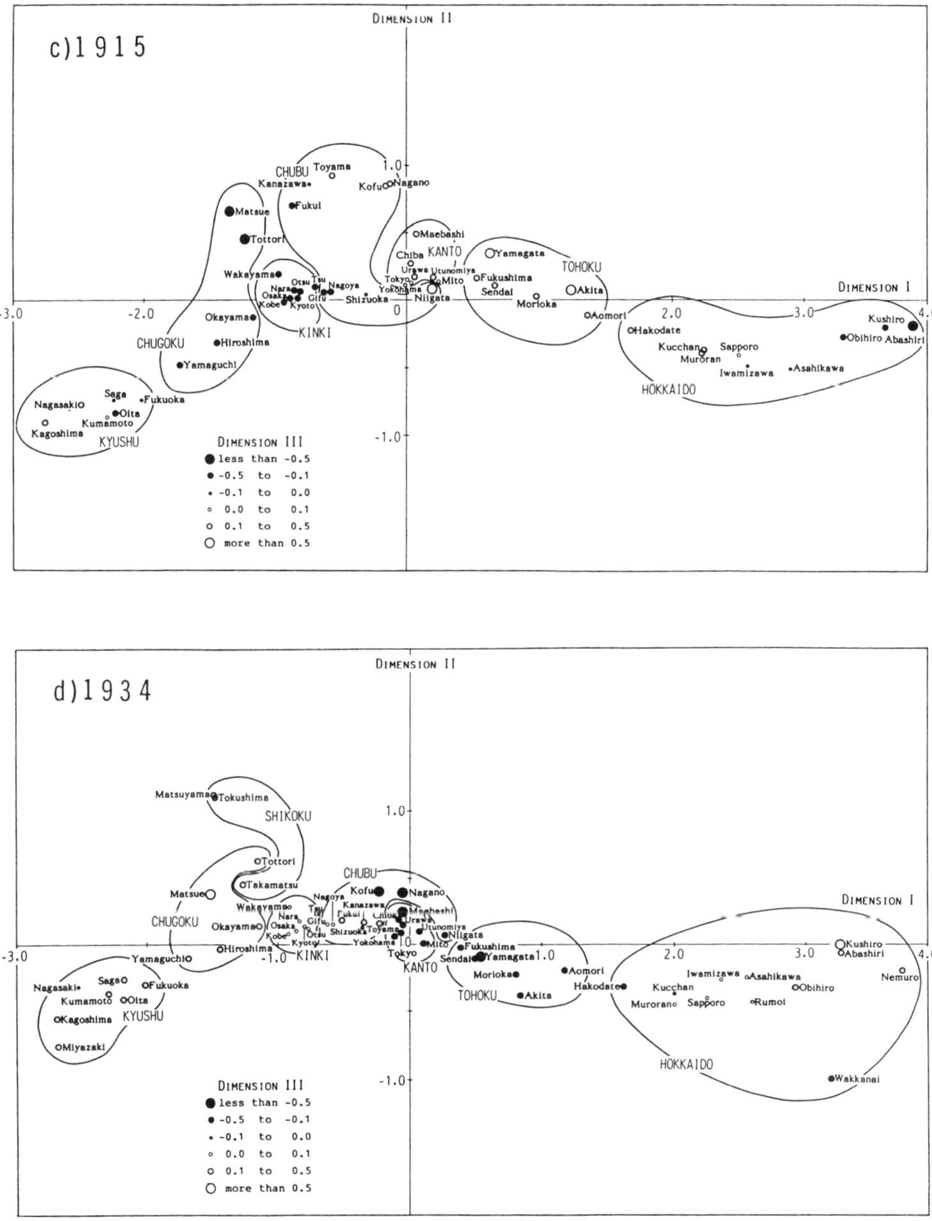

Figure 6.5. Location of cities as derived from ALSCAL (c, d) *(continued).*

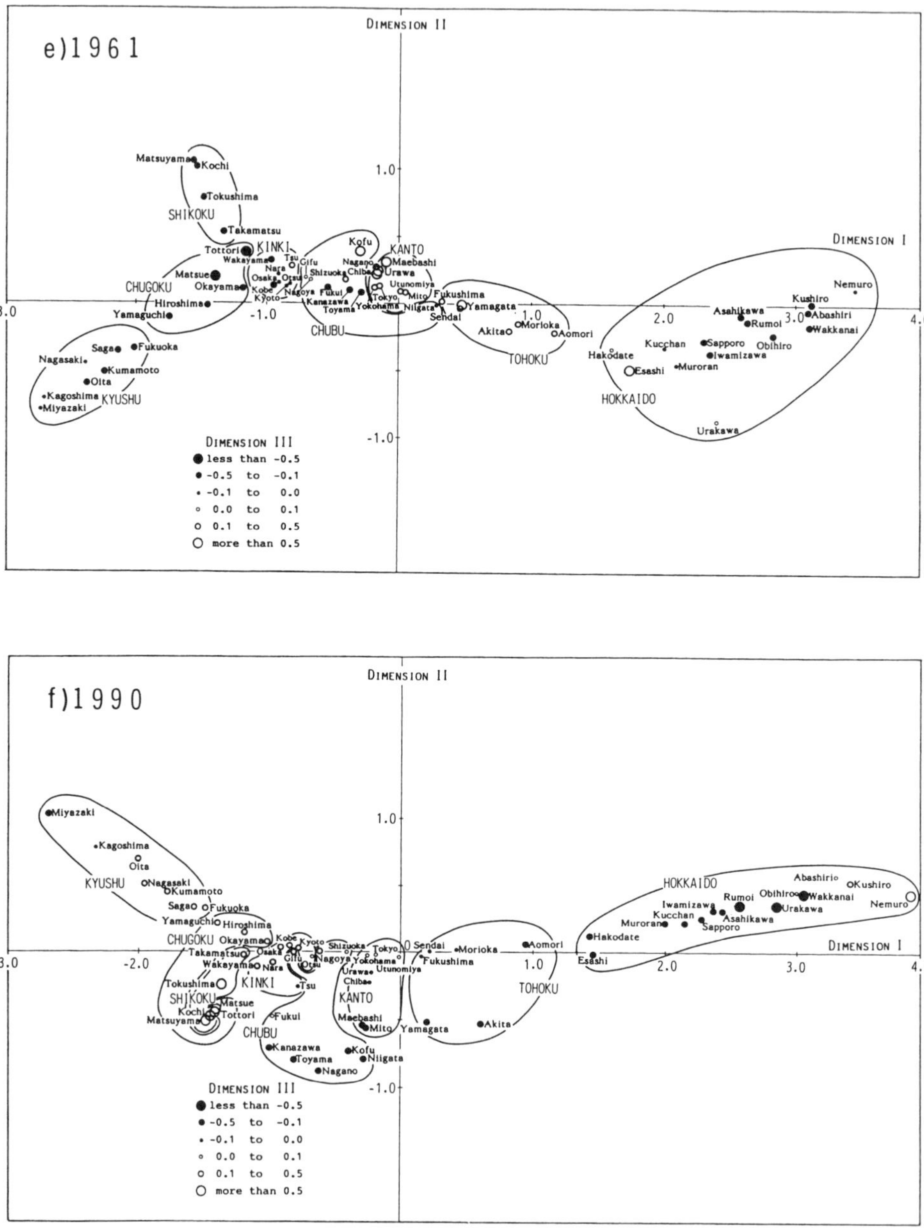

Figure 6.5. Location of cities as derived from ALSCAL (e, f) *(continued).*

The stress value in 1990 is the largest (0.054) of the six time periods (see Table 6.3). This suggests that the 1990 MDS space is very complicated because the three dimensional solution is least appropriate in 1990.

Overall, two characteristic patterns in the time-series change during this century can be noted. First, the locations of cities in Kanto and Kinki were virtually constant through the six periods because of their strong and stable inter-connectivity, while cities in Chubu, Chugoku and Shikoku moved their relative locations here and there because of their weak connectivity. Second, the location of Tokyo, the largest city in Japan, gradually moved toward the west, reflecting more rapid increase of accessibility in western Japan than in eastern Japan. At the same time, Sapporo, Fukuoka, Sendai and Hiroshima have converged to the origin of the graph. This convergence reveals that these major cities have enjoyed reduced accessibility differentials due to the impact of the Shinkansen (Figure 6.6).

6.7. Conclusion

The increase in intercity accessibility has encouraged and strengthened the unity of the Japanese urban system. Over the course of the century, travel time in Japan has been reduced to one-fifth.

In the earliest period before the 1910s, the Japanese urban system was loosely unified by the few cities linearly arranged from east to west. The original form of the present urban system was established in the 1920s when most of the major cities were incorporated as members of the system. Between 1910 and 1960, the concentrated investment in rail construction and the gradual reduction of travel time had contributed to the balanced development of the Japanese urban system. But the emergence of Shinkansen, the high-speed transit system in the 1960s, destroyed this spatial structure. The Shinkansen cities immediately gained the highest locational advantage, while non-Shinkansen cities in remote regions moved further toward the periphery as the MDS space shows. As a result, the shape of the Japanese urban system has been transformed into a distorted configuration caused by the convergence of Shinkansen-related cities and the divergence of other cities.

These findings pose a new question: how have these accessibility changes influenced the economic growth/stagnation and the interdependency of cities? The answer lies in a quantitative examination of the relationships between accessibility changes and socio-economic trends of city attributes. This theme will be the subject of future research.

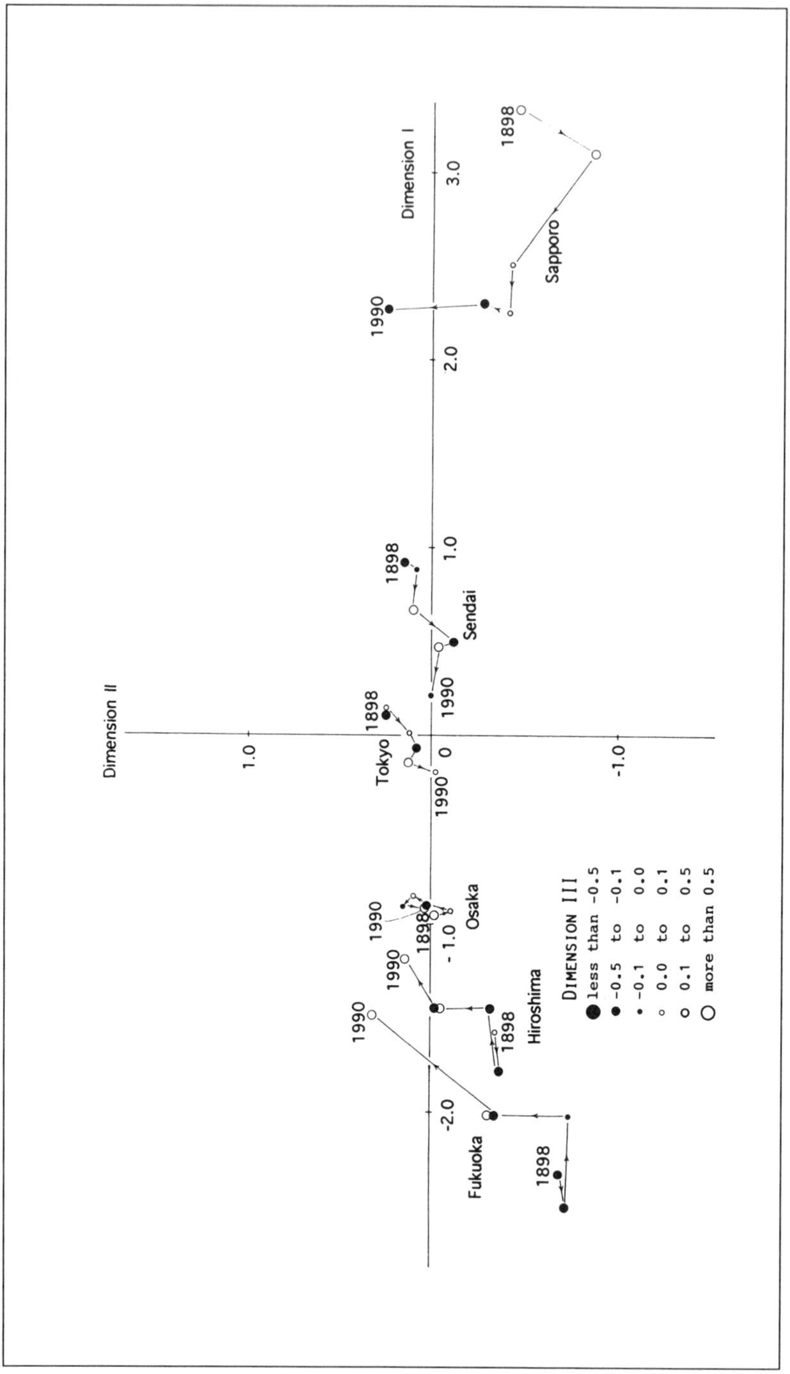

Figure 6.6. Changes in the location of cities as derived from ALSCAL, 1898-1990

Note

1. The Japanese National Railways (JNR) were disbanded and divided into six private companies in 1987. To simplify the discussion, however, these companies are considered here to be national railways.

2. In this analysis, a link is drawn only when a network satisfies the principle of triangle law in which the sum of any two sides of a triangle exceeds that of the third side. Namely, a minimum time route which is not necessarily a direct route is selected in the triangular network. For example, the figure below shows that the direct route between A and C is ignored because the indirect route via B is faster.

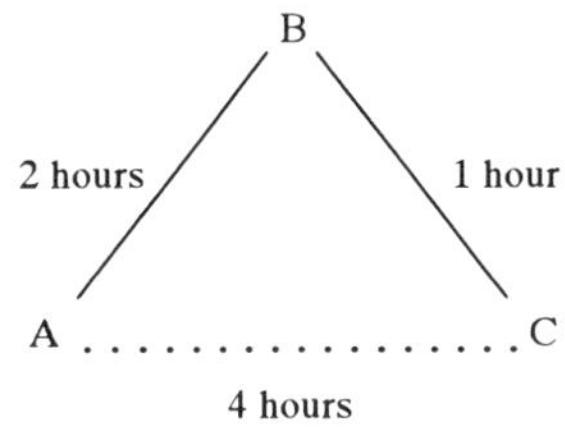

3. see note 2.

References

Kruskal, J.B. (1964) Multidimensional scaling by optimizing goodness of fit to a nonmetric hypothesis, *Psychometrika* **29**, 1-28.

Murayama, Y. (1992) Japanese commodity flows in the mid-Taisho period, *Ann. Rep., Inst. Geosci.*, Univ. of Tsukuba **17**, 13-19.

Shimbel, A. (1953) Structural parameters of communication networks, *Bull. of Math. Biophys.* **15**, 501-507.

Young, F.W., Takane, Y. and Lewyckyj, R.(1978) Three notes on ALSCAL, *Psychometrika* **43**, 433-435.

CHAPTER 7

SPATIAL DYNAMICS OF MIGRATION IN THE JAPANESE URBAN SYSTEM

7.1. Introduction

In this chapter, in order to clarify the origin and destination characteristics of internal migration in Japan, an attempt is taken to elucidate how the Japanese migration patterns changed spatially in the period between 1954 and 1987. The long-term transitional trend of the migration movement at the national level has not been fully examined yet by Japanese geographers. The O-D matrices of 34 years with 46 rows and 46 columns showing the inter-prefectural migration are employed here as the analytical data which are issued by Statistics Bureau, Management and Coordination Agency, Japan. Okinawa Prefecture is excluded from the analysis due to the unavailability of data before 1962. Figure 7.1 shows the study area and the names of prefectures.

The study of human migration forms an interdisciplinary research field, which is related with many disciplines such as geography, economics, sociology and anthropology. Geographers attempt to clarify the mechanism of the migration movement from a spatial point of view by focusing on the flow elements of volume, direction and distance.

7.2. Transition of Migration Patterns in Japan

Before analysis, I would like to outline the transition of migration patterns in Japan briefly. Total migrants[1] including both intra- and inter-prefectural migration in Japan are 6,536,880 in 1987, and among them the inter-prefectural migrants reached 3,171,170. This number corresponds to 5.38 per cent of the total Japanese population 122,263,735 in 1987. It means that 5 out of 100 people move in a year. Figure 7.2 indicates the number of inter-prefectural migrants, their annual increase rate and their sex ratio. From the late 1950s till the early 1970s, the number of migrants rapidly increased and reached the peak in 1971 (4,256,605 migrants). After that year, the number of migrants gradually decreased until 1985, but since 1986 the line of graph has shown the tendency of the slight increase.

The changing tendency of the increase rates suggests that migration was active in the high economic growth period, and was inactive in the low economic growth period. It is estimated that acceleration of economic growth promoted the strong stream of migration. By contrast, economic stagnation weakened the mobility of migration stream. The sex ratio indicates that the male ratio was constantly high through the whole periods. In the high economic growth period, however, the difference between the

Figure 7.1. Study area

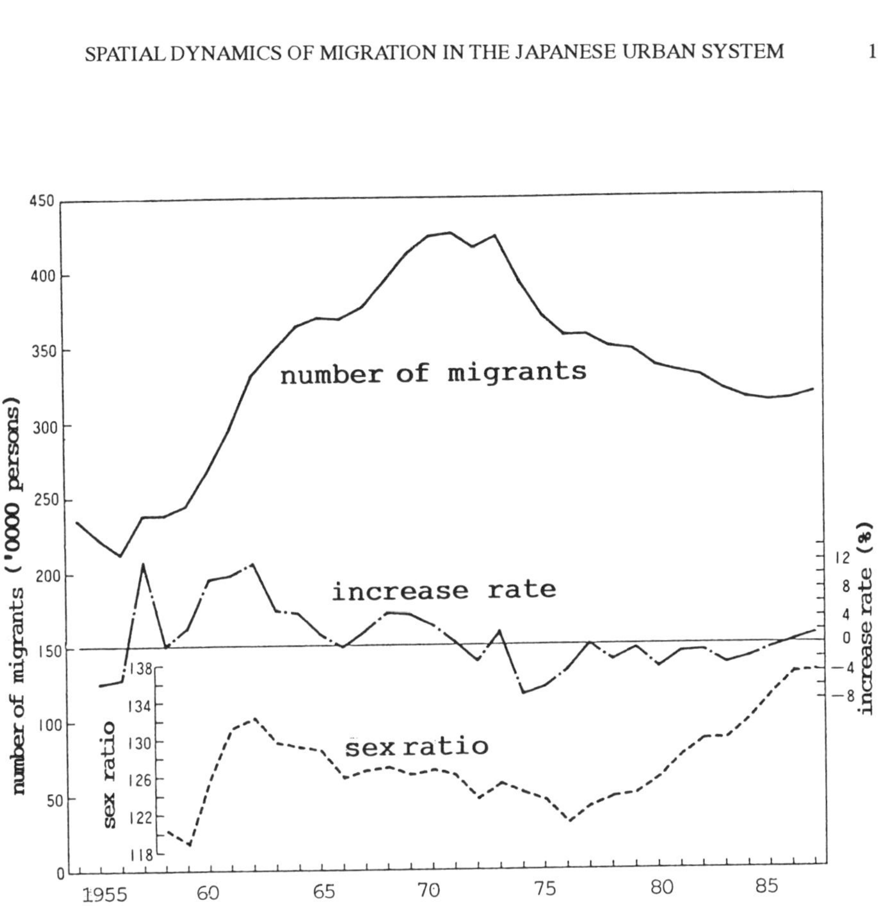

Figure 7.2. Changes in interregional migration flows

number of the male and female migrants became smaller. It suggests that, in the high economic growth period, many male migrants who sought for new employment did not move alone but with their families.

Figure 7.3 reveals the transitional trend of migration rate which is the ratio of migrants to the total Japanese population. If comparing the inter-prefectural migration rate with the intra-prefectural one, we understand that the latter was constantly higher than the former before 1962 and after 1973. In the period between 1963 and 1972, however, the both inter- and intraprefectural migration showed almost the same rates. The interprefectural migration generally grew in the period of high mobility when long distant migration occurred more frequently than that of the short distant.

Table 7.1 indicates the coefficients of variation of 34 years in terms of 2,070 (46 × 45) interprefectural pairs. The larger coefficient means the larger dispersion of flows. In other words, large coefficient indicates that flows are not dispersed uniformly to 2,070 pairs, but concentrate on the particular interprefectural pairs. The coefficients of variation had increased from 1958 to 1973 when the first oil crisis broke out. Therefore the larger coefficients in the high economic growth period signify the greater increase of the number of migrants in the certain interprefectural pairs connecting non-urbanized areas with urbanized areas.

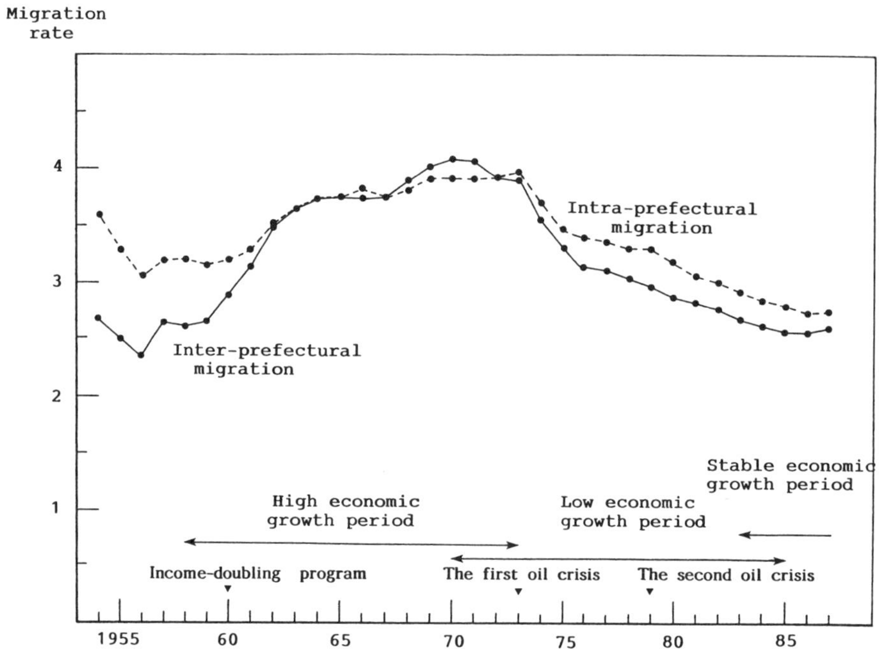

Figure 7.3. Changes in migration rates

Table 7.1. Coefficients of variation of migration flows

Year	Mean	Standard deviation	Coefficient of variation
1954	1136.6	3645.3	3.21
1955	1072.4	3451.0	3.22
1956	1024.9	3313.6	3.24
1957	1149.8	3654.3	3.18
1958	1150.0	3723.2	3.24
1959	1179.9	3930.5	3.33
1960	1294.5	4234.8	3.27
1961	1426.2	4726.1	3.31
1962	1595.5	5423.0	3.40
1963	1678.9	5873.6	3.50
1964	1755.7	6364.1	3.63
1965	1783.4	6411.7	3.60
1966	1779.3	6535.4	3.67
1967	1817.0	6626.5	3.65
1968	1902.0	6922.0	3.64
1969	1988.1	7173.5	3.61
1970	2056.8	7439.6	3.62
1971	2056.3	7326.0	3.56
1972	2005.6	7347.3	3.66
1973	2017.1	7480.4	3.71
1974	1864.0	6690.9	3.59
1975	1758.5	6150.2	3.50
1976	1694.9	6017.7	3.55
1977	1697.2	5869.9	3.46
1978	1657.4	5807.5	3.50
1979	1647.4	5826.2	3.54
1980	1596.9	5529.8	3.46
1981	1573.8	5227.2	3.32
1982	1560.2	5072.8	3.25
1983	1516.9	4914.1	3.24
1984	1488.7	4790.1	3.22
1985	1480.0	4773.5	3.23
1986	1488.1	4874.2	3.28
1987	1506.7	5249.9	3.49

7.3. Change in the Spatial Migration Pattern

In order to grasp the temporal and spatial variation of Japanese internal migration, the analytical framework shown in Figure 7.4 is built. The matrix cube consisted of 46 origins, 46 destinations and 34 years is transformed into two dimensional matrix of 2,070 rows and 34 columns. Then the factor analysis is practiced for this matrix to reduce 34 years into a small number of time dimensions. The result of the analysis is shown in Table 7.2. Two factors with eigenvalues over 1.0 are extracted to explain 99.0 per cent of the total variance.

Factor I having 25 factor loadings over 0.7 indicates the spatial migration pattern after 1963. Figure 7.5-a shows the factor score distribution. Flow lines are classified into four stages based upon the magnitude of factor scores. The larger the factor score is, the more the volume of migration is. Broken lines show negative factor scores less than -1.0. The negative large score appears, when the remarkable difference of migration volume yields between factors I and II. The largest positive flow is the one from Tokyo to Saitama whose factor score is 27.4. There exist 15 flow lines with factor scores over 3.0. Origins and destinations of these 15 flow lines are all included within the Tokyo Metropolitan Area or within the Osaka Metropolitan Area. In the Tokyo Metropolitan Area, migration movements among four regions of Tokyo, Saitama, Chiba and Kanagawa are dominant. Within this metropolitan area, the flows show two-way directional movement, but to be more precise, the migrants from Tokyo to its surround-

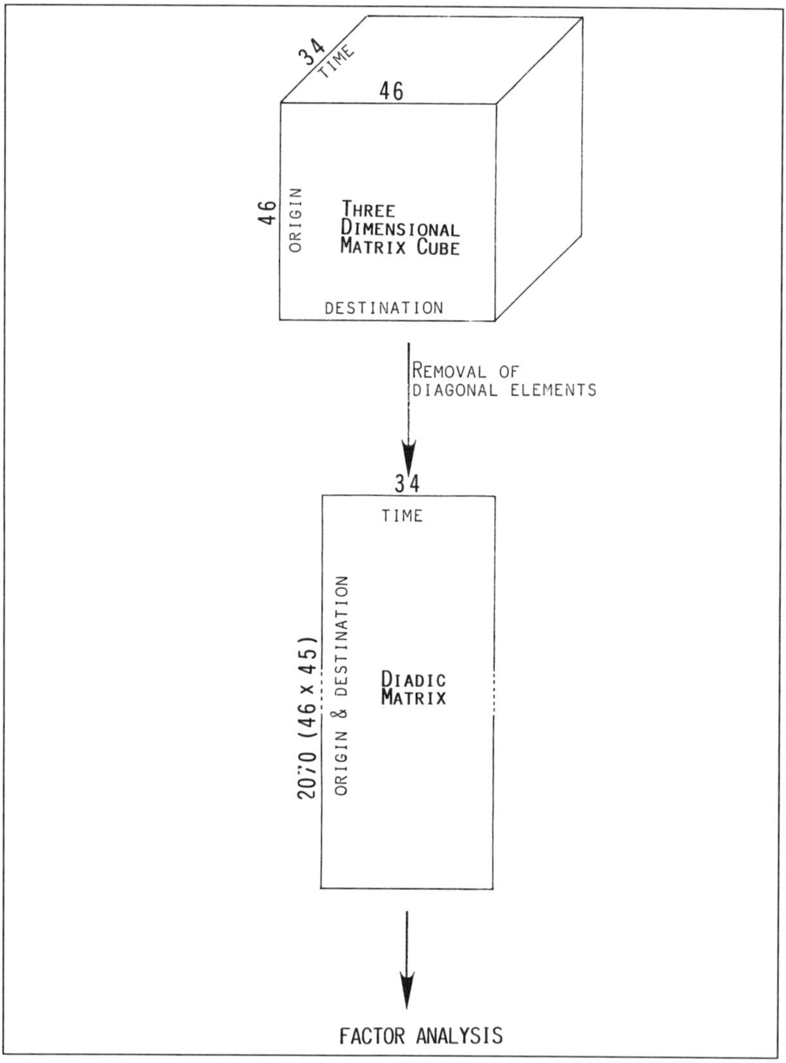

Figure 7.4. Procedure of factor analysis

-ings are a little bit more in number than the opposite flow, that is, the migrants from the Tokyo's surroundings to Tokyo. In the Osaka Metropolitan Area also, the migration movement from Osaka to its surroundings is striking.

Table 7.2. Factor loading matrix of gross-migration flows

Year	Communality	Factor I	Factor II
1954	0.996	0.457	0.878
1955	0.993	0.477	0.866
1956	0.998	0.470	0.877
1957	0.997	0.471	0.878
1958	0.998	0.510	0.857
1959	0.998	0.542	0.838
1960	0.998	0.578	0.809
1961	0.998	0.627	0.768
1962	0.999	0.683	0.718
1963	0.998	0.722	0.676
1964	0.999	0.752	0.642
1965	0.999	0.765	0.633
1966	0.999	0.786	0.606
1967	0.999	0.806	0.582
1968	0.999	0.814	0.571
1969	0.999	0.826	0.553
1970	0.994	0.841	0.523
1971	0.999	0.843	0.529
1972	0.999	0.862	0.501
1973	0.999	0.876	0.475
1974	0.998	0.874	0.481
1975	0.999	0.869	0.489
1976	0.999	0.869	0.490
1977	0.999	0.863	0.500
1978	0.999	0.867	0.490
1979	0.999	0.868	0.487
1980	0.999	0.866	0.490
1981	0.999	0.853	0.511
1982	0.999	0.839	0.530
1983	0.999	0.828	0.545
1984	0.999	0.822	0.550
1985	0.999	0.824	0.545
1986	0.999	0.831	0.537
1987	0.998	0.863	0.492
Percent of eigenvalue		32.293	1.343
Variance (%)		95.0	4.0
Cumulative variance (%)		95.0	99.0

Note: Factor loadings of over 0.7 are enclosed within the rectangles.

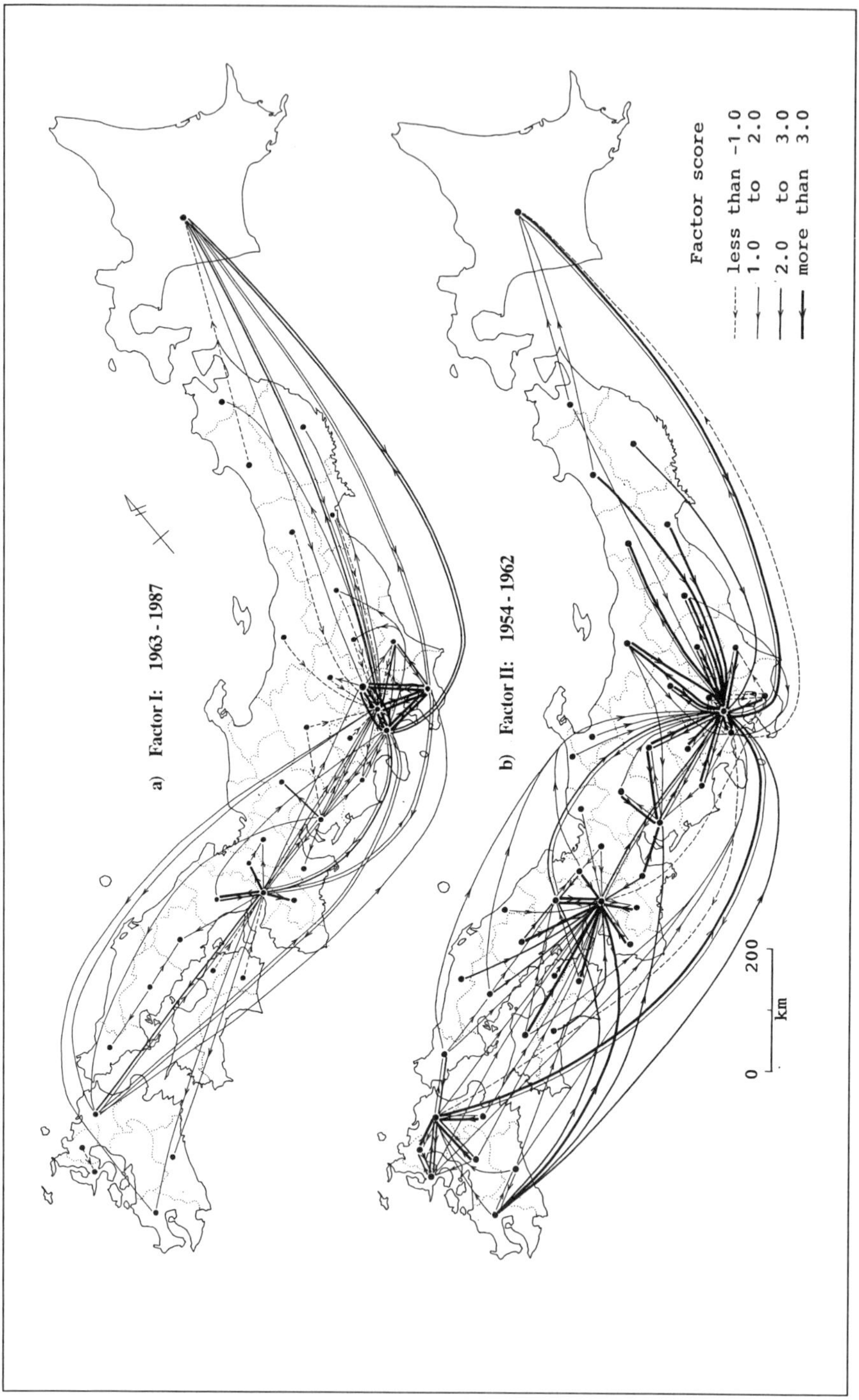

Figure 7.5. Spatial gross-migration patterns

Factor II having nine loadings over 0.7 indicates the spatial migration pattern until 1962. Figure 7.5-b shows the factor score distribution. The largest score is the flow from Niigata to Tokyo whose score value is 11.58. The flow lines from non-urbanized areas to urbanized areas are widely distributed in the map. Tokyo, Osaka, Aichi and Fukuoka play important roles as the destinated places of migration. Tokyo attracts a great number of migrants from the central and eastern Japan. Osaka also draws a great number of migrants from the western Japan, especially from the Kinki and Shikoku regions. Aichi gathers a considerable number of migrants from the Chubu region and Fukuoka from the Kyushu region. It is characteristic that there exist two large migration fields centering on Tokyo and Osaka, and two small sub migration fields centering on Aichi and Fukuoka.

If comparing the map patterns of these two periods, we can say that the centripetal nodal structure focusing on Tokyo, Osaka, Aichi and Fukuoka as the nucleuses of in-migration until 1962 turned into the weak centrifugal structure dominated by the migration from the large metropolitan areas to their surroundings or to distant regional capital cities after 1963. In short, the dominance of in-migration has shifted gradually into the dominance of out-migration since 1963 in the large metropolitan areas. Factor II has 32 factor scores more than 3.0, whereas Factor I has only 15 scores over 3.0. The fact that the largest scores over 3.0 are a few in number means that a small number of inter-prefectural pairs occupy a great number of migrants. Therefore in the period from 1963 to 1987, migration movement within the Tokyo Metropolitan Area are extremely active beyond our expectation.

7.4. Change in the Spatial Net-Migration Pattern

Net-migration data are employed here to grasp the more drastic change of Japanese migration flows for 34 years. Net-migration is computed as the difference between in- and out-migrations for each inter-prefectural pair. After the same procedure as Figure 7.4 is practiced using the matrix of net-migration,[2] the result shown in Table 7.3 is obtained. Extracted four factors whose eigenvalues are over 1.0 explain 74.9 per cent of the total variance altogether. Neighboring years got together and four clusters of time periods are obtained. Consequently 34 spatial migration patterns for 34 years are reduced into the four representative patterns.

Factor I accounting for 54.1 per cent of the total variance possesses high factor loadings over 0.5 for 15 years after 1954. It indicates that the spatial net-migration pattern had been almost similar during these 15 years. This period corresponds to the term from the pre-High Economic Growth to the mid-High Economic Growth in Japan.

Factor II accounting for 13.7 per cent of the total variance possesses high factor loadings over 0.5 during 12 years from 1962 to 1973. It can be said that the spatial migration pattern had been stable during these 12 years. The High Economic Growth period in Japan ended in 1973 when the first oil crisis occurred. Factor II is interpreted as the spatial net-migration pattern in the High Economic Growth period in Japan.

Table 7.3. Factor loadings matrix of net-migration flows

Year	Communality	Factor I	Factor II	Factor III	Factor IV
1954	0.704	0.804	0.207	0.091	0.098
1955	0.688	0.800	0.190	0.084	0.096
1956	0.756	0.842	0.202	0.097	0.133
1957	0.756	* 0.852	0.178	0.063	0.122
1958	0.725	0.827	0.189	0.091	0.106
1959	0.801	0.830	0.307	0.090	0.135
1960	0.822	0.750	0.431	0.090	0.190
1961	0.822	0.699	0.473	0.078	0.229
1962	0.821	0.663	0.530	0.094	0.234
1963	0.836	0.642	0.568	0.103	0.245
1964	0.840	0.628	0.589	0.120	0.226
1965	0.832	0.615	0.608	0.141	0.194
1966	0.804	0.597	0.610	0.149	0.181
1967	0.828	0.566	0.660	0.189	0.186
1968	0.822	0.557	0.669	0.196	0.167
1969	0.809	0.481	* 0.702	0.206	0.198
1970	0.804	0.432	0.689	0.236	0.268
1971	0.780	0.404	0.690	0.264	0.267
1972	0.727	0.374	0.658	0.311	0.233
1973	0.660	0.251	0.597	0.407	0.229
1974	0.594	0.162	0.447	0.583	0.154
1975	0.515	0.115	0.239	0.677	0.022
1976	0.486	-0.062	0.044	0.727	-0.013
1977	0.564	-0.001	0.079	0.777	0.089
1978	0.557	0.120	0.018	0.745	0.146
1979	0.586	0.054	0.075	* 0.781	0.132
1980	0.546	0.083	0.155	0.679	0.251
1981	0.555	0.147	0.249	0.610	0.293
1982	0.600	0.236	0.268	0.534	0.455
1983	0.677	0.266	0.305	0.479	0.567
1984	0.672	0.266	0.326	0.445	0.573
1985	0.699	0.240	0.395	0.384	* 0.618
1986	0.718	0.342	0.425	0.346	0.550
1987	0.726	0.366	0.439	0.396	0.480
Eigenvalue		18.395	4.660	1.468	1.001
Variance (%)		54.1	13.7	4.3	2.8
Cumulative variance (%)		54.1	67.8	72.1	74.9

Note: 1) Factor loadings of over 0.5 are enclosed within the rectangles.
2) The year with largest factor loading is marked by * for each factor.

Possessing high loadings for 9 years after 1974, Factor III has the highest loading in 1979 when the second oil crisis broke out, and since then the loadings have gradually lowered. This factor reveals the spatial net-migration pattern in the Low Economic Growth period.

Factor IV, which possesses high loadings for four years after 1983, indicates the spatial net-migration pattern in the stable economic growth period which follows the Low Economic Growth period. As mentioned above, the change of Japanese migration patterns during 34 years is divided into four periods, based upon the interregional similarity of the net-migration. It is considered that a certain year with the highest loading represents the spatial pattern of the factor in each period. Accordingly, the transition from the first to the fourth period will be discussed here by mapping the flow patterns of four representative years, that is 1957 (Factor I loading of 0.852), 1969 (Factor II loading of 0.702), 1979 (Factor III loading of 0.781) and 1985 (Factor IV loading of 0.618).

7.4.1. THE FIRST PERIOD: 1957

Figure 7.6-a shows the spatial net-migration pattern in 1957. The largest net-migration is from Niigata to Tokyo whose excess number of migrants is 19,627, and the second largest from Fukushima to Tokyo whose excess number of migrants is 19,598. The enormous excess of in-migration of Tokyo from the eastern Japan stands out remarkably. There exist ten flow lines whose net-migrants exceed 10,000 in number. Nine out of the ten flow lines direct their courses toward Tokyo. This was the golden period when Tokyo had strong absorbing force gathering a great number of migrants from all over the country.

The pulling force is not so strong as that of Tokyo though, Osaka collects many flow lines of in-migration from the western Japan, preferentially from the Kinki, Chugoku and Shikoku region. Aichi draws four flow lines of in-migration from neighboring prefectures such as Shizuoka, Mie, Nagano and Gifu. Though Fukuoka collects six flow lines from the whole prefectures in the Kyushu region, the pulling force itself is not so strong.

This corresponds to the period when many industrial enterprises began to require mass labor force in the large metropolitan areas such as Tokyo, Osaka, Aichi and Fukuoka where many factories were rapidly established. Qualitative change of Japanese economy from agriculture to industry brought about the increase of labor migrants from rural to urban areas. The simple flow structure characterized by the strong stream of migration from the remote rural areas to the central urban areas, especially to Tokyo, is noted for attention in this period. However, strong inflow to Tokyo's neighboring areas from the distant rural areas was not yet appeared in this period.

7.4.2. THE SECOND PERIOD: 1969

Figure 7.6-b indicates the spatial net-migration pattern in 1969. The largest net-migration is the flow from Tokyo to Saitama whose excess number of migrants is 80,371, and then comes the flow from Tokyo to Chiba whose excess number of migrants is 67,301. Tokyo still continued to absorb many migrants from the eastern Japan and Osaka from the western Japan, but the pulling force itself tended to weaken in comparison with the first period. Both Tokyo and Osaka decreased the number of in-migrants from the neighboring areas. Tokyo decreased the number of in-migrants from the Kanto region, and on the contrary increased considerably the number of out-migrants to the Kanto region. It is also noticeable that there appeared the flow lines from rural areas to Tokyo's surroundings such as Saitama, Kanagawa and Chiba. This phenomenon was not seen in the first period.

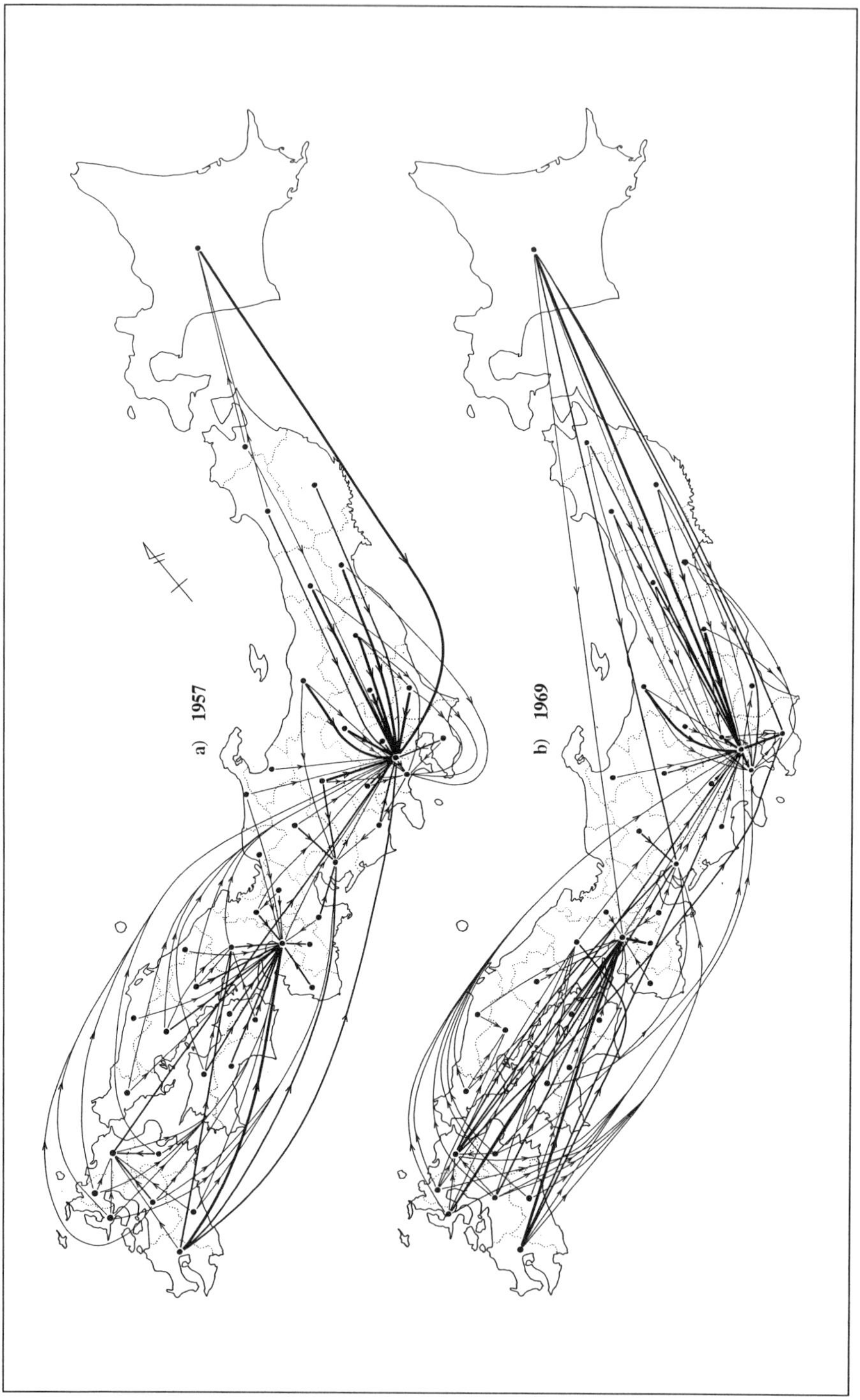

Figure 7.6. Spatial net-migration patterns (a, b)

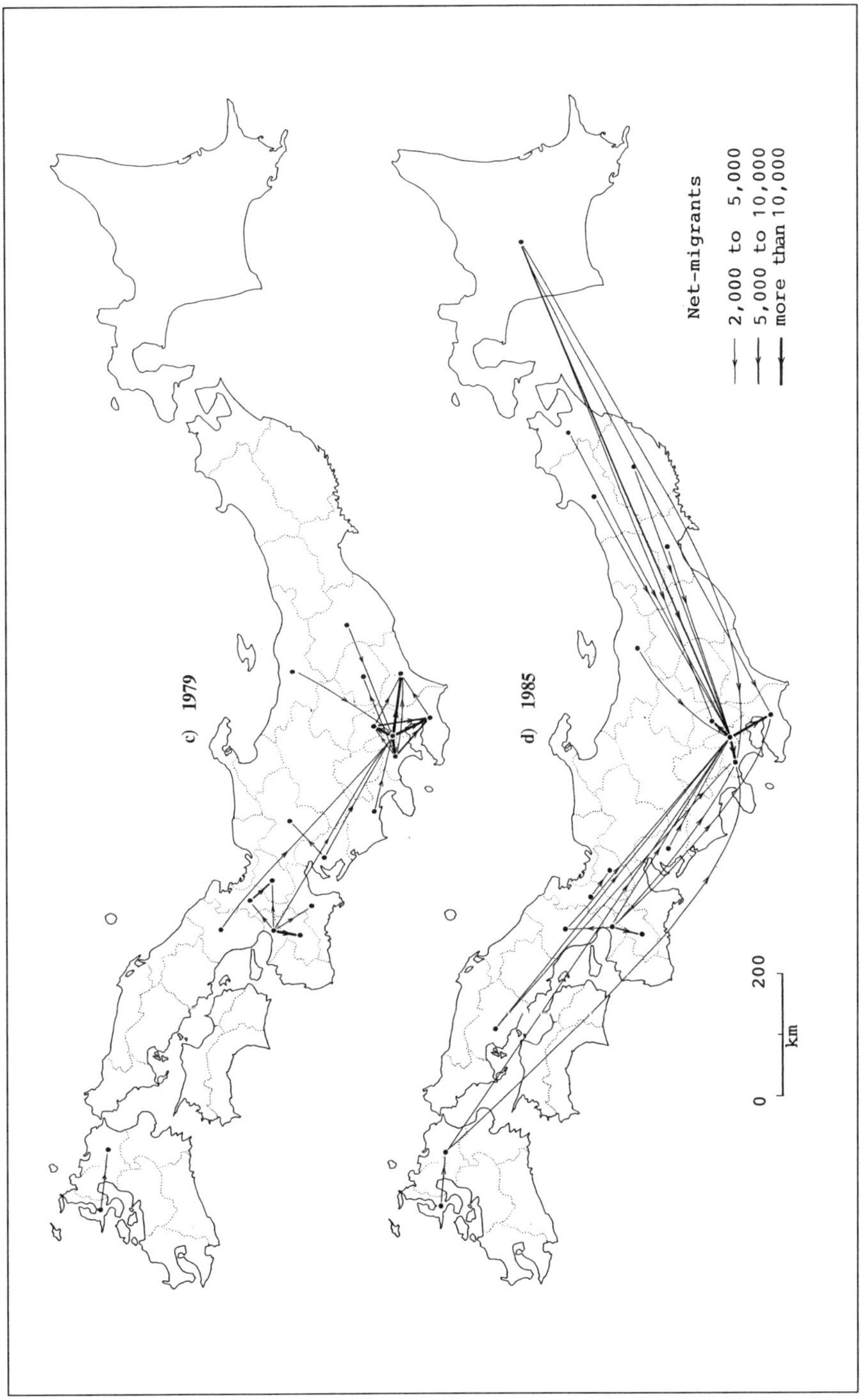

Figure 7.6. Spatial net-migration patterns (c, d) *(continued)*.

7.4.3. THE THIRD PERIOD: 1979

Figure 7.6-c indicates the spatial net-migration pattern in 1979. The largest net-migration is the flow from Tokyo to Saitama whose excess number of migrants is 55,499 and comes next the flow from Tokyo to Chiba whose excess number of migrants is 51,917. Many flow lines seen in the previous periods disappeared almost in this period. Long distant migrants to Tokyo and Osaka were a few in number and especially Tokyo's pulling force of migration became weak this year. The active movements within the Tokyo Metropolitan Area were only worth to point out. Tokyo's substantial outflow to its surroundings were still noticeable, although the number of out-migrants decreased to some extent. Though the flow scale was not so large, limited in-migration to Tokyo from the Kinki region remained unchanged even in this period. Osaka's substantial increase of out-migrants to its surroundings was also remarkable. Net-migrants from Osaka to Nara reached 19,969 in number, indicating that Nara became to encourage the satellite function as the bedroom town for Osaka as time passed.

On the whole, this year is characterized by the existence of a small number of flow lines whose lengths are comparatively short. This fact suggests that out- and in-migrants keep the balance in number. The migration rates are also low in this period as Figure 7.3 shows. The migration rate which was 4.1 per cent in 1970, decreased to 3.0 per cent in 1979. Thus we may say that pushing and pulling forces were at the weakest level in this period out of the established four periods. There exist many prefectures having no flow lines outside the Tokaido Megalopolis. The balance of out- and in-migration in the non-urbanized areas outside the Megalopolis is supposed to have been brought about by the increase of returned migration from the areas within the Tokaido Megalopolis to these non-urbanized areas. This returned migration was not recognized in the previous periods.

7.4.4. THE FOURTH PERIOD: 1985

Figure 7.6-d indicates the spatial net-migration pattern in 1985. The largest net-migration is the flow from Tokyo to Saitama whose excess number of migrants is 18,409. Though the outflow from Tokyo to its neighboring areas was still prevailing, the total number decreased to one third of 1979. Tokyo weakened the pushing force, on the contrary, strengthened again the pulling force particularly from the eastern Japan. Flow lines originating from the western Japan directed toward Osaka in the first and second periods, but they changed their destinations from toward Osaka to Tokyo by this period. In this manner, Osaka lowered pulling function, by contrast Tokyo heightened it. At the same time, due to the intensity of Tokyo's pulling power, its neighboring prefectures also became to intensify the pulling force.

7.5. Conclusion

The purpose of this chapter was to reveal how many clusters of periods the 34 years after 1954 on Japanese internal migration was divided into and to clarify what kind of spatial pattern each derived period indicates. Main findings are summarized in the following.

Only two factors were obtained by the practice of factor analysis to the total migration flow data. This result suggests that drastic structural changes have not occurred, and that the spatial migration pattern has fundamentally changed with stability for 34 years. From the temporal point of view, Japanese domestic migration was represented roughly by the two spatial patterns, that is, the patterns until and after 1962. The spatial pattern until 1962 showed the simple bipolar nodal structure dominated by the strong migration stream from the eastern Japan to Tokyo and from the western Japan to Osaka. By contrast, the spatial pattern after 1962 was characterized by the dominant complicated internal movement within the Tokyo Metropolitan Area, not showing the distinct nodal structures centering on Tokyo and Osaka as the nucleuses. The pulling function of Osaka lowered after 1963.

In terms of net-migration, 34 spatial patterns on 34 years were reduced into four major patterns, that is, the patterns 1954 to 1964 (the first period), 1962 to 1973 (the second period), 1974 to 1982 (the third period) and from 1983 to 1986 (the fourth period). These four periods which were closely related with the trend of Japanese economy were named here 1) pre- and early High Economic Growth period, 2) High Economic Growth period, 3) Low Economic Growth period and 4) Stable Economic Growth period. Strong streams of migration from the eastern Japan to Tokyo and from the western Japan to Osaka in the first period demonstrate that in the large metropolitan areas enterprises needed a great deal of labor force from non-urbanized areas with fewer job opportunities. The spatial pattern in the second period was very similar to that of the first period. But the pulling force of Tokyo and Osaka lowered to some extent in the second period. It is also characteristic that the excess tendency of out-migrants from Tokyo to its neighboring areas became prominent in this period. In the third period, out-migrants from Tokyo to its surroundings were still a great number but the volume itself decreased. Dominance of one-way directional migration stream was stopped by this period and the out- and in-migration came in balance. People did not prefer to move in the third period. In the fourth period, the similar spatial pattern with the first and second periods reappeared. In short, the movement from the non-metropolitan to metropolitan areas became distinct again. By encouraging the pulling function, Tokyo became to attract many migrants again not only from the eastern Japan but also from the western Japan. We have to note that Osaka which had had the strong pulling force could not recover the former power unlike Tokyo until 1987.

Finally the following two points became evident as conclusions of this analysis. Firstly, one-way movement from rural to urban areas was prevailing in the first and second periods, but the strong stream of migration calmed down due to the rise of two-way movement and the decrease of the migration rate in the third and fourth periods. Secondly, the migration pattern was transformed from the bipolar structure centering on Tokyo and Osaka in the 1950s into the unipolar structure centering on only Tokyo in the 1980s.

Note

1. Migrants refer to those persons who in changing their addresses crossed municipality boundaries, excluding those who moved within a municipality, those who immigrated from or emigrated to foreign countries, those who were exempted from the application of the Basic Resident Register Law, that is, those who did not have Japanese nationality, and those whose previous addresses were not known.

2. The logarithmic transformation to the net-migration data is employed to bring the data near to the normal distribution.

PART IV

REGIONAL URBAN SYSTEM

CHAPTER 8

POPULATION MOVEMENT

8.1. Introduction

Since around 1970, the structural changes becoming evident in metropolitan areas of the United States and Europe have been characterized by the reduction of population differentials between the central city and its suburbs.[1] In other words, the urban functions which had previously been concentrated in the central city are now being dispersed to the suburbs. This phenomenon has been recognized in the United States and Europe by Vining and Pallone (1982) and Champion *et al.* (1989), among others. Gordon (1979) identified the causes of functional dispersion as being changes in corporate location factors. In other words, companies which seek external economies but are located in central cities are seeking economies of scale which accompany this growth. Substantial areas of inexpensive land are needed for expansion, so industries move to the suburbs. As a result, employment opportunities increase in the suburbs, which no longer have to depend on the central cities for jobs.[2] Gordon placed this restructuring of the metropolitan area within the expansion process of the urban region.

The increase in employment opportunities in the suburbs, assuming it is the cause of population in-migration there, can be thought to have a great effect on the characteristics of the in-migrants because gender and age will greatly influence their ability to find work or accept a transfer (Takahashi ed., 1990). On this point, the work of Zelinsky (1978) has attracted considerable attention. Zelinsky, in considering the changes in population structure in the United States, revealed that population fluctuations in the suburbs differ according to gender and age group.[3] As Takahashi (1987), Shinohara *et al.* (1992) and others have reported, there are several types of movement mechanisms in the life stage. They suggest that the potential for each of these mechanisms differs according to gender and age groups. Therefore, the focus on population movement patterns by gender and age group is extremely useful in helping us to understand the socio-economic structure of the metropolitan region. Taking such a perspective, this chapter will consider the characteristics of in- and out-migrants in an effort to provide an understanding of the spatial structure of population movement in the Aichi regional system.

8.2. Study Area and Method

8.2.1. MATERIALS AND DESCRIPTION OF THE REGION

The materials used in this chapter included the 1988 Survey of Population Movement in Aichi Prefecture published by the Statistical Division of Aichi Prefectural Planning Department. This report is a statistical record of the population movement by gender and age groups among the various administrative areas of Aichi Prefecture from October 1987 to September 1988. There were five age groups in the survey: 0-14, 15-19, 20-29, 30-64, and 65-and-older. The division of these groups was not equal, but, as Takahashi ed. (1990) have indicated, these age classifications are probably suitable for considering work, marriage, childbirth, and retirement as mechanisms for increasing population movement potential.

Looking at the Nagoya Metropolitan Area, which accounts for most of the relocations in Aichi Prefecture, we can see that population movement was very active during the period of high economic growth, with very high rates of in- and out-migration and a large continuous influx of population. However, with the Oil Crisis of 1973, out-migration began to surpass in-migration, with the rates for both falling to around the 2% level. Beginning 1984, there was once again greater in-migration, although the two population movements have roughly been in equilibrium since 1990 (Figure 8.1).

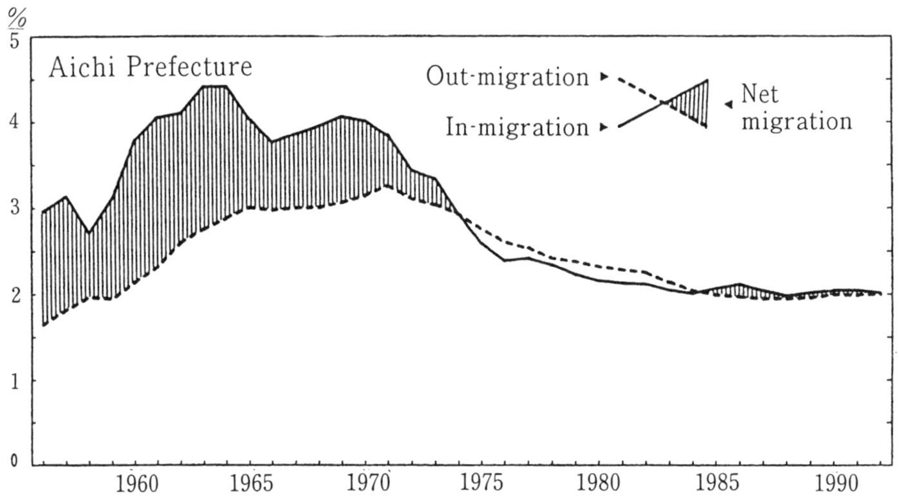

Figure 8.1. Absolute changes in rates of in- and out-migration in Aichi Prefecture

Source: The Annual and Quarterly Reports on the Internal Migration in Japan compiled by the Statistics Bureau, Management and Coordination Agency.

There are 103 administrative districts in Aichi Prefecture, including 88 municipalities (Figure 8.2). Most of the cities are in the western part of the prefecture.[4] Looking at the population movement from October 1987 to September 1988, we can see that 136,539 people moved out of Aichi into other prefectures while 137,807 people migrated into the prefecture from elsewhere in Japan, resulting in a negligible increase of in-migration. There were also 196,609 people who moved within Aichi Prefecture. Of all the people who relocated during that time, 48.8% were from 15 to 29 years of age, showing a very large potential for the movement of the youthful population.

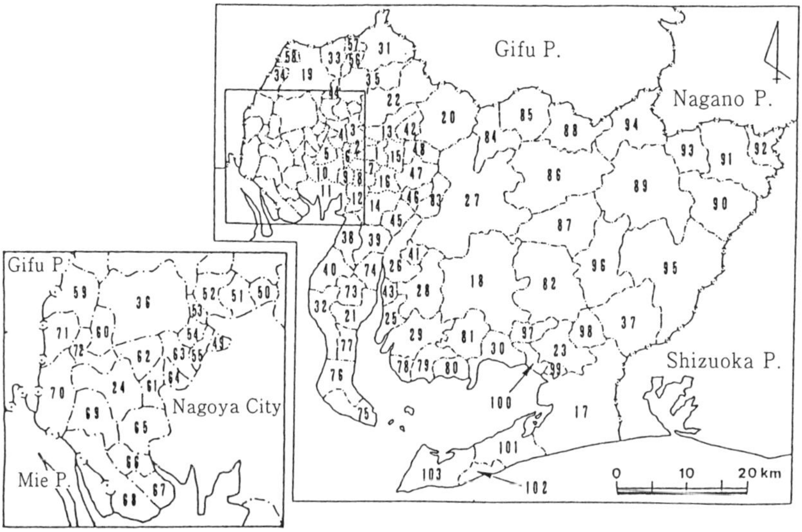

Figure 8.2. **Municipalities in Aichi Prefecture**

Nagoya City (1. Chikusa Ward, 2. Higashi Ward, 3, Kita Ward, 4. Nishi Ward, 5. Nakamura Ward, 6. Naka Ward, 7. Showa Ward, 8. Mizuho Ward, 9. Atsuta Ward, 10. Nakagawa Ward, 11. Minato Ward,12. Minami Ward, 13. Moriyama Ward, 14. Midori Ward, 15. Meito Ward, 16. Tempaku Ward), 17. Toyohashi City, 18. Okazaki City, 19. Ichinomiya City, 20. Seto City, 21. Handa City 22. Kasugai City, 23. Toyokawa City, 24. Tsushima City, 25. Hekinan City, 26. Kariya City, 27. Toyota City, 28. Anjo City, 29. Nishio City, 30. Gamagori City, 31. Inuyama City, 32. Tokoname City, 33. Konan City, 34. Bisai City, 35. Komaki City, 36. Inazawa City, 37. Shinshiro City, 38. Tokai City, 39. Obu City, 40. Chita City, 41. Chiryu City, 42. Owariasahi City, 43. Takahama City, 44. Iwakura City, 45. Toyoake City, 46. Togo Town, 47. Nissin Town, 48. Nagakute Town, 49 Nishibiwajima Town, 50. Toyoyama Town, 51. Shikatsu Town, 52. Nishiharu Town, 53. Haruhi Village, 54. Kiyosu Town, 55. Shinkawa Town, 56. Oguchi Town, 57. Fuso Town, 58. Kisogawa Town, 59. Sobue Town, 60. Heiwa Town, 61. Shippo Town, 62. Miwa Town, 63. Jimokuji Town, 64. Oharu Town, 65. Kanie Town, 66. Jushiyama Town, 67. Tobishima Village, 68. Yatomi Town, 69. Saya Town, 70. Tatsuta Town, 71. Hachikai Village, 72. Saori Town, 73. Agui Town, 74. Higashiura Town, 75. Minamichita Town, 76. Mihama Town, 77. Taketori Town, 78. Isshiki Town, 79. Kira Town, 80. Hazu Town, 81. Koda Town, 82. Nukata Town, 83. Miyoshi Town, 84. Fujioka Town, 85. Obara Town, 86. Asuke Town, 87. Shimoyama Village, 88. Asahi Town, 89. Shitara Town, 90. Toei Town, 91. Toyone Village, 92. Tomiyama Village, 93. Tsugu Village, 94. Inabu Town, 95. Horai Town, 96. Tsukude Town, 97. Otowa Town, 98. Ichinomiya Town, 99. Kozakai Town, 100. Mito Town, 101. Tahara Town, 102. Akabane Town, 103. Atsumi Town

Looking at demographic trends during this period in cities and counties, we can see that urban population declined by 0.02% in contrast to county population which increased by 0.30%. On a municipality basis (Figure 8.3), we can see that the population of 61 municipalities (26 cities, 30 towns, 5 villages), including the cities of Okazaki and Toyota, increased. This trend was especially noticeable in the hilly region east of Nagoya where residential development has been proceeding in Owariasahi City and the towns of Nagakute and Nisshin. It was also noteworthy in the Toyota area where population increases were particularly high in Fujioka Town and Shimoyama Village. On the other hand, 27 municipalities, including the cities of Gamagori, Bisai and Tokoname, experienced population decline, a trend which was particularly great in

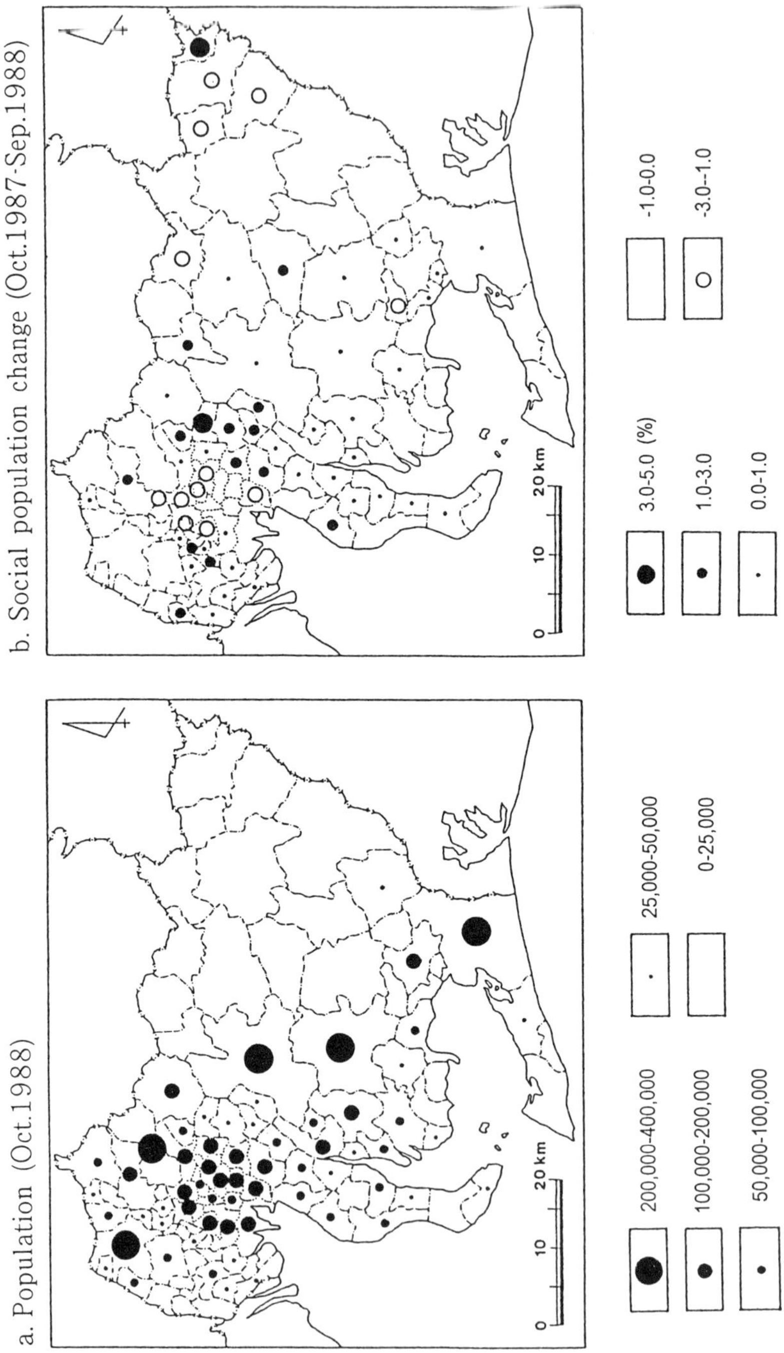

Figure 8.3. Population and social population change

Source: 1988 Population Movement Data for Aichi Prefecture, Aichi Prefecture Department of Planning and Statistics.

Asahi Town, Toei Town, Tsugu Village, and other places far distant from Nagoya and Toyohashi. In Nagoya itself, population decline was low in Naka and Atsuta wards in the relatively sparsely populated central area while it was high in the highly populated Chikusa ward and Nishi ward on the outskirts of the city. Finally, the hilly areas such as Midori and Tempaku wards in the southeastern part of the city, witnessed the same type of high population increase that the suburban areas did.

8.2.2. IN- AND OUT-MIGRATION PATTERNS BY GENDER AND AGE GROUPS

First, let us examine the in-and out-migration rates for gender and age groups in the entire Aichi Prefecture (Figure 8.4). Here, the in-and out-migration rate is the ratio of migrants to resident population.[5] In this analysis, Aichi Prefecture is treated as "within the metropolitan area," neighboring Gifu, Shizuoka, and Mie prefectures being as "outside the metropolitan area," outside that is "other regions". In both the in-migration rate (a) and the out-migration rate (b), there are large numbers in both the 20-29 and 30-64 age groups, with most of these people located in the metropolitan area (Figure 8.4). While there was almost no difference between males and females in the 20-29 age group, males accounted for more in- and out-migrants in the 30-64 age group. Among this latter group, females accounted for more of the in- and out-migrants within the metropolitan area (Aichi Prefecture), strongly suggesting that females tend to move shorter distances. Both males and females in the 65-and-over group showed the least tendency to relocate, with movement mostly occurring within the metropolitan area. Females in this group tended to move more than males, possibly because females, who on average live longer than males, are often forced to relocate due to the deaths of their husbands.

Compared with other age groups, 15-19 year-olds, especially males, showed the least tendency to move within the metropolitan area and had a wider range of destinations. This is probably the result of high school graduates leaving the metropolitan area to work or seek higher education in Tokyo, Osaka, and other cities. Although more males relocated than females in both the 15-19 and 30-64 age groups, females in the 20-29 age group tended to move as much as their male counterparts, possibly because of marriage.

Examining the in- and out-migration rates in Aichi Prefecture on an administrative area basis (Figure 8.5), we can see that, for all age groups, the rates were highest in the wards, indicating a high degree of mobility in the core of the metropolitan area. Compared to other administrative areas, the wards experienced more population movement to and from other prefectures, with short-distance moves accounting for little of the total. The difference between male and female trends in the wards was quite striking, with males accounting for the vast majority of relocations.

Compared to cities and wards, mobility in towns was low, and much lower in villages. The low mobility of village residents is a result of a high farming population. In towns and village, the proportion of intra-prefectural relocation was high. Furthermore, the in-migration rate for towns was higher than the out-migration rate for all age groups. This implies that housing developments in the towns are attracting population.

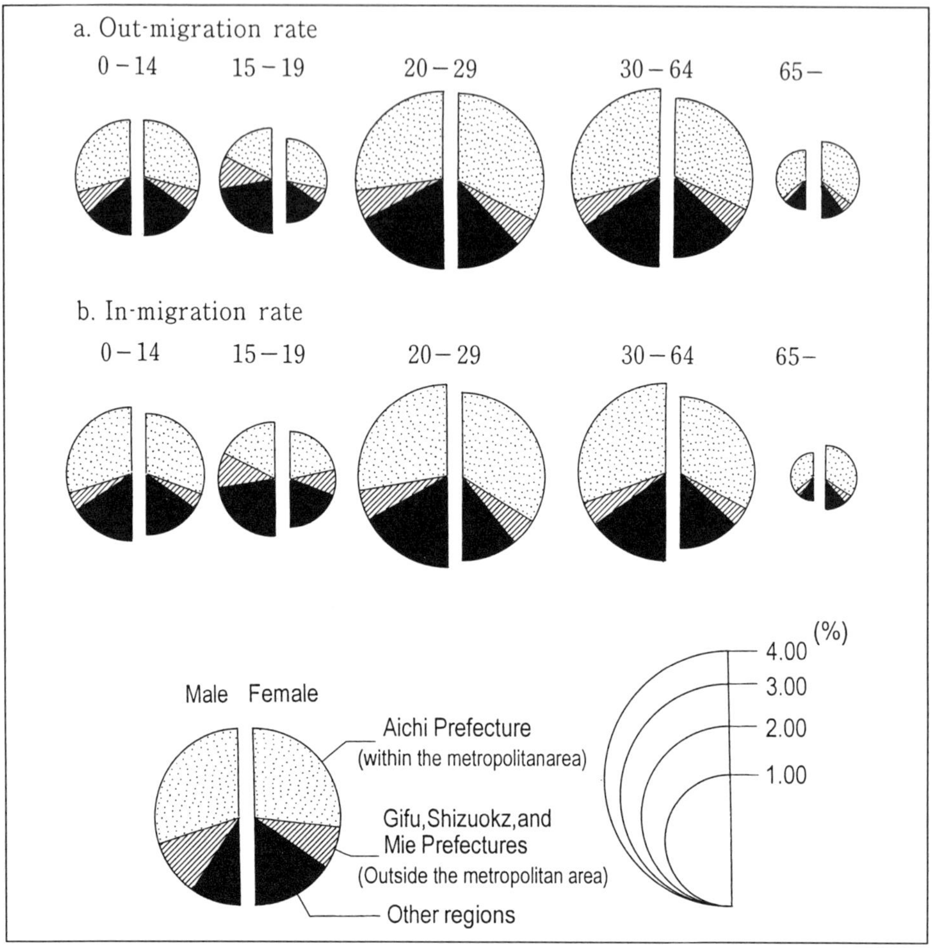

Figure 8.4. Overall population movement in Aichi Prefecture by age and gender (1987-88)

Figure 8.6 shows out-migration rates by sex and distance zones in Aichi Prefecture. The out-migration rate for 0-14 year-olds in the 0-50km zone decreases with distance from the central city. Relocation to other places within Aichi is high, with little movement to points outside the prefecture. The closer people live to the metropolitan center, the greater the tendency to move far away.

Compared with the 0-14 year-olds, 15-19 year-olds show a much lower out-migration rate overall. Their rate of out-migration to cities is lower than all other age groups for all distance zones. Movement to Tokyo and Osaka is particularly noteworthy.

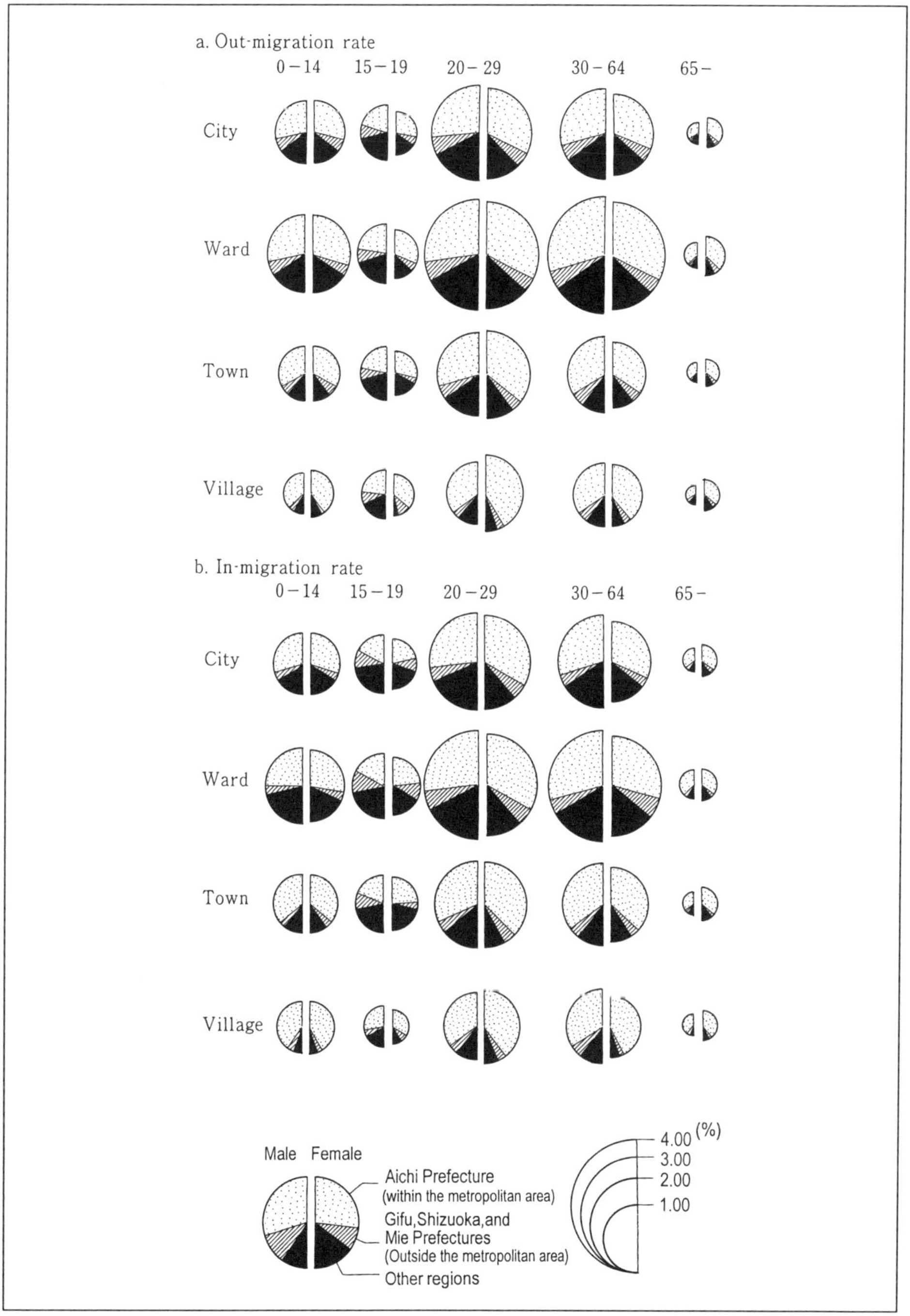

Figure 8.5. Population movement in administrative areas of Aichi Prefecture by age and gender (1987-88)

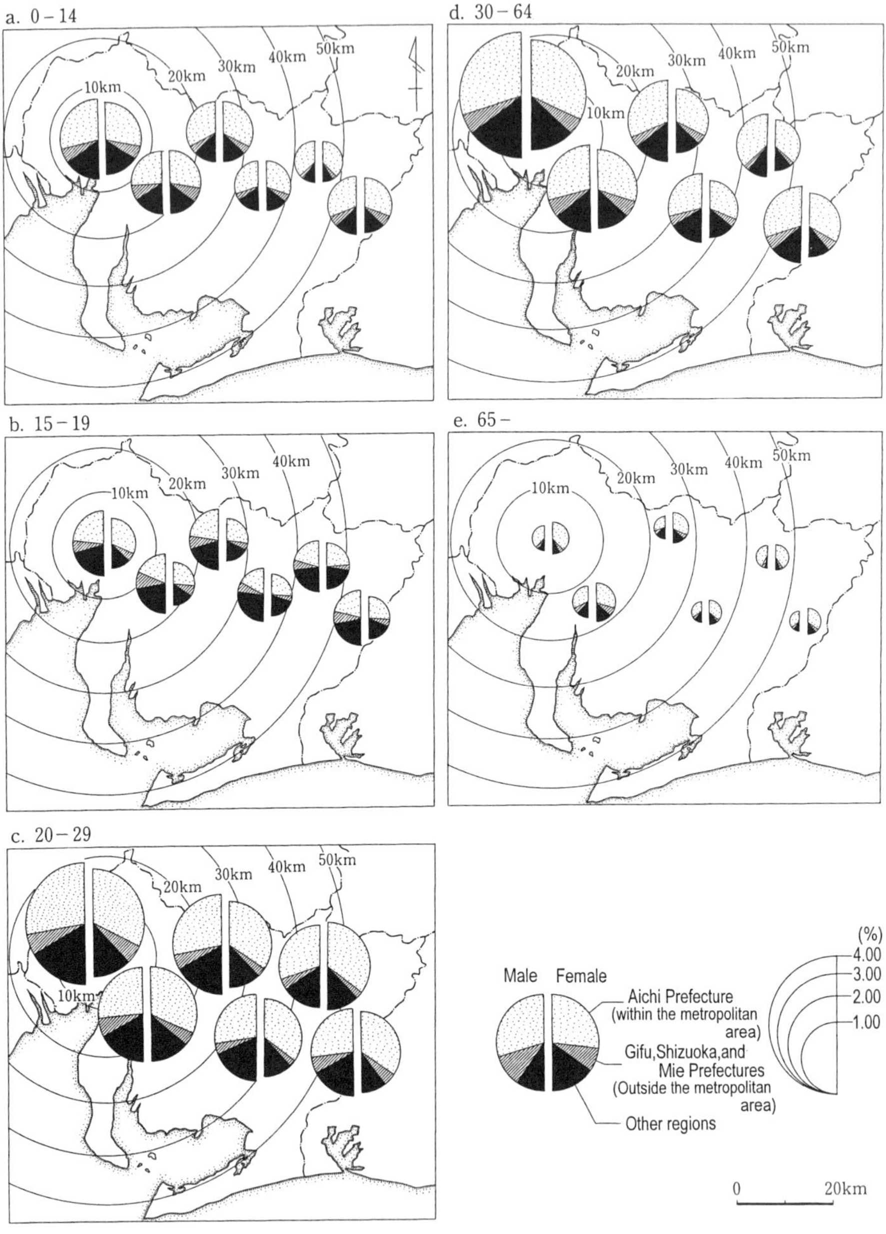

Figure 8.6. Rates of out-migration from Aichi Prefecture by age and gender (1987-88)

Relocating more often were 20-29 year-olds. With the exception of moves of 0-10 km, their out-migration was the highest among all age groups. Out-migration decreased with distance from the central cities. For all distance zones, females showed a greater tendency than males to move within the prefecture.

Among the 30-64 year-olds, relocation was extremely high for moves up to 10 km. As with the other age groups, out-migration tended to decrease as distance from the central city increased. There was an exceptional difference between males and females in moves of 20-50 km, yet a comparison of out-migration destinations shows no gender-related differences for all distance zones.

For those aged 65 and older, there was little movement for all distance zones, with relocations within Aichi Prefecture constituting the vast majority of moves.

Figure 8.7 shows in-migration rates by gender and distance zone for Aichi Prefecture. For 0-14 year-olds, we can see that the in-migration rate gradually declines with distance from the central city in moves of 0-50 km, although the proportion moving into metropolitan areas is high. However, members of this age groups moving to metropolitan areas tended to move outside the prefecture, especially for long distances.

Among the 15-19 year-olds, the in-migration rate of males is much higher than that of females for moves of 0-40 km, especially 20-30 km. In the 0-30 km range, in-migration from the neighboring prefectures of Mie, Shizuoka, and Gifu is very conspicuous.

Among 20-29 year-olds, there was less in-migration for all distance zones among males than females. This trend was especially evident in the 0-10 and 20-30 km distance zones. Males were prone to come from other regions, suggesting a concentration of male in-migrants coming from Tokyo, Osaka, and other metropolitan areas.

The overwhelming majority of in-migrants 30-64 years of age moved within a radius of only 0-10 km from their former homes. As with the other age groups, in-migration of 0-50 km tended to decrease with increasing distance from the central city. In all distance zones, male in-migrants outnumbered females, with such differentials becoming particularly noticeable toward the central city. A structural comparison of destinations shows no major differences between males and females.

Finally, the in-migration rates of the 65 and over group were low for all distance zones; furthermore, most of these moves were within Aichi Prefecture.

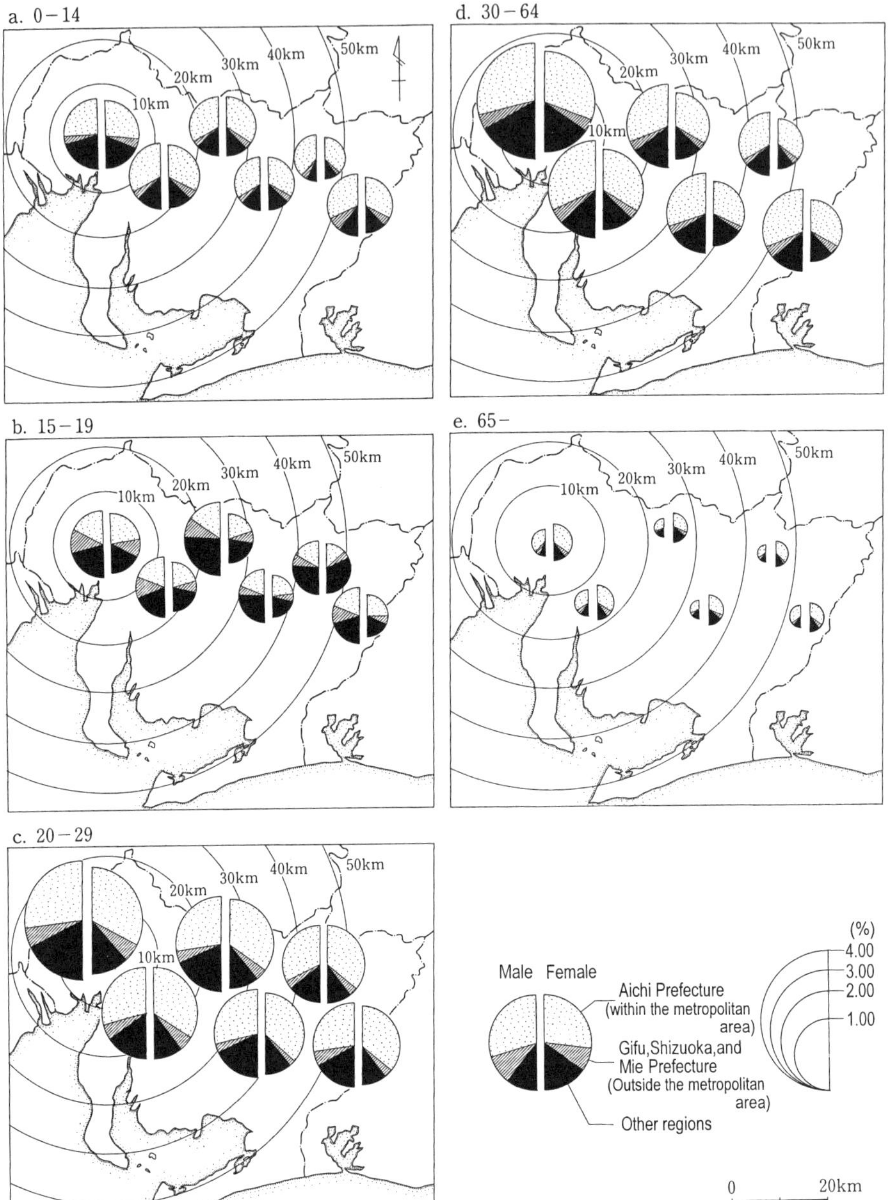

Figure 8.7. Rates of in-migration into Aichi Prefecture by age and gender (1987-88)

8.3. Population Movement by Age and Gender

The preceding analyses have shown that most of the population movement in Aichi Prefecture is mostly confined to within the metropolitan areas. Although established patterns for the direction of this movement have been identified, the details are quite complex. Therefore, this section attempts to simplify patterns of movement between points of origination and destination in order to facilitate the interpretation of the patterns. The following section will note the similarities between movement by gender and that by age group to identify the fundamental structure of population movement. In order to achieve these goals, this study employs three-mode factor analysis (Tucker, 1966; Yano, 1989; Hashimoto and Murayama, 1991) (Figure 8.8). In this analysis, the square sum for each factor of each gender and age group was maintained at a standard 1.0.[6] This helped us to clarify any differences in movement patterns among gender and age groups.

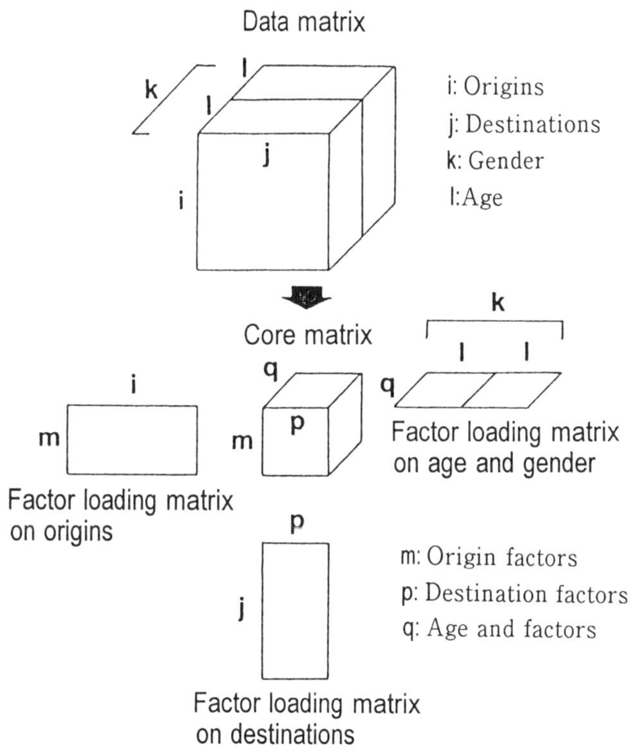

Figure 8.8. The framework of three-mode factor analysis

Note: *i*: Origins, *j*: Destinations, *k*: Gender, *l*: Age, *m*: Origin factors, *p*: Destination factors, *q*: Age and gender factors.

8.3.1. FACTORIZING THE DESTINATION ZONES

To start the analysis, the destination zones were factorized. From standardized four-dimensional data, a matrix Z_1 was formulated with 1030 rows, representing gender × age group × origination zone, and 103 columns representing destination zones.[7] Varimax rotation was used to facilitate interpretation. From the results, 8 factors were extracted to explain 71.6% of the total variance (Table 8.1).[8] Factors I and VI showed a high factor loading in the Nishi Mikawa region of central Aichi. In contrast to Factor I which showed a high loading only in urban areas, Factor VI also showed a high loading in counties. Since all factors displayed a high factor loading in Okazaki, we can assume that Okazaki formed the core for the complex population movements of the Nishi Mikawa region.

Table 8.1. Factor loadings for destination zones

Factor I	(39.8%)
18 Okazaki City	0.65
27 Toyota City	0.79
28 Anjo City	0.83
29 Nishio City	0.72
Factor II	(8.6%)
5 Nakamura Ward	0.72
9 Atsuta Ward	0.70
11 Minato Ward	0.74
Factor III	(6.5%)
1 Chikusa Ward	0.76
16 Tempaku Ward	0.72
42 Owariasahi City	0.66
47 Nisshin Town	0.69
48 Nagakute Town	0.70
Factor IV	(4.4%)
35 Komaki City	0.82
Factor V	(3.3%)
14 Midori Ward	0.87
Factor VI	(3.3%)
18 Okazaki City	0.72
Factor VII	(2.7%)
15 Meito Ward	0.74
Factor VIII	(2.6%)
10 Nakagawa Ward	0.65

Note: 1) The numbers correspond to those of unit areas shown in Figure 8.2.
2) The values in parenthesis represent % of total variance.

Factors II, III, V, VII, and VIII were all located in the Nagoya area and all had high factor loadings in the surrounding municipalities. The regions explained by these factors differed according to the direction away from the central Nagoya, with Factor II showing high loading in SW Nagoya and Factor III showing high loading in the NE part of the city. There were numerous instances in which destinations were locations in suburban Nagoya, with patterns dependent on the direction of movement. Finally, Factor IV, which explains the urban region in northern Nagoya, showed an especially high factor loading in Komaki City.

Table 8.2. Factor loadings for origination zones

Factor I	(14.1%)
18 Okazaki City	0.95
26 Kariya City	0.87
41 Chiryu City	0.86
43 Takahama City	0.86
83 Miyoshi Town	0.91
84 Fujioka Town	0.83
85 Obara Town	0.84
86 Asuke Town	0.82
Factor II	(12.2%)
19 Ichinomiya City	0.81
22 Kasugai City	0.92
31 Inuyama City	0.91
33 Konan City	0.90
36 Inazawa City	0.81
44 Iwakura City	0.93
50 Toyoyama Town	0.90
51 Shikatsu Town	0.93
52 Nishiharu Town	0.83
56 Oguchi Town	0.88
Factor III	(11.9%)
27 Toyota City	0.97
28 Anjo City	0.89
81 Koda Town	0.92
82 Nukata Town	0.91
Factor IV	(11.6%)
5 Nakamura Ward	0.94
9 Atsuta Ward	0.83
11 Minato Ward	0.95
65 Kanie Town	0.85
Factor V	(11.5%)
2 Higashi Ward	0.92
13 Moriyama Ward	0.84
15 Meito Ward	0.88
Factor VI	(10.9%)
10 Nakagawa Ward	0.98
Factor VII	(10.5%)
16 Tempaku Ward	0.92
45 Toyoake City	0.90
Factor VIII	(9.9%)
1 Chikusa Ward	0.99
Factor IX	(1.1%)
6 Naka Ward	0.82

Note: 1) The numbers correspond to those of unit areas shown in Figure 8.2.
2) The values in parenthesis represent % of total variance.

8.3.2. FACTORIZING THE ORIGINATION ZONES

The origination zones were factorized by reconducting the 1-mode factor analysis of matrix S_1 derived from the data matrix Z_1 of the destination zone.[9] From this analysis, 9 factors explaining 93.7% of the total dispersion were obtained (Table 8.2).[10]

Factors I and III showed high factor loadings in the Nishi Mikawa region of central Aichi. The distribution shows a loading for Factor I in most municipalities, whereas Factor III shows high loading only in the municipalities surrounding Okazaki. This signifies that, as was previously indicated by the destination zone factors, the population movement patterns in the Nishi Mikawa region are quite complex.

High factor loadings for Factor II appear in the communities in the northern part of Nagoya, especially in cities such as Inuyama, Kasugai, and Ichinomiya in the Komaki City area.

Factors IV through IX all show high loadings in the wards of northern Nagoya and their surrounding areas. Among these six factors, IV through VII are located in wards on the Nagoya fringe, i.e., Factor IV shows high loading in southwest Nagoya; Factor V, in the city's northeast; Factor VI, in western Nagoya; and Factor VII shows high loadings in the southeastern part of the city. Finally, Factors VIII and IX explain central Nagoya, and show a high loading in only one ward apiece. These results suggest the existence of complex movement patterns with the City of Nagoya and surrounding municipalities acting as origination zones.

8.3.3. POPULATION MOVEMENT PATTERNS BY AGE AND GENDER

From the preceding analyses, 8 destination zone factors and 9 origination zone factors were obtained. The size/proportions of the interfactor movement are shown in the 2-mode factor matrix S_2 derived from the factorization of the input data Z_1 of the destination and origination zones.[11] This matrix S_2 was standardized by gender and age group and movements judged to significantly show absolute values of 2.0 or greater were mapped.

First, the movement of males was investigated (Figure 8.9). Looking at the relocations occurring in suburban Nagoya, we can see high values for intermovement between central Nagoya and its surrounding area, and between the peripheral wards of the city and their surrounding areas for all age groups except the 15-19 year-olds. Movement from central Nagoya to its periphery is noteworthy in eastern Nagoya but is not particularly evident elsewhere. This is in contrast to the 15-19 year-olds, who only showed high values for movement between the central city and its surrounding area, especially in eastern and southern Nagoya. Finally, there were high values shown by all age groups for movement into Komaki City, north of Nagoya, from the surrounding communities.

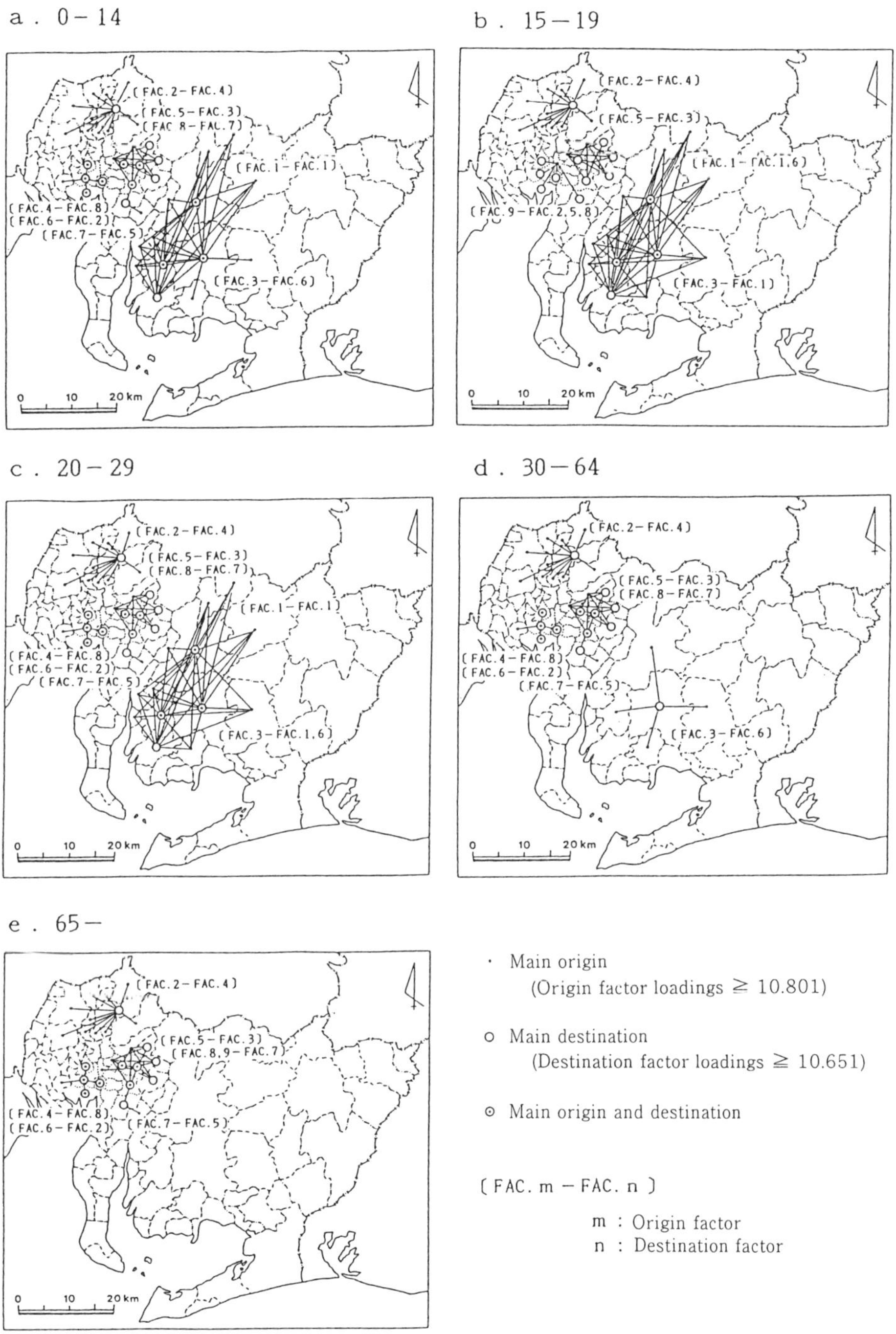

Figure 8.9. Male migrations by age in Aichi Prefecture (1987-88)

Note: The straight lines show the flows with factor scores over |1.5|.

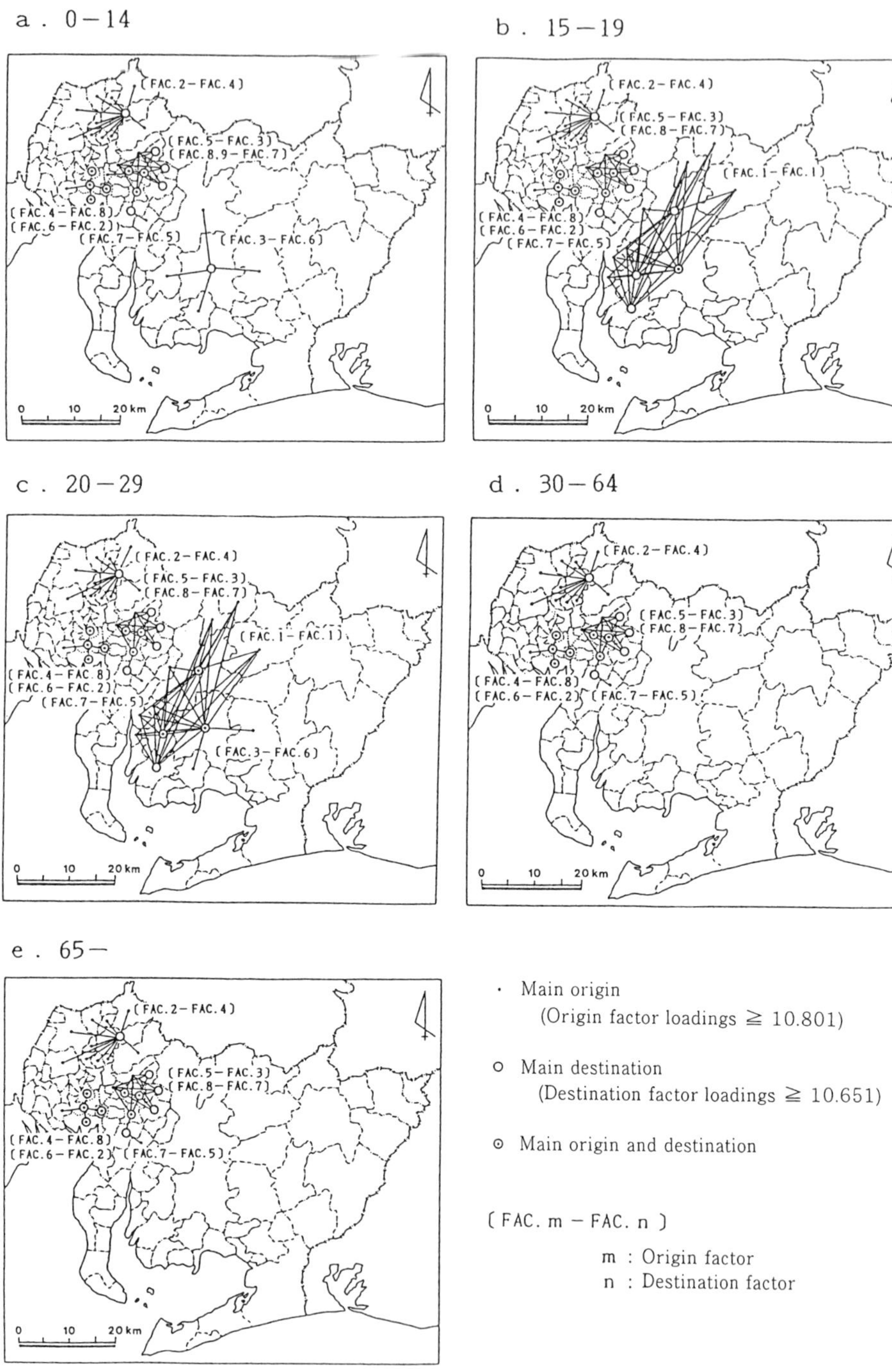

Figure 8.10. Female migrations by age in Aichi Prefecture (1987-88)
Note: The straight lines show the flows with factor scores over |1.5|.

In the Nishi Mikawa region, there are high values shown for county-to-city and intercity migration for the 0-14, 15-19, and 20-29 age groups, indicating complex patterns of movement. In contrast, the only high values shown by the 30-64 age group are for movement into Okazaki City from the surrounding communities. Finally, no high values were seen for the 65-and-over group.

Next, female population movement was examined (Figure 8.10). Looking at relocations occurring in suburban Nagoya, we can see high values for intermovement between central Nagoya and its surrounding area, and between the peripheral wards of the city and their surrounding areas for all age groups. Female movement, like that of males, from central Nagoya to the surrounding area was conspicuous in the eastern part of the city, with little in evidence elsewhere. Finally, there were high values shown by all age groups for movement into Komaki City from the surrounding communities.

In the Nishi Mikawa region, there are high values shown for county-to-city and intercity migration for the 15-19 and 20-29 age groups, indicating complex patterns of movement. In contrast, the only high values shown by the 0-14 age group were for migration into Okazaki City from the surrounding communities. Finally, no high values were seen for the 30-64 and 65-and-over groups.

8.4. Basic Structure of Population Movement in the Nagoya Metropolitan Area

The previous section clarified population movement in the Nagoya Metropolitan Area by age and gender. As a result, considerable differences could be seen in the extent and direction of movement according to different characteristics. To further clarify these differences, this section will factorize gender and age groups. In addition, gender and age groups will be used as one element in the analysis.[12]

Table 8.3. Factor loadings by age and gender

Gender	Age	Factor I	Factor II	Factor III
Male	0-14	0.54	0.33	0.23
	15-19	0.20	0.14	0.63
	20-29	0.48	0.32	0.32
	30-64	0.50	0.38	0.23
	65-	0.30	0.57	0.17
Female	0-14	0.55	0.30	0.22
	15-19	0.42	0.33	0.38
	20-29	0.49	0.34	0.29
	30-64	0.47	0.40	0.24
	65-	0.36	0.54	0.17
Variance (%)		88.55	6.08	2.49
Cumulative variance (%)		88.55	94.63	97.12

Note: Underlined figures represent absolute values of 0.4 or above.

To begin the analysis, a new matrix Z_3 was created from the matrix S_2 derived by the two-mode factor analysis in section 8.3 by multiplying destination zone factors by origination zone factors in the rows, and multiplying gender by age in the columns. Factor analysis on this matrix provided us with three factors.[13] The factor loading after varimax rotation is shown in Table 8.3.

Factor I shows high factor loading for males in the 0-14, 20-29, and 30-64 age groups, and for females in the 0-14, 15-19, and 30-64 age groups. Since movement of the 0-14 year-olds is believed to accompany that of their parents, Factor I is believed to reflect the movement of parents with children.

Factor II shows a high factor loading for both males and females 65 and over; therefore, it is interpreted to be a reflection of the movement of the elderly. However, because it also shows slightly high loadings for both females and males 30-64, it is also believed to include middle-aged adults.

Factor III shows a high factor loading for males 15-19 years in age, thus reflecting the movement of male youths. However, since the factor loading here for females in the age group 15-19 is high, it is also believed to include the movement of older teenage girls.

Comparing each factor loading of male movement with each factor loading of female movement, we can see that there is little difference in inter-factor values. This indicates that movement patterns by age cannot be as easily distinguished as those for males. Furthermore, females in all but the 65-and-over age group showed the greatest factor loading in Factor I, suggesting that much of this can probably be explained by family-related moves such as marriage.

In order to clarify the 3 movement patterns derived from the gender and age group factors, a 3-mode factor matrix S_3 was calculated. This 3-mode factor matrix S_3 had been factorized from all the origins of Matrix Z_1, that is, 103 origination zones, 103 destination zones, and the two sexes and five age groups.[14] This matrix, which has been derived from the 3-mode analysis of 9 origination zones factors, 8 destination zone factors, and 3 gender-age factors, is the core matrix of Z_1 and its values illustrate the size of the regional inter-factor population movement for each gender and age factor.

Factor I shows movement of parents with children and young teenagers (Figure 8.11-a). The highest values are shown in movement from the wards of NE Nagoya to the surrounding communities, movement from Tempaku Ward and Toyoake City to Midori Ward, and movement from Naka Ward of central Nagoya to the wards of Nakamura, Atsuta, Minato, and the eastern part of Meito, all in southwestern Nagoya. In addition, there is a high value for movement into Komaki City from the surrounding communities, and in the Nishi Mikawa region, we can see a high value for movement into Okazaki City from neighboring communities.

Factor II represents movement of the elderly (Figure 8.11-b). High values are seen only for movement in central Nagoya and surrounding areas, especially movement between the wards of SW Nagoya and surrounding communities, from Chikusa and Naka wards of central Nagoya to Meito and Tempaku wards and surrounding communities in the east, and from Chikusa Ward to Nakagawa Ward in the western part of the city.[15] The movement of the elderly is particularly striking in areas undergoing urbanization, with little in- or out-migration occurring in farming communities. Furthermore, movement is usually of a very short distance, with most of it occurring between neighboring regions.

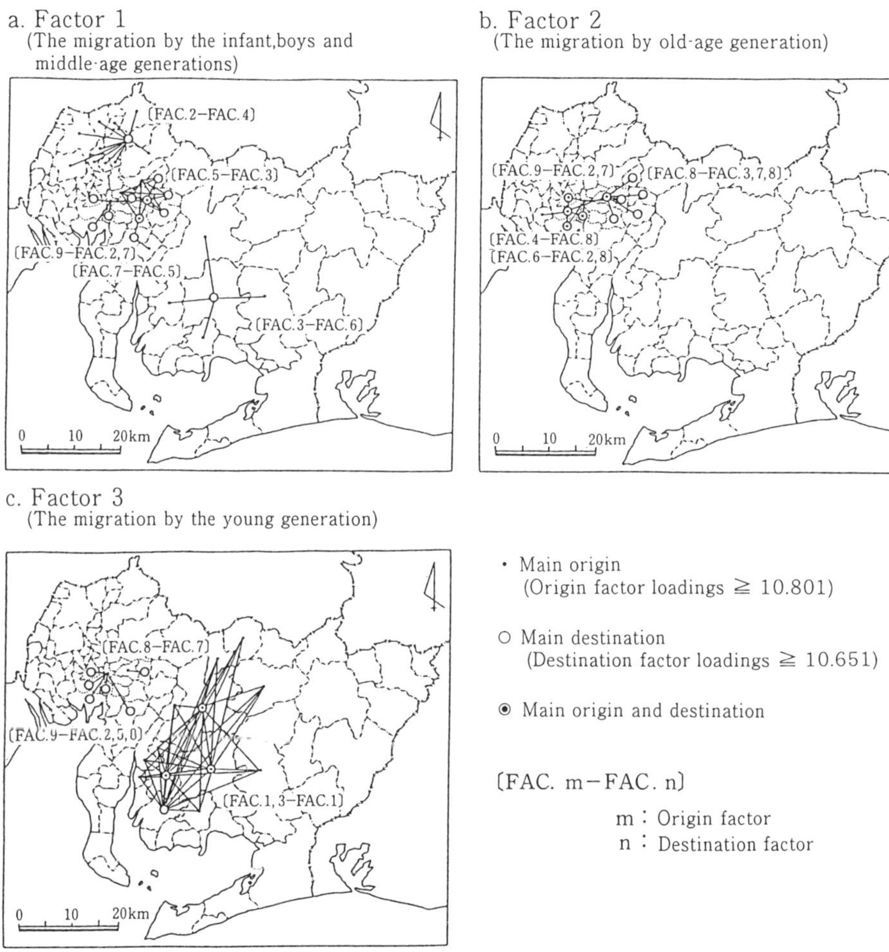

Figure 8.11. Migrations by age in Aichi Prefecture (1987-88)
Note: The straight lines show the flows with factor scores over |1.5|.

Factor III represents the movement of older teenagers (Figure 8.11-c). High values are shown by movement from Chikusa Ward in central Nagoya to Meito Ward in the east, as well as from Naka Ward to Midori Ward in the SE and the SW wards of Nakamura, Atsuta, Minato, and Nakagawa. In the Nishi Mikawa region, high values are shown by movement into the cities of Okazaki, Toyota, Anjo, and Nishio from the surrounding area.[16] The distance moved is relatively long.

From these results, we can make the following conclusions: First, all age and gender factors indicate considerable movement from central Nagoya to surrounding wards and communities which can be attributed to sectorality. However, parents, children and young teens (Factor I) show high values for movement to the hilly region of eastern Nagoya, in contrast to the older teen group (Factor III), which shows considerable movement to S and SW Nagoya. This movement of parents with children is believed to be a result of improvement and purchase of housing, while the movement of older teens is thought to reflect continuing education or commencement of employment.[17] Finally, the elderly (Factor II) show considerable movement into E and SW Nagoya, as well as among the wards of the SW. Since characteristics other than age and gender are believed to be responsible for movement of the elderly, further analysis is required.

Next, looking at high movement values outside the Nagoya area, we can see simple movement of parents with children into single cities of Okazaki, Komaki, etc., from their respective neighboring areas. This is in contrast to the more complex movement of older teens, who not only move into a number of cities from the hinterlands, but also exhibit a great tendency to move from one city into another. Finally, little movement is seen among the elderly outside the Nagoya area; most of it involves short distance relocation within central Nagoya and the surrounding area which are served by a good public transportation network.

8.5. Conclusion

This chapter has been a discussion of population movement in Aichi Prefecture from October 1987 to September 1988. It has revealed that most of the population movement by both gender and age in the Nagoya urban belt occurs from within Aichi Prefecture. Therefore, further analyses were conducted on population movement patterns by gender and age groups, with the following results:

From the linkages between origination zone and destination zone factors, population movement patterns by age and gender were identified. The resulting comparison showed significant movement from central Nagoya to neighboring municipalities on the east, and movement into Komaki City from surrounding communities, for all gender and age groups. Furthermore, with the exception of the 15-19 year-old males, there is considerable short distance movement among the wards of central Nagoya and surrounding wards, and clear sectors have been detected in this inter-ward movement. In the Nishi Mikawa region, the movement of both males and females in the 15-19 and 20-29 age groups is occurring among municipalities covering a wide area, in contrast to the 30 and over groups, which show almost no movement.

By factorizing gender and age groups, three types of factors were derived: one to explain parents with children, another to explain the elderly, and the third to explain older teens. Looking at the movement patterns corresponding to the respective factors, we can see that in all age and gender groups, movement is in evidence from the wards of central Nagoya to surrounding wards and municipalities. However, the direction of this movement depends on the factor, with the main movement of the parents-children group to eastern Nagoya, older teens moving to S and SW Nagoya, and the elderly moving between the central wards and those in the E and SW. Looking at movement occurring in the rest of Aichi Prefecture, we can see simple movement of the parents-children group into the cities of Komaki, Okazaki, etc., from their respective hinterlands. This is in contrast to the more complex movement of older teens into cities from various outlying areas. Finally, movement of the elderly is mostly confined to the Nagoya area.[18]

This research has partially explained the structure of Aichi Prefecture from the perspective of population movement. However, since it shows population movement for only one 12-month period, it is merely a cross-sectional type of analysis. In the future, materials should be prepared for a time-series analysis to serve as a mechanism for understanding structural changes in the metropolitan area.

Notes

1. This phenomenon has been discussed in detail by Fujii (1983) and Morikawa (1988).

2. Using his analysis of the retail industry, Fujii (1983) indicated that this phenomenon is progressing in Japan.

3. For example, the number of 15-34 year-olds in the suburbs from 1940-50 decreased in all regions, but increased in areas of close proximity to large cities from 1960-70. In contrast, the number of people 55 and over in suburbs increased throughout the country from 1940-50, but decreased in areas of low proximity to major cities from 1960-70.

4. In the *Fifth Aichi Regional Plan*, the prefecture was divided into three regions: Higashi Mikawa in the east; Nishi Mikawa in the central area; and the Owari region in the west. This study is following these regional boundaries.

5. The source for this information is the *Statistical Tables of Households and Population* from the Basic Resident Registers.

6. By setting the volume of movement in origination zone i (i=1,2,3...n_O), destination zone j (j=1,2,3...n_D), sex k (k=1,2,3...n_S), age group l (l=1,2,3...n_A) at x_{ijkl}, the element z_{ijkl} of the standardized matrix can be obtained. In other words, for all k=1, n_S and l=1,2,3...n_A,

$$z_{ijkl} = x_{ijkl} \Big/ \sqrt{\sum_i \sum_j x_{ijkl}^2} \qquad (1)$$

Here, n_O is the total number of origination zones (n_O=103), n_D is the total number of destination zones (n_D=103) n_S is sex (n_S=2), and n_A is the number of age groups (n_A=5). This operation allows us to standardize the square sum for each gender and age group at a uniform 1.0.

7. This analysis was conducted as follows: First, from standardized 4-dimensional data, a 1,030-row, 103-

column matrix Z_1 was created by multiplying gender × age × origination zone for the rows, with the columns being comprised of destination zones. From this matrix Z_1, plus its transposed matrix ${}^tZ_1$, Equation 2 can be used to derive a product sum matrix M_1 of n_D rows and n_D columns, i.e., 103 rows and 103 columns.

$$M_1 = (1/n_O n_S n_A)^t Z_1 Z_1 \qquad (2)$$

Next, from the analysis of the eigen value of the product sum matrix M_1, the initial factor loading matrix F_1 was derived for the destination zone. Finally, in this study, varimax rotation was used to simplify the interpretation of these results.

8. The standard number of destination zone factors was determined to be all factors in which a large loading for the absolute value appeared.

9. This analysis was conducted as follows: First, a 1-mode factorized matrix S_1 was calculated from the factorized destination data matrix Z_1 in note 7. The factor loading matrix F'_1 was established for the destination zone (after varimax rotation) and the 1-mode factorized matrix S_1 was derived from Equation 3.

$$S_1 = Z_1 F'_1 ({}^tF'_1 F'_1)^{-1} \qquad (3)$$

From S_1, a matrix Z_2 was created by multiplying gender × age × destination zone for the rows, with columns consisting of origination zones. From this matrix Z_2, plus its transformation matrix ${}^tZ_2$, a product sum matrix M_2 of 103 rows and 103 columns was obtained using equation 4.

$$M_2 = (1/n_{DF} n_S n_A)^t Z_2 Z_2 \qquad (4)$$

Here, n_{DF} was the number of destination zone factors. Next, eigen value analysis of the product sum matrix M_2 provided the initial factor loading F_2 for the origination zones. Finally, this study employed varimax rotation to simplify the interpretation of the data.

10. The number of origination zone factors was derived in the same way as the number of destination factors, that is, it was determined to be all factors in which a large loading for the absolute value appeared.

11. Calculations for the 2-mode factorized matrix S_2, which was used to factorize the data matrix Z_2 for destination zone in note 9, were performed as follows: Giving the factor loading matrix for origination zones (after varimax rotation) as F'_2, a 2-mode factorized matrix S_2 was obtained from Equation 5.

$$S_2 = Z_2 F'_2 ({}^tF'_2 F'_2)^{-1} \qquad (5)$$

12. According to Dai (1982), "element" is the formal data and "mode" signifies the collection of concepts which is the objective of the research. In this study, however, "element" is the dimension of the data while "mode" represents the factorized element.

13. This analysis was conducted as follows: First a matrix Z_3 was created from the two-mode factorized matrix S_2 derived in note 11 by multiplying origination zone factors by destination zone factors for the rows, and by multiplying gender by age group for the columns. From this matrix Z_3 and the transposed matrix ${}^tZ_3$, a product sum matrix M_3 of 10 rows × 10 columns ($n_S \times n_A$ rows, $n_S \times n_A$ columns) was obtained using Equation 6.

$$M_3 = (1/n_{DF} n_{OF})^t Z_3 Z_3 \qquad (6)$$

Where n_{OF} is the number of origination zone factors. Next, using eigen value analysis of the product sum matrix M_3, an initial factor loading matrix F_3 was derived for gender and age groups. In this study, varimax rotation was performed to facilitate interpretation of the data. Finally, the number of factors was determined in the same way they were for origination and destination zones, i.e., by standardizing the large absolute values which appeared in the all factors.

14. The calculations of gender for the data matrix M_3 in note 13 and age groups from the factorized 3-mode factorized matrix S_3 were performed as follows. Giving the factor loading matrix for gender and age groups (after varimax rotation) as F'_3, the 3-mode factorized matrix S_3 could be derived from Equation 7 as follows:

$$S_3 = Z_3 F'_3 (F'_3 F'_3)^{-1} \tag{7}$$

15. This trend corresponds to the fact discovered by Saino (1989) in his analysis of areas of high average age in the city of Nagoya.

16. In Yano's (1989) research, older teens and the elderly were considered as one factor, which is believed to be the reason that only interurban movement was dealt with.

17. In this study, these opportunities for movement are presumed to be represented by gender and age groups, though further research, including written questionnaires and personal interviews, will be needed to perform a more detailed analysis of this relationship. Furthermore, once this relationship is made clear, it will become possible to analyze the movement patterns during various life stages.

18. Comparing the 2-mode factorized matrix obtained from the factorized origination and destination zones with the 3-mode factorized matrix derived from all factorized elements, we can see that differences in movement patterns by gender and age groups are more evident in the three-mode factorized matrix. This causes the complex movement patterns of one gender and age group. In this study, the underlying factors for movement were restricted to age and gender; analysis was conducted based on the assumption that these two factors determined movement patterns. However, looking at the factor loadings of age and gender, we can see that this assumption is inadequate since complex movement patterns have been revealed for a single gender/age group. On this point, the 3-mode factor analysis conducted in the latter half of this study may be an effective method for further clarifying the differences in movement patterns by age and gender.

References

Champion, A.G. ed. (1989) *Counterurbanization: The Changing Pace and Nature of Population Deconcentration*, Edward Arnold, London, 266p.

Dai, K. (1982) Application of the three mode factor analysis to industrial design of chair styles, *Jap. Psych. Rev.* **25**, 91-103. (J with E)

Fujii, T. (1983) The locational change of retail functions in the Keihanshin metropolitan area: a phase in the development process of the metropolitan area, *Hum. Geogr.* **35**, 210-232. (J with E)

Gordon, P. (1979) Deconcentration without a clear break, *Env. and Plann.* **11**, 281-290.

Hashimoto, Y. and Murayama,Y. (1991) Regional characteristics of the migration by sex and age classes in Aichi Prefecture, *Ann. Tohoku Geogr. Assoc.* **43**, 122-138. (J with E)

Morikawa, H. (1988) On the phenomenon of population turnaround or "conuterurbanization", *Geog. Rev. Jap.* **61**A, 685-705. (J with E)

Saino, T. (1989) On the spatial pattern of aging in Nagoya City, *Ann. Tohoku Geogr. Assoc.* **41**, 110-119. (J)

Shinohara, S., Hashimoto, Y., Kashima, H. and Taira, A. (1992) The spatial characteristics of population in Tsukuba City, *Area. Res. Pap.* **14**, 9-23. (J)

Takahashi, N. (1987) A geographical study on life space in Japan, *Hum. Geogr.* **39**, 295-318. (J with E)

Takahashi, N. ed. (1990) *Life Space in Japan*, Kokon Shoin, Tokyo, 259p. (J)

Tucker, L.R. (1966) Some mathematical notes on three mode factor analysis, *Psychometrika* **31**, 279-311.

Vining, D.R., Jr. and Pallone, R. (1982) Migration between core and peripheral regions: a description and tentative explanation of the patterns in twenty-two countries, *Geoforum* **13**, 339-410.

Yano, K. (1989) Age-disaggregated intrametropolitan migration in Tokyo metropolitan area, *Geogr. Rev. Jap.*

62A, 269-288. (J with E)

Zelinsky, W. (1978) Is nonmetropolitan America being repopulated? The evidence from Pennsylvania's minor civil divisions, *Demography* **15**, 13-39.

(J = in Japanese; J with E = in Japanese with English abstract)

PART V

INTRA-URBAN SYSTEM

CHAPTER 9

COMMERCIAL ACTIVITIES IN A SMALL CITY

9.1. Introduction

According to Census of Commerce in Japan, the number of retail stores in Japan has declined steadily since reaching their peak in 1982, falling by 12.8% from then until 1994. As we can see in Figure 9.1, annual sales and the number of employees have been growing. Thus, the scale of stores in Japan has been increasing, while family-owned shops are under pressure to close. As is well known, most family-owned small shops are located in the traditional central commercial district (from now CCD).

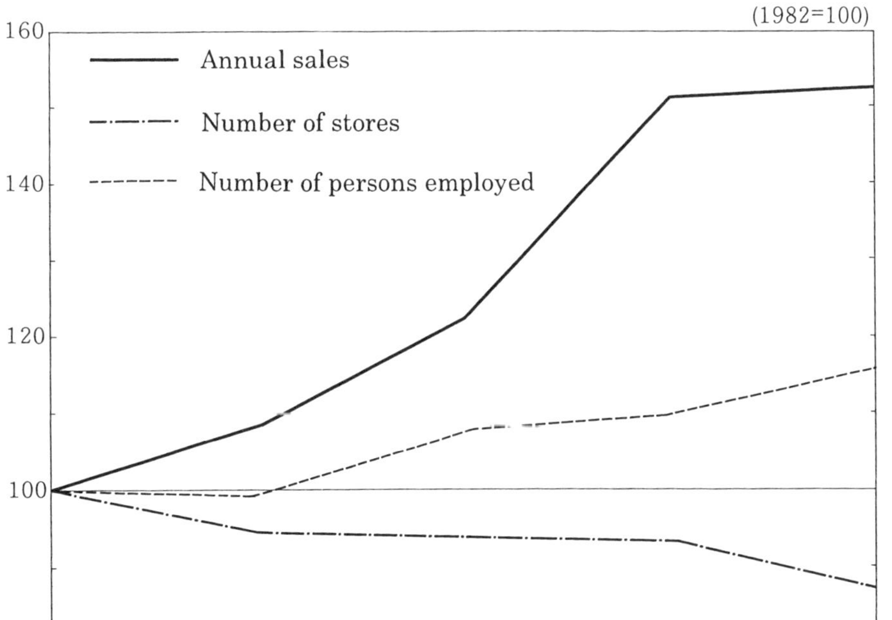

Figure 9.1. Change in the retail activity in Japan
Source: Census of Commerce in Japan.

According to the "Survey of the State of Retail Areas" conducted by the Small and Medium Enterprises Agency in 1996, more than 90% of all CCDs in Japan are either stagnating or declining. This trend is especially noticeable in smaller regional cities. Non-durable (low order) goods are purchased in suburban or exurban areas, while higher-order durable goods fall into the hinterlands of larger cities nearby. Thus, the CCDs of small regional cities are in danger of going out of existence.

Given the above conditions, this chapter will examine how once thriving CCDs have suddenly gone into decline by revealing the factors and process involved. It will also consider what economically devastated CCDs should do to survive into the future.

9.2. Motorization and Changes in Urban Structure

The modernization of Japanese cities began with the opening of railroads. Beginning in about the mid-1880's, railroads began to spread out into the countryside, and the areas around train stations became urban centers that were soon surrounded by retailers (Figure 9.2). Housing in these central areas was dense and compact. The only transportation from CCDs to residential areas was by bicycle, by foot, or by bus (which was usually very expensive), and it was difficult for local residents to commute long distances to work.

Due to these restrictions on movement and the high population density of the central area, the commercial areas around the railroad stations in smaller cities prospered. Because CCDs offered dense conglomerations of a variety of retail establishments, consumers were able to effectively combine several shopping trips into one, without the use of a motorized vehicle. The ability to buy both durable and non-durable goods was very convenient for shoppers who had limited access to long-distance transportation. For their part, the CCDs that came to dominate commercial activity, were blessed with such effective concentration of retailers that customers often came from outside the local urban area, even from outlying farming communities, to purchase durable goods there.

However, with the advent of motorization came dramatic changes to this compact centripetal urban spatial structure. Family cars now began to liberate local residents who were squeezed into small living space in central areas, from the restrictions placed on their movement by distance. The numbers of people moving out of the central area to the suburbs and exurbs increased, and the population became more dispersed. This phenomenon started to become obvious as motorization became entrenched and the modern road network was developed, in the 1970s. At the same time, motorization initiated fierce competition among cities as shoppers began venturing even farther from home. The hinterlands for durable goods in smaller cities were swallowed up by larger neighboring cities, and the CCDs in the smaller cities began to lose their commercial vitality. The driving force behind this trend was the development of exurban retail areas along bypasses that were far removed from city centers.

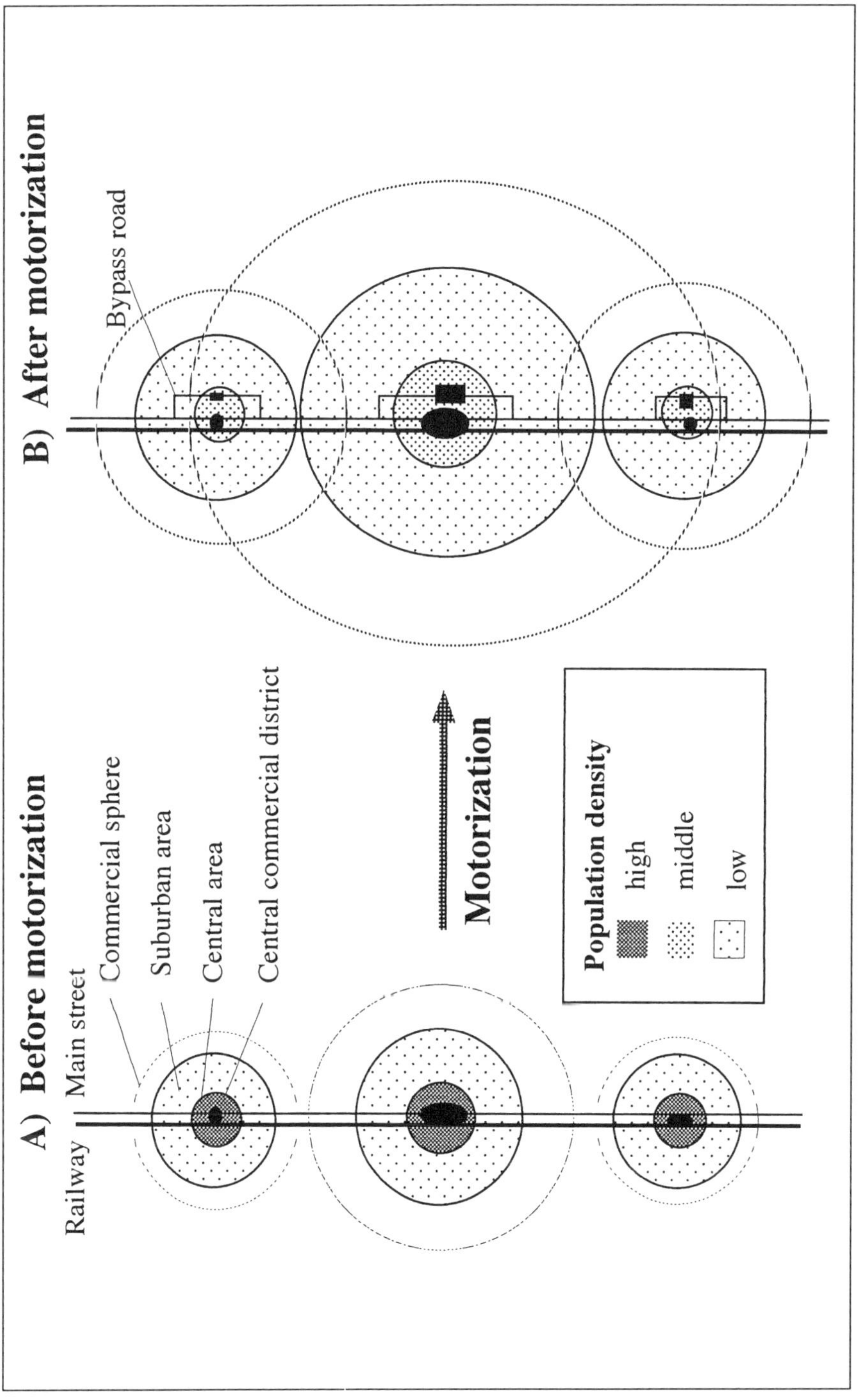

Figure 9.2. Change in the urban structure

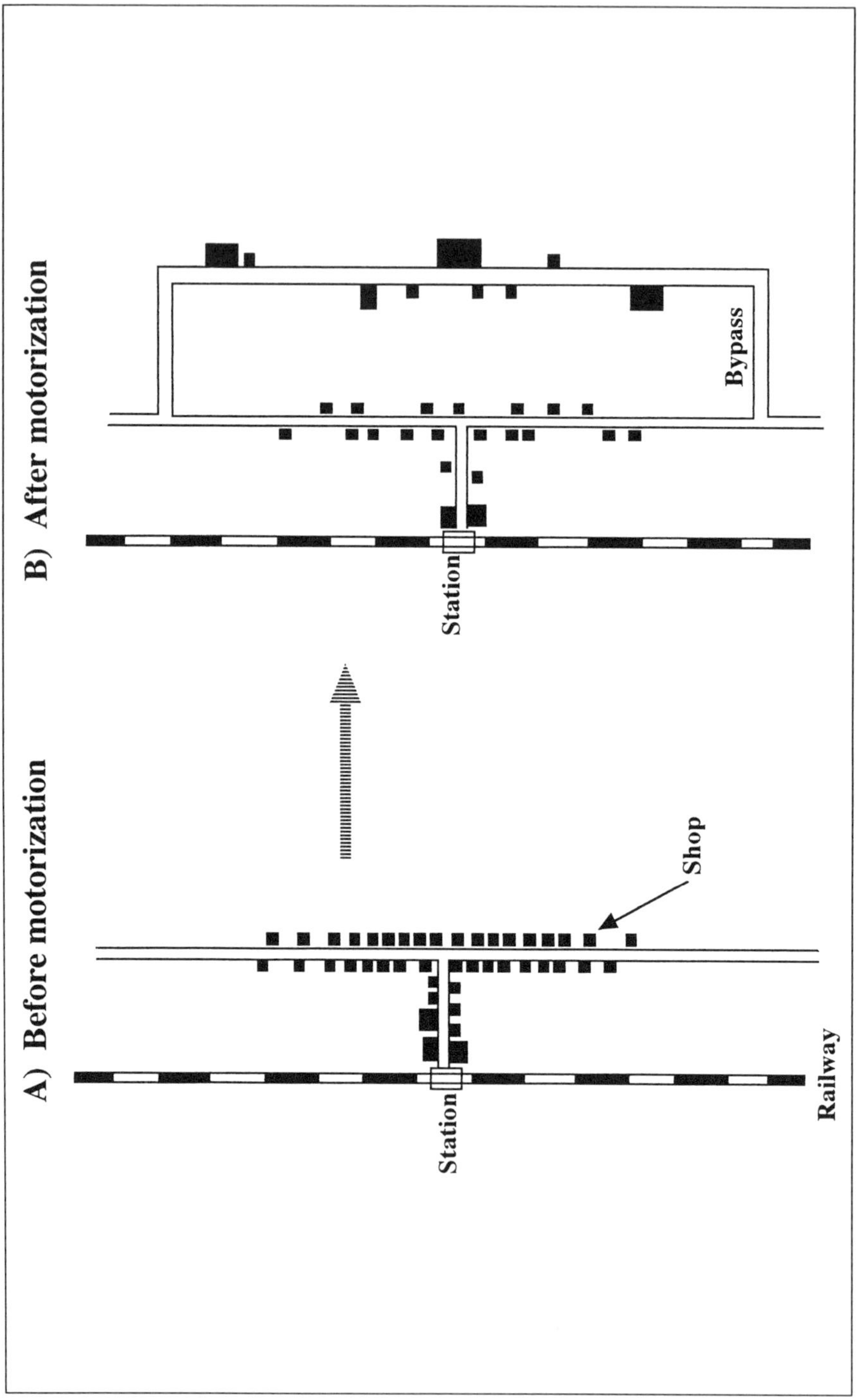

Figure 9.3. Locational change in shops in a small city

9.3. The Decline of CCDs and Development of Exurban Retail Areas

Main Streets in the CCDs were laid out before the advent of the automobile, so they are usually narrow, and pedestrians often have to walk on the same road with motorized traffic. This led to chronic congestion in city centers. To provide relief from this congestion, city leaders began to plan bypasses (Figure 9.3). Naturally, the intent of these bypasses was to channel large trucks and other unrelated traffic out of the city center to facilitate the flow of vehicles, rather than block their movement into the CCD. However, the opening of bypasses wrought major changes in consumer behavior. Large retailers and entertainment facilities started to move in, targeting users of private automobiles. Not only was movement easier along bypasses, but land was more spacious and cheaper, attracting the attention of chain stores and other non-local business owners.

Changes in lifestyles also had quite an effect on store location. During Japan's period of *High Economic Growth*, workers in tertiary industries began to earn more than farm workers, and more job opportunities began opening up for women, changing the behavioral patterns of consumers who started to make mass purchases on their days off. Shoppers converged on "megastores" in exurban commercial areas, while struggling shops in the CCD were largely ignored.

The first group of CCD stores to sustain damage were the sellers of non-durable goods with the frequent visit, such as grocery stores, which had serious problems trying to compete with the large exurban stores that offered larger selections, fresher goods, the ability to buy in bulk, and more spacious parking. As exurban commercial areas grew, the sales of not only non-durable goods shop but also clothing and appliance stores declined. Some of these shops aimed at survival by closing their downtown stores and re-opening in exurban retail areas, or by becoming tenants of larger establishments along bypasses. As a result, bypasses became retail "ribbons" of large and small stores owned by out-of-towners and locals alike. Eventually, the only shops remaining in the CCD were those that had lost the will to survive, or those that had no successor to carry on the business.

The decline of the CCD began on its edge, but gradually spread to the central area. Today, there are many empty shops, even around railway stations. For CCDs, which had made optimal use of the density effect, this loss of continuity and unity has been a major blow to their vitality. In recent years, not only vacant shops but also vacant houses have become noticeable (Photo 9.1). The number of people who move out (either to outlying areas or large cities) when they close their stores is increasing. Many if not most of the remaining residents are elderly former shopkeepers, but when they die, their houses often remain vacant. Once a house is no longer inhabited, there are few people who will take out a loan to buy it. This is because of the undesirable living environment in the city center. The relationship between retail stores and the period of store closings in the CCD is summarized in Table 9.1.

Photo 9.1. Typical landscape of the declining CCD in a small city (1999)

Table 9.1. Relationship between characteristics of small retailers and sustainability in the CCD

Period of closing	Type of business	Type of ownership	Ownership of land/bldg.	Employees	Location
Early	Low-order goods	Corporation	Lease land/ bldg.	Many	Surrounding area
Intermediate	High-order goods	Individual (Successor available)	Only bldg. leased	Few	Central area
Late	Specialty goods	Individual (No successor)	Own land and bldg.	Non	Station area

9.4. A Case Study of Hitachiota City

Located 120 km northeast of Tokyo with a population of around 40,000, the small regional city of Hitachiota will be used here to illustrate the decline of the CCD and the development of exurban retail areas. Based on the local retail system, Table 9.2 classifies 115 shops located in the CCD and in the exurbs into 5 categories.

Table 9.2. Number of shops by type in Hitachiota (1997)

Type of the stores	Central commercial district		Exurban commercial area		Total	
	numbers	%	numbers	%	numbers	%
Type A (corporate management)	5	7.5	19	39.6	24	20.9
Type B (semi-corporate management)	10	14.9	4	8.3	14	12.2
Type C (subsistence management, having successor)	15	22.4	6	12.5	21	18.3
Type D (subsistence management, successor undecided)	15	22.4	10	20.8	25	21.7
Type E (subsistence management, no successor)	22	32.8	9	18.8	31	27.0
Total	67	100.0	48	100.0	115	100.0

Source: Kawase *et al.* (1998).

After its opening in the 1970s to route traffic around the city center, the local bypass began to attract retail establishments. This had a "snowball effect" so that by the time of the author's field survey (1997), there were no less than 48 bypass stores. In contrast, there were close to 100 shops in the CCD in the early 1980's, but by 1997 that number had dwindled to 67. Furthermore, there is a danger that in the near future, nearly half of the remaining shops will close. This is because there is a large proportion of Type E shops (having no successor) and Type D shops (having no definite successor) in the CCD. The CCD is quickly losing its centripetal ability so that the only establishments that have an apparent chance of surviving are those that target elderly residents living in the area, shops that have no local competitors, and/or traditional service businesses such those that deliver rice and liquor, and bicycle repair shops. It should be noted that there are many "mom and pop" operations in the CCD that are run as a hobby rather than for profit. These are not "stores" in the truest sense, and they are fated to disappear soon.

In comparison, there is an overwhelming number of Type A stores (having many employees) in the exurban commercial area, with very few Type C, D and E shops that are owned and operated by individuals or families. Unlike in the CCD, there are almost no struggling shops where the owners live on the premises. Therefore, it is thought that the number of retail establishments there will continue to increase. Naturally, there are also retailers along the bypass who go out of business due to declining sales. However, due to the immense locational advantage of the exurban retail area, there is almost always a ready tenant for a vacant establishment. In stark contrast, there is almost never a new tenant that can be found for a CCD shop once it closes. This sort of commercial polarization between CCD and exurbs is becoming a very serious problem in Hitachiota.

9.5. Conclusion: Toward the Sustainable Development of Central Commercial Districts

CCDs in small cities everywhere are declining, and various programs are being undertaken to restore some sort of vitality to them. Thus, the big question is, are these programs going to succeed? Unfortunately, the answer is no. There is still more than the optimal number of shops in the CCD, and the number of closings is going to increase. As lifestyles and consumer behavior become more diversified and motorization becomes more entrenched, the CCDs of smaller regional cities are going to completely lose whatever locational advantages they still have. We might say that they have already fulfilled their historical role.[1]

Given this situation, what kinds of steps can be taken to give CCDs sustainable viability and stop their devastation? The author believes that so-called "community service businesses" would probably be the most effective way to give downtown sustainable growth, or at least much vitality. With the maturation of society, there undoubtedly will be increasing demand for specialty services such as bars and restaurants, entertainment and leisure establishments, educational facilities for adults, and health-related services. Looking at this in terms of the future society, we can see that the potential for all these businesses is very high. If suitable measures can be implemented, then CCD shops may be able to survive. If vacant shops are converted into these new types of businesses, and the central streets offer a mixture of traditional and "community service" establishments as residential functions are added, then maybe the population exodus can be stopped and vacant houses will become a thing of the past. Although the population of city centers is rapidly aging, if community service establishments are concentrated there, they should be able to meet the needs of elderly residents who do not own automobiles and have trouble traveling long distances.

The author also believes that it is important for the CCD to become a meeting and leisure place for local residents. In fact, in Hitachiota, empty shops have been converted into a hall for public events, much to the approval of local residents, who can gather and intermingle at the exhibits, speeches, etc., that are held there (Photo 9.2).

We should emphasize not only the economic aspects of CCDs, but also their social and environmental sides. It is safe to say that their future depends on their ability to become comfortable districts for supporting and enriching the lives of local residents.

Notes

1. There are still many shops in the CCD that are still in business but are not profitable. Since the same kinds of shops disappeared from European and American CCDs about 20 years ago, it appears that Japanese shopkeepers have had greater survivability. This is in large part due to favorable laws that restrict the location of large stores and to the traditional division of labor in which the wife tended the shop while the husband held an outside job. However, if satisfactory measures are not taken soon, most of these enterprises will be forced out of business, and the number of vacant shops and houses in the CCD will continue to increase.

Photo 9.2. Newly established community center in the CCD in Hitachiota (1998)

References

Kawase, M., Murayama, Y., Iwama, N. and Kaneko, J. (1998) Changes in the commercial structure of a small city during the transition to motorization: A case study of Hitachiota, Ibaraki, Japan, *Ann. Rep., Inst. Geosci., Univ. Tsukuba*, **24**, 1-5.

CHAPTER 10

MIGRATION CHAIN IN A MEDIUM CITY

10.1. Introduction

10.1.1. PREVIOUS RESEARCH AND PURPOSE

Generally, the process of relocation involves four types of changes in social status or life event: entering a higher level of school, obtaining or changing occupation (including transfer to another office), marriage, and retirement (Lewis, 1982; Clark, 1987). These four sequential movements account for most of an individual's migration history. Some sort of spatial regularity is believed to exist in these sequential movements which respond to the socio-economic characteristics of individuals and families. In discussing these spatial chains, we must consider whether movements are confined to one urban/metropolitan area or not, that is, whether movements are recognized as intra-urban migration or interregional migration. Most interregional migration is due to one of the four changes in social status mentioned above. On the other hand, as Wolpert (1965) has suggested, most intra-urban migration is made to improve one's so-called "place utility" (Brown and Longbrake, 1970).

The improvement of place utility is a concept similar to "housing adjustment" which Clark and Onaka (1983) have thought to be a significant factor of migration. "Housing adjustment" refers to changes in types of housing, neighborhood, and/or accessibility. Specifically, it entails movements to new housing, movements to improve one's neighborhood environment (such as better public facilities), and movements to be closer to one's job or school. Thus, migration to improve place utility generally is not induced by changes in social status. This migration, however, is often simultaneous with a change in life cycle. For example, perhaps a young person finds work close to his or her parents' home but moves from the home at the same time to begin an independent life. This movement is not essential to the change in his or her social status, that is, finding work. We should perhaps treat the social change not as a reason but as an "opportunity" for the movement. It is believed that similar cases occur in short-distance movements such as intra-urban migration. On the other hand, intra-urban migration resulting purely from changes in social status is, with the exception of marriage, limited in number.

From the above discussion, we can see that the spatial chains shaped by intra-urban migration are going to exhibit more complex patterns than those by interregional migration. In a number of intra-urban migration studies, geographers and demographers have referred to the relationship between migration chains and life courses by classifying intra-urban movements according to several attributes of movers (e.g.,

McCarthy, 1976; Clark and Onaka, 1983; Gober *et al.*, 1991). There have also been studies that, in consideration of directional bias of migration, mention the relationship between migration patterns and life cycle changes (Adams, 1969; Abler *et al.*, 1971; Donaldson, 1973; Hartshorn, 1980). Thus the spatial chains (or channels) of intra-urban migration have attracted the attention of many geographers and demographers. However, the migration chains which have been handled in these studies are not channels traced by individual migration histories. Rather they are channels analogically inferred from the aggregate of "one-time migration," i.e., non-chain migration of numerous residents. To the best knowledge of the authors, no research has undertaken systematic or detailed corroborative analysis of spatial chain patterns by tracing individual movements of numerous residents. The greatest reason for this is the lack of sufficient Origin-Destination data for the routes of individual movements.

Therefore, this chapter will quantitatively analyze the track of individual chain-type migration of many residents to determine how they move in response to changes in their lives, i.e., whether there is spatial regularity on the chain patterns of intra-urban migration. The study area for this research will be the industrial city of Yokkaichi, Japan. The study should help us to understand the state of intra-urban migration at the non-aggregate level for all residents during a 15-year period. Special attention will be given to the age at which a resident relocates, and to the interval between two successive movements, i.e., the stationary period.

10.1.2. STUDY AREA AND MATERIALS

Yokkaichi is an industrial city of 274,530 people (1989) located in the Nagoya Metropolitan Area, one of the three largest urban areas in Japan. It is situated approximately 40km SW of Nagoya along Ise Bay and is topographically comprised of coastal lowlands and interior uplands. Numerous industrial complexes in the coastal lowlands have been constructed next to pre-existing built-up areas. Large-scale housing complexes were built in various locations in the upland areas mainly from 1965-1980.

Yokkaichi is divided into 23 administrative districts (or "ku") (see Figure 10.1). These divisions are roughly equivalent to the old "choson" (i.e., town and village) boundaries of the city. Of these 23 districts, Chubu (District No.1) and Kyohoku (No.23) comprised Yokkaichi when it was first designated as an official city in 1897. The CBD is located in Chubu.

The raw data used in this chapter are from the 15-year period 1975-89 and were supplied by the Yokkaichi City Office from individual change-of-address notices. The data sets consist of a personal ID code and age and sex of the relocating party(s), as well as origin and destination codes. Here, the origin and destination will both be within the 23 administrative districts. The raw data were realigned by individual ID code to gain a thorough understanding of the migration chains of residents. From this realigned data, we can see that during the 15-year period, there were 2,862 people making 4 consecutive intra-urban movements (i.e., all movements were made within the city); 941 people making 5 successive movements; 331 people with 6 successive movements; 127 people with 7 successive movements; 51 people with 8 successive movements; 18 people with 9 successive movements; and 31 people moving within Yokkaichi at least 10 consecutive times. These raw data indicate a great degree of mobility in Yokkaichi. The authors have previously touched upon the demographic features of Yokkaichi as well as intra-urban migration patterns in this city (Murayama and Inoue, 1989).

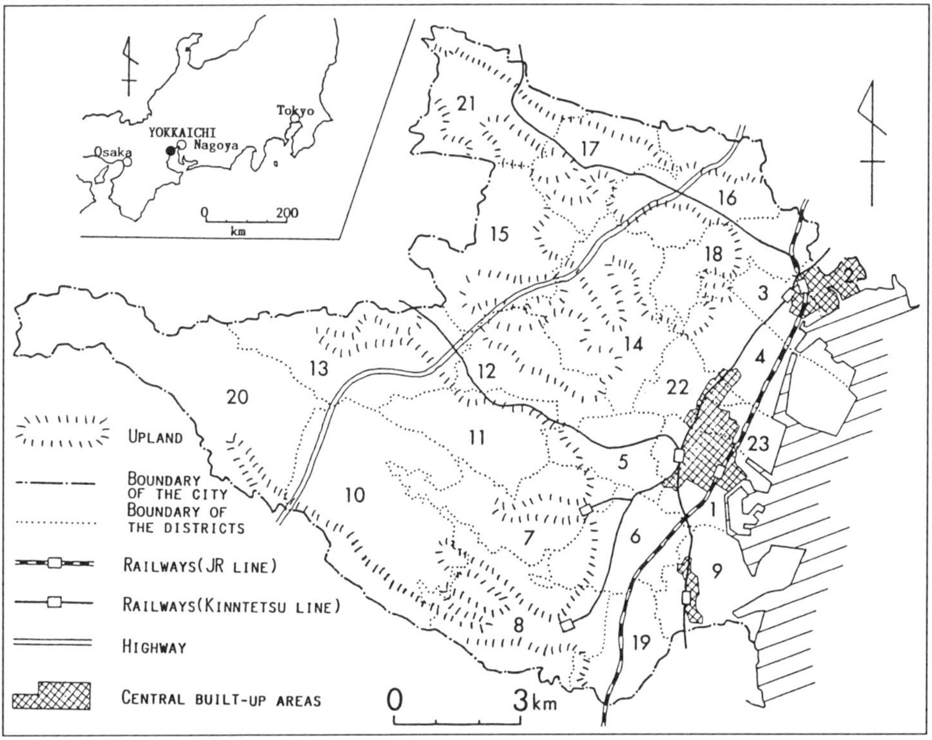

Figure 10.1. Study area

1. Chubu, 2. Tomisuhara, 3. Tomida, 4. Hazu, 5. Tokiwa, 6. Hinaga, 7. Yogo, 8. Uchibe, 9. Shinohara, 10. Oyamada, 11. Kawashima, 12. Kanzaki, 13. Sakura, 14. Mie, 15. Agata, 16. Yasato, 17. Shimono, 18. Ooyachi, 19. Kawarada, 20. Suizawa, 21. Hobo, 22. Kaizo, 23. Kyohoku

10.2. Methodology

10.2.1. UNIT REGIONS

To facilitate visual analysis from quantitative analysis and mapped data, the 23 districts were grouped into several analytical unit regions using Ward's method of cluster analysis. These regions were established as follows. First, the data used in the cluster

$$MO_i = \left(\sum_j MM_{ij}\right) \Big/ MP_i$$

$$MI_i = \left(\sum_j MM_{ji}\right) \Big/ MP_i$$

$$FO_i = \left(\sum_j FM_{ij}\right) \Big/ FP_i$$

$$FI_i = \left(\sum_j FM_{ji}\right) \Big/ FP_i$$

Where, MO_i: male out-migration ratio in district i,
MI_i: male in-migration ratio in district i,
FO_i: female out-migration ratio in district i,
FI_i: female in-migration ratio in district i,
MM_{ij}: male migration from district i to district j,
FM_{ij}: female migration from district i to district j,
MP_i: male population in district i, and
FP_i: female population in district i.

Figure 10.2. Formulas for calculating migration rates

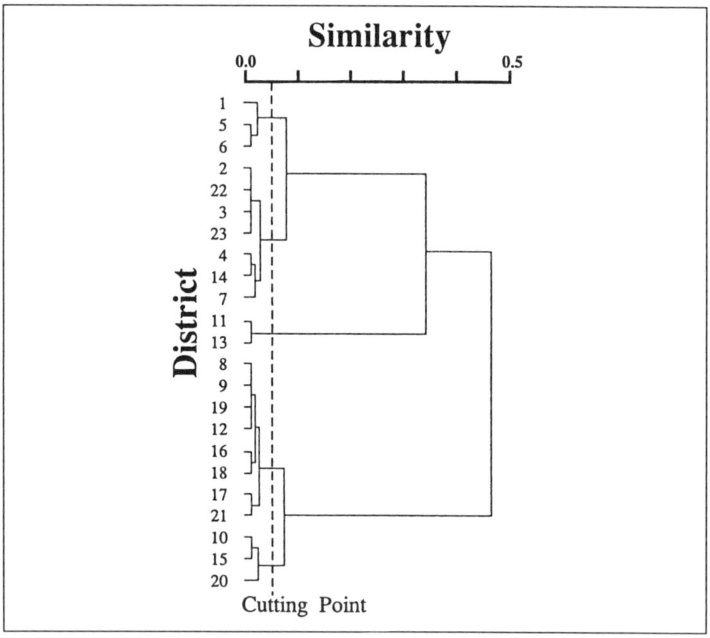

Figure 10.3. Dendrogram

analysis were composed of the rates of male in-migration, male out-migration, female in-migration, and female out-migration per district in 1989. In other words, a 23-row, 4-column geographical matrix was created using these four migration variables as the columns and the 23 districts as the rows. The formulas for calculating the migration rate for each district are shown in Figure 10.2. The results of the analysis, in which the 23 districts were grouped into five unit regions, are shown by the dendrogram in Figure 10.3. Figure 10.4 shows the five new aggregate regions.

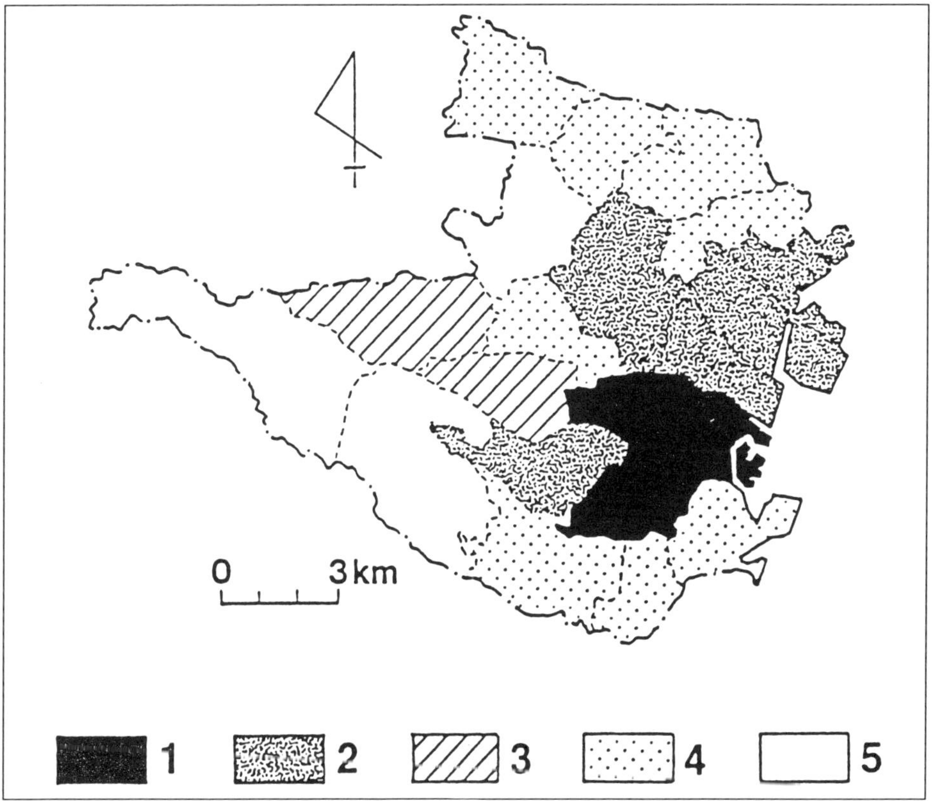

Figure 10.4. Classification of the 23 districts of Yokkaichi

Note: Region numbers correspond to those in Table 10.1.

As we can see by the average values of migration rates in Table 10.1, each of the unit regions thus established maintained its homogeneity, i.e., high degree of similarity of migration characteristics. Region numbers correspond to the ones in Figure 10.4. We can define the characteristics of these regions as follows:

Region 1 is the urban core, the central region of high population density. Here considerable migration is observed, with high rates of in- and out-migration for both males and females. Region 2 is a transition area surrounding Region 1, showing somewhat lower mobility than the central region. This is an area of comparatively large numbers of apartment houses. Though Regions 3 and 4 are both located on the outskirts of Yokkaichi, Region 3, an "area of increasing suburban population," is witnessing rapid in-migration by both males and females. The region is a new housing area of numerous single-family homes located along the relatively convenient Kintetsu Railway line. Region 4, on the other hand, is an "area of stable suburban population" where in- and out-migration are in equilibrium and the mobility of residents is low. Finally, Region 5 is a rural area with even lower mobility.

Table 10.1. Average values of migration rates

Region	Out		In	
	Male	Female	Male	Female
1	6.47	6.40	6.00	5.70
2	5.34	4.96	4.41	4.33
3	3.05	3.25	8.90	8.45
4	3.48	3.69	3.58	3.71
5	2.03	2.60	2.47	3.10

10.2.2. MIGRATION CHAIN UNITS

This analysis involved migration chains, that is, consecutive intra-urban movements (i.e., two or more consecutive movements within the Yokkaichi city limits) during the 15-year period from 1975 to 1989. Naturally, this 15-year period allows us to understand only a fraction of an individual migration chain. Nevertheless, we can estimate the overall individual migration chain for the life course by connecting migration chains which different age groups made during this period.

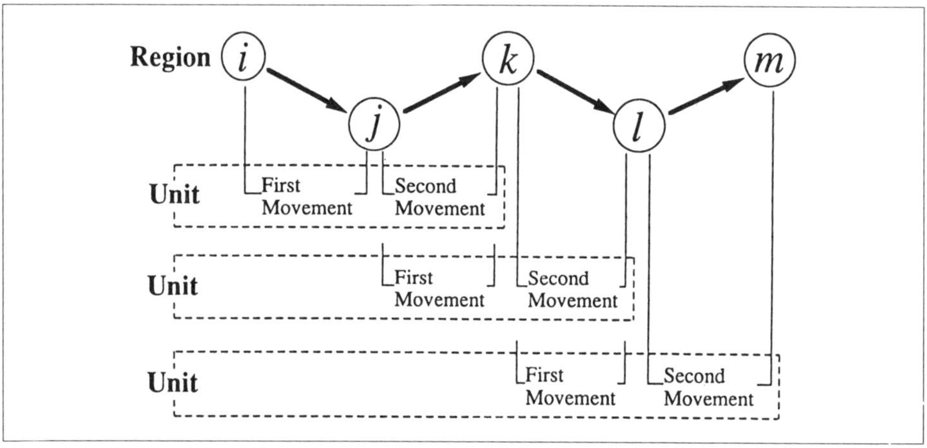

Figure 10.5. Units of migration chains

Obviously, the migration chains observed during this period comprise various numbers of movements, that is, they vary in size. The following problem arises in treating various-size migration chains in their entirety; that is, the total stationary period for an observed migration chain is shorter on average than that for the other non-observed chains of the same size, and furthermore, this tendency to be shorter increases as the size increases. Therefore, in this analysis, migration chains have been divided into minimum-size linkages, the basic units for data formulation. For example, Figure 10.5 shows that a migration chain from Region *i* through Regions *j*, *k*, and *l* to Region *m* consists of four successive movements, that is, three linkages (*i-j-k*, *j-k-l*, *k-l-m*). We can consider each of these three linkages as a migration chain unit.

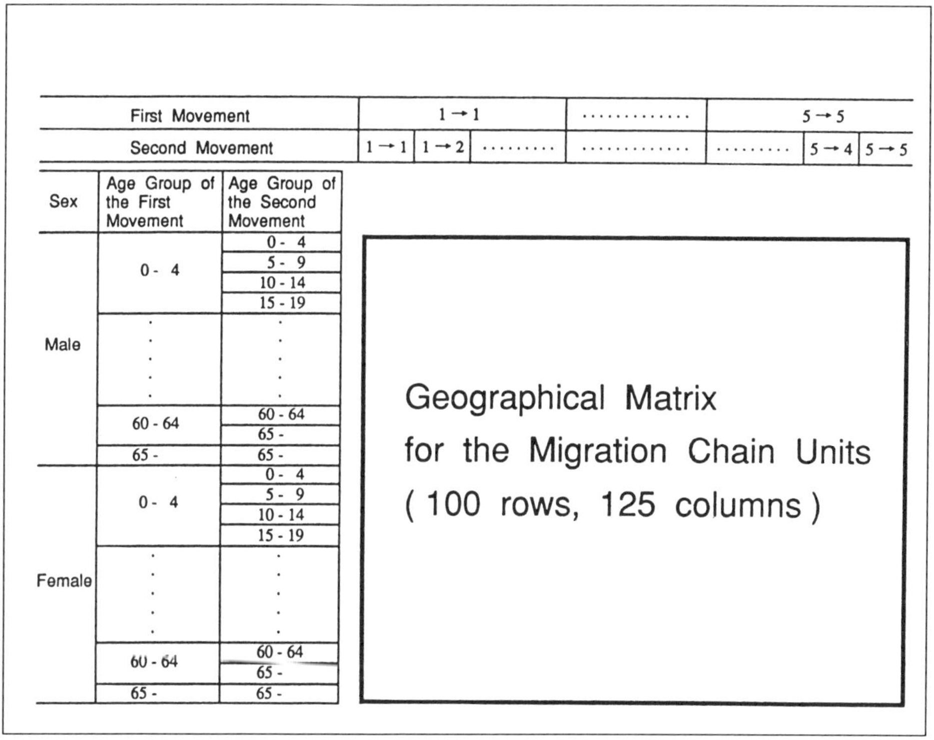

Figure 10.6. Data matrix

10.2.3. CREATION OF A GEOGRAPHICAL MATRIX

The migration chain unit, as defined in the previous section, can be thought of as a pair of intra-urban movements. In other words, the most fundamental migration attributes of one unit are the relocator's sex, age at which the first movement was made, age at which the second movement was made, the original residence (i.e., the first origin), the first destination, and the second destination.[1] Considering these six fundamental attributes, we quantified the migration chain units and created a geographical matrix which was directly subjected to mathematical analysis.

First, the ages at the times of the first and second movements were classified into 14 age groups of five-year intervals: 0-4, 5-9, 11-14, etc., up to 65-and-over. Consequently, the relocation's stationary period between the first and second movements could be determined by a shift between two age groups. The shifts, however, were restricted to those having not more than three steps (e.g., a shift from 20-24 to 35-39), because this study was concerned with data from a 15-year period. Therefore, 50 possible pairs of age groups were generated each for males and females, for a total of 100 pairs. These 100 pairs of age groups were distributed in the rows of the matrix.

Next, combining the three regions of the original residence, the first destination, and the second destination, we expressed the spatial patterns of migration chain units. Since there were five unit regions in this study, there were 125 (= 5 × 5 × 5) of these regional combinations (hereafter called O-D combinations). These 125 O-D combinations were distributed in the columns of the matrix. Naturally, the components of the matrix represent the number of intra-urban migrants. From the above procedure, we constructed a 100-row, 125-column geographical matrix for the migration chain units (Figure 10.6).

10.2.4. STANDARDIZING THE MATRIX

Geographical matrices are usually standardized by means of standard deviations in application of factor analysis. This method, however, is not appropriate for standardizing the matrices which, like the above geographical matrix, include many zero-value components. This is because the method may not maintain the contrast between zero-value components and non-zero components. Thus we adopted the method by which the square sum of each column (i.e., each of the 125 O-D combinations) of a geographical matrix was standardized at 1.0. The conversion of every zero-value component into itself is a distinctive characteristic of this standardization.[2]

If the *ij*-component of the geographical matrix constructed in the preceding section is represented by x_{ij} (i = 1,2, ... , 100; j = 1,2, ... , 125), the standardized value z_{ij} for all i, j is derived from the following.

$$z_{ij} = x_{ij} \Big/ \sqrt{\sum_i x_{ij}^2} \tag{1}$$

Using this operation, we obtained a matrix, Z, whose component was given by the standardized value z_{ij}, and where the square sum was 1.0 for each column.

10.2.5. CALCULATING FACTOR LOADINGS

Matrix Z is composed of 100 rows (age group pair × sex) and 125 columns (origin × first destination × second destination). Matrix Z and its transposed matrix ${}^tZ$ provided a 125-row, 125-column product sum matrix M (Equation 2).

$$M = (1/100)\,{}^tZZ \tag{2}$$

Next, eigenvalue analysis was performed on the product sum matrix M,

$$M = EU\,{}^tE \tag{3}$$

giving us matrices E and U, where E is an eigen vector matrix of 125 rows and n columns, U is the diagonal matrix with the eigen value as a diagonal component, ${}^tE$ is the transposed matrix of the eigen vector matrix E, and n is the number of factors related to O-D combinations. In addition, giving the diagonal components of matrix U as $(\lambda_1, \lambda_2, \dots, \lambda_n)$, then $(\sqrt{\lambda_1}, \sqrt{\lambda_2}, \dots, \sqrt{\lambda_n})$ provides us with diagonal matrix V. Multiplying this matrix V by the eigen vector matrix E gives us the initial loading factor matrix F (Equation 4).

$$F = EV \tag{4}$$

The rows of this initial loading factor matrix F are composed of origin × first × second destination, while the columns contain the factors related to O-D combinations. In addition, the number of factors was determined as the number of eigenvalues greater than the average value calculated from the standard deviation of each column of the standardized square sum matrix Z. To simplify interpretation, factors were subjected to varimax rotation.

$$L = FT \tag{5}$$

Where L signifies the factor loading matrix after rotation and T is the rotated matrix of the initial factor loading matrix F.

Finally, factor scores were calculated. To do this, weight matrix W was obtained from the following equation:

$$W = L({}^tLL)^{-1} \tag{6}$$

As shown in Equation 7, factor score matrix S was generated by multiplying the weight matrix W by the standardized data matrix Z.

$$S = ZW \tag{7}$$

The factor score matrix S is composed of 100 rows (age group pair × sex) and n columns (factors related to O-D combinations). Although S was obtained in the form of a standardized square sum matrix, to facilitate interpretation, we further standardized the matrix S using standard deviations, that is, conventional factor analysis method.[3]

10.3. Results of the Analysis

As a result of factor analysis, the migration chain patterns within Yokkaichi could be broken down into six factors. Table 10.2 shows the variance and the ratio for each factor. In addition, Table 10.3 and Figure 10.7 depict the distribution patterns of standardized factor scores and factor loadings, respectively. Table 10.3 shows the age group pairs (i.e., ages at the time of the first and second movements) with a factor score over 1.0. Figure 10.7 illustrates factor loading patterns in the form of migration chain units divided according to residents' first destinations. The thin arrows indicate routes giving factor loadings of 0.5-0.7, while the thick ones denote routes giving factor loadings over 0.7. Using Table 10.3 and Figure 10.7 as a base, we interpreted the migration chain patterns of the six factors which had been identified.

Table 10.2. Variance of each factor

Factor	Variance	Proportion (%)
I	0.6992	55.94
II	0.0831	6.65
III	0.0455	3.64
IV	0.0355	2.84
V	0.0297	2.38
VI	0.0258	2.07

10.3.1. FACTOR I

The variance of Factor I accounts for 55.94 percent of the total variance. Accordingly, a substantial portion of chain migration within Yokkaichi can be explained by Factor I. Every age group pair with a factor score of more than 1.0 is a combination of the same age group. Furthermore, with one exception, every pair is a combination of youthful age groups from 15-19 to 25-29 (see Table 10.3). Factor I, therefore, shows a period of less than five years between the first and second movements for these young groups, indicating a rather short stationary period. Finally, we should note that female factor scores show much higher values than male ones.

The most obvious loading pattern of Factor I in Figure 10.7 is that which provides several routes through Region 1 toward Region 2. When the regional characteristics of Regions 1 and 2, as well as the migration attributes of Factor I (age level of mover, length of stationary period) derived from the factor score are taken into consideration, this migration chain pattern is believed to represent the following scenario: Young people move from all areas of the city into the central area, then head toward apartment houses and other such housing in the surrounding area almost immediately thereafter. Hence, we can see that this migration chain pattern is basically a reflection of taking action to obtain higher education or employment for the first movement, and to improve place utility for the second.

Table 10.3. Factor scores

Factor I

Sex	Age group		Score
	First movement	Second movement	
Male	20-24	20-24	3.70
	25-29	25-29	2.84
Female	0-4	0-4	1.33
	15-19	15-19	1.77
	20-24	20-24	5.89
	25-29	25-29	3.35

Factor II

Sex	Age group		Score
	First movement	Second movement	
Male	20-24	25-29	3.75
	25-29	30-34	2.81
	30-34	35-39	1.99
Female	20-24	25-29	6.84
	25-29	30-34	1.28

Factor III

Sex	Age group		Score
	First movement	Second movement	
Male	65-	65-	4.67
Female	65-	65-	7.82

Factor IV

Sex	Age group		Score
	First movement	Second movement	
Male	0-4	0-4	4.71
	0-4	5-9	1.89
	5-9	5-9	1.32
	25-29	35-39	1.23
	30-34	30-34	2.62
	30-34	35-39	1.95
	35-39	35-39	1.31
	35-39	40-44	1.17
Female	0-4	5-9	1.52
	5-9	5-9	1.01
	25-29	25-29	2.82
	25-29	30-34	1.28
	30-34	30-34	20.60

Factor V

Sex	Age group		Score
	First movement	Second movement	
Male	0-4	10-14	1.81
	20-24	20-24	1.27
	25-29	30-34	4.98
	30-34	40-44	1.01
Female	20-24	20-24	2.33
	25-29	30-34	2.30
	25-29	35-39	1.70
	30-34	40-44	1.63

Factor VI

Sex	Age group		Score
	First movement	Second movement	
Male	10-14	15-19	1.10
Female	10-14	15-19	1.43
	15-19	15-19	1.35
	15-19	20-24	5.39
	25-29	30-34	1.56

Note: Standardized factor scores over 1.0 are shown.

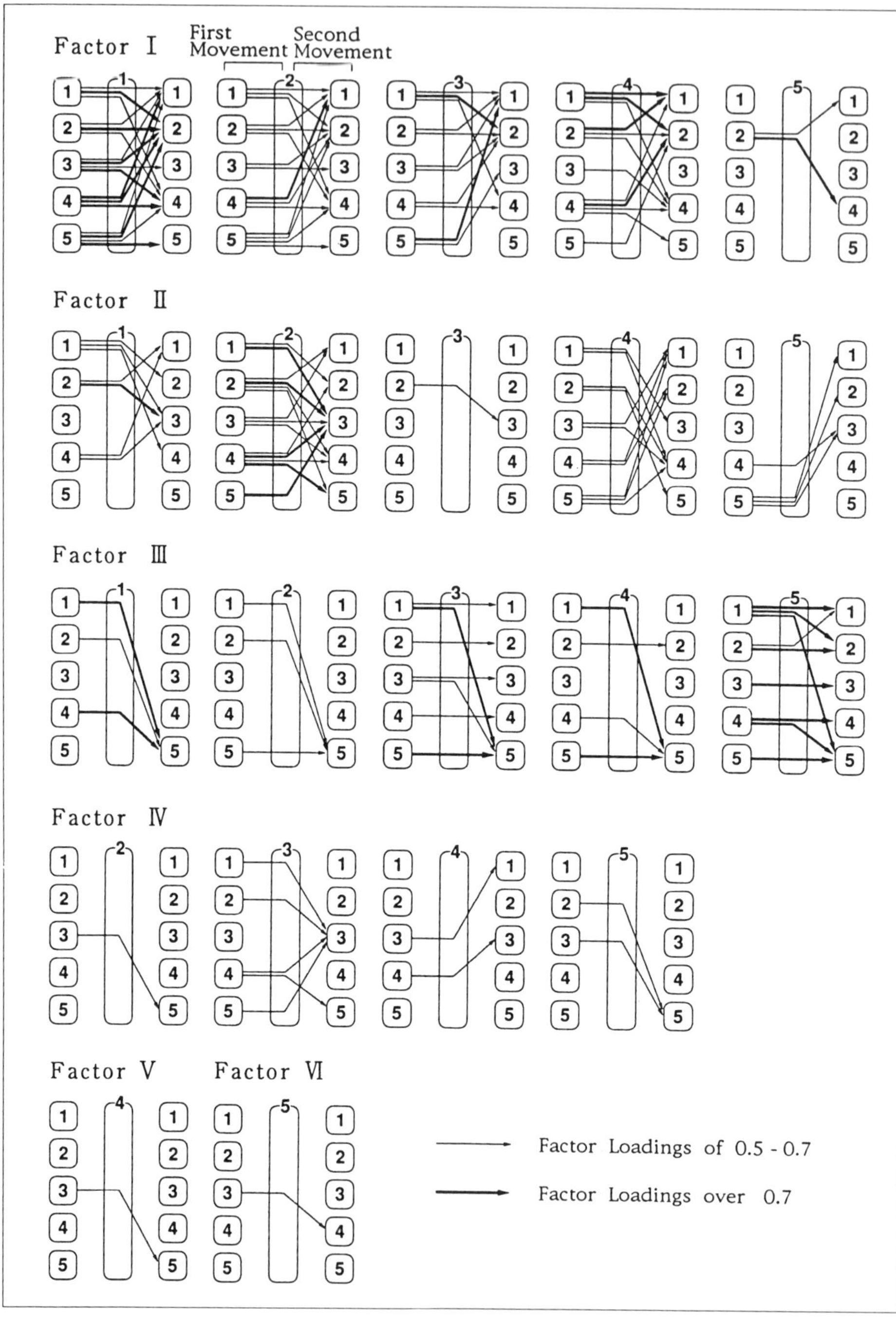

Figure 10.7. Factor loading patterns

Note: Factor loadings over |0.5| are shown.

Additionally, in the loading patterns of Factor I, there is a considerable number of cases in which people return to their original residences after having spent time in Region 1. In other words, all patterns in which people move from Regions 2, 3, 4, and 5 to Region 1, then return to their original residences show, with the exception of Region 3, high factor loadings above 0.7. These migration chain patterns indicate that young people were moving from other regions in the city to the central area, then almost immediately returning to their original residences. Here, the first movement, like the pattern previously reported, appears to be connected with securing higher education or employment. We can think of the second movement as being a kind of U-turn migration, even though "U-turns" are usually observed in interregional migration. These U-turns naturally result from life events such as commencement of new or different employment. Nevertheless, the likelihood that the second movement is necessary to life cycle changes is rather low, since the movement is small-scale migration within one urban area. We consider, therefore, the movement as an effort to improve place utility. Finally, because the factor scores for females are higher than those for males, the second movement may be effected by female marriage or preparation for marriage. In factor I, there is very little movement through Region 5 (farming areas) to other regions.

10.3.2. FACTOR II

In comparison with Factor I, the variance of Factor II is low, accounting for only 6.65% of the total. According to Table 10.3, the average stationary period is about five years longer than that of Factor I. Moreover, the age group pair in Factor II, being a combination of age groups from 20-24 to 35-39, is somewhat older than that in Factor I. In the scores of Factor II, female values are strongly biased toward the young age group pair of 20-24 and 25-29 in contrast to male values, which not only contain such young pairs but also the older pair of 30-34 and 35-39. These male-female differences in the score distribution of age group pairs correspond well to those in the age distribution of marriage in Japan.[4] This indicates that either the first or second movement is connected with marriage.

Figure 10.7 shows that there is a very conspicuous pattern of migration through Region 2 to the outlying Region 3. This reflects the migration of young people into a rapidly growing peripheral area of mostly single-family homes after living some time near the city center. Although there are various reasons for making the first of such movements, the second possibly results from marriage, or to improve place utility in response to childbirth or both. As a result, the improvement of place utility is believed to involve mostly housing, especially the acquisition of single-family homes. Movement chains going through Region 3 are distinctively few in number. Since this region is experiencing rapidly increasing population and new housing construction, it is not a temporary housing area but a permanent destination.

10.3.3. FACTOR III

The variance of Factor III at 3.64%, is somewhat lower than that of Factor II. As Table 10.3 clearly illustrates, this factor represents a migration chain of the 65-and-over age group. Furthermore, considering the different life expectancies of males and females, women naturally have a higher factor score.

The most distinguishing characteristic of factor loadings is the emergence of migration patterns of elderly people from all parts of the city to the mostly agricultural Region 5 or back to their hometowns (see Figure 10.7). Given the characteristics of Region 5 and the movers in this age group, we can assume that the first movement is very likely made after retirement, although it is somewhat difficult to make a judgment concerning the second movement.

10.3.4. FACTOR IV

The variance proportion of Factor IV, at 2.84%, is low. Age group pairs with a factor score exceeding 1.0 are clearly divided into two generations: combinations of the youngest groups under 10 years of age and combinations of groups from 25-29 to 40-44 (see Table 10.3). Obviously, Factor IV represents migration of parents accompanied by their children. Moreover, we can conclude that the stationary period is comparatively short. The distribution of these age group pairs shows males with a wider range in the upper levels, suggesting that this factor is the result of household relocation.

Factor loading patterns here are much simpler than those up to and including Factor III; furthermore, there are no migration chains giving a factor loading over 0.7. This pattern indicates that the first movement is directed toward the new housing areas of Region 3, while the second movement tends to be made within Region 3. Consequently, this migration appears to be made by young families who seek to improve their place utility by changing residences. There is apparently the movement back to the urban core (Region 1).

10.3.5. FACTORS V AND VI

The variance proportions of both Factors V and VI, at the 2% level, are quite low. Moreover, since there is only one O-D combination which can be obtained from each factor loading, the two factors will be treated together in this section.

Factor V, like Factor IV, shows migration of parents accompanied by their children, though the factor is dissimilar from Factor IV in that it shows a wider age difference in each pair of age groups, that is, a longer stationary period. Factor loadings greater than 0.5 are found only in the migration chain from Region 3 through 4 to 5.

Factor VI represents migration of young women within a ten-year period; the score for the age group pair of 15-19 and 20-24, at 5.39, is particularly striking. Factor loadings in excess of 0.5 are found only in the migration chain from Region 3 through 5 to 4.

10.4. Discussion: Life-Course Patterns

In the previous section, six factors were identified from migration chain patterns within the city of Yokkaichi, then the distinguishing characteristics of each factor for this migration were investigated. Specifically, from the score and loading of each factor, we could gain an understanding of the migration attributes (movers' age groups, stationary periods, movement routes, and so forth) which in turn allowed us to suggest reasons for relocating. The concentration of movement chain patterns into 6 analyzable factors suggests a strong interrelationship among age, stationary period, route of movement, etc. The matrix was derived from the data of minimum-size migration chains, that is, migration chain units. We must, therefore, not only take the unique patterns of each factor into account, but also understand the various patterns from a holistic standpoint in order to discuss the lifelong chain migration of individuals. Thus, this section will compare and investigate the migration chain patterns of all factors together. It will also attempt to find the typical life course of the residents of Yokkaichi.

Examining the factor loading patterns in Figure 10.7, we can see that several age groups tend to move along their own peculiar channels. For example, Factor I shows that there is a notable tendency for young people in their late teens and early twenties to migrate from some region through Region 1 to Region 2. According to Factor II, people in their twenties and thirties show a strong inclination to relocate through Region 2 to Region 3. In Factor IV, many residents in their late twenties to early forties, after moving into Region 3, relocate within Region 3. These spatial patterns indicate that intra-urban chain migration in Yokkaichi has "spatial directionality." Furthermore, when these spatial chain patterns are exhibited not simultaneously by some age groups but sequentially by one cohort, we can combine the several chains into one long chain. This allows us to derive the following typical migration history.

First, people in their late teens and early twenties move into the central area from all other regions in order to receive higher education or find employment, then move to neighboring regions in an effort to improve their place utility (Factor I). Next, residents in their late twenties and thirties mostly use marriage and/or childbirth as an opportunity to move into new housing areas in the surrounding regions (Factor II). Subsequently, people in their thirties and early forties move with their children within these new housing developments (Factor IV).

As mentioned above, we could find that the spatial directionality appears in chain migration and then, by developing this idea, we expressed our views on the relationship between chain migration patterns and life histories. In addition to these inclinations, the loading patterns of Factors I and III show that "spatial regression" occurs in chain migration within Yokkaichi. In other words, even if young people in their late teens and twenties move to Region 1, they return to their original residences (Factor I); while people aged 65 and over eventually return to their hometowns, even if they first spend some time in Region 5 (Factor III). Such spatial regression suggests that many migration routes of young and elderly people, before turning to their origins, converge on the central area and outlying agricultural areas, respectively.

10.5. Conclusion

This chapter has attempted to confirm spatial chain patterns of intra-urban migration corresponding to the life cycle changes of residents, i.e., the spatial regulation of migration chains within a city. To achieve this objective, the industrial city of Yokkaichi, which offered 15-year data on the intra-urban movements of all residents which could be understood at the non-aggregate level, was selected as the study area, and the track of individual chain migration of numerous residents was quantitatively analyzed.

This chapter integrated the several channels of migration having spatial directionality into the following typical migration chain. First, young people in their late teens and early twenties move into the city center from all other regions to receive higher education or secure employment, then move into adjacent areas to improve their place utility. Next, in their late twenties and thirties, they move from the adjacent areas into new housing developments, mainly as a result of marriage and/or childbirth. Thereafter, as adults in their thirties and early forties, they relocate with their children within these housing developments.

This chapter was also able to verify a trend toward spatial regression in the above mentioned migration chain patterns, in which young and elderly movers tend to return to their original residences after staying in the central area and outlying agricultural areas, respectively.

From the above results, we can see that two of the distinguishing features of chain patterns for migration within the city of Yokkaichi are spatial directionality and spatial regression. Moreover, we know that these two tendencies manifest themselves in their own peculiar age group pairs and O-D combinations. This chapter has discussed the reasons behind the appearance of these distinguishing migration characteristics chiefly in terms of place utility and changes in social status; however, since no data were available from movers concerning their incentives to relocate, part of the discussion has admittedly been based on conjecture. Therefore, in order to conduct a more thorough discussion of the spatial regularity of migration chains, we should obtain migration data which include movers' incentives at the non-aggregate level, and develop methods for quantitatively analyzing these migration data.

Notes

1. Naturally, this corresponds to the point of origination of the second movement.

2. Please refer to Hashimoto (1992) regarding the effectiveness of the square sum standardization.

3. This standardization gave us a factor score matrix where the average value and standard deviation of each column was 0.0 and 1.0, respectively.

4. The average age of marriage in Japan in 1988 was 29.7 for males and 26.8 for females.

References

Abler, R., Adams, J.S. and Gould, P. (1971) *Spatial Organization: The Geographer's View of the World*, Prentice-hall, Englewood Cliffs, N. J., 587p.

Adams, J.S. (1969) Directional bias in intra-urban migration, *Economic Geography* **45**, 302-323.

Brown, L.A. and Longbrake, D.B. (1970) Migration flows in intra-urban space: place utility considerations, *Annals of the Association of American Geographers* **60**, 368-384.

Clark, W.A.V. (1987) *Human Migration*, Scientific Geography Series No. 7, Sage Publications, Beverly Hills, 96p.

Clark, W.A.V., and Onaka, J. (1983) Life cycle and housing adjustment as explanations of residential mobility, *Urban Studies* **20**, 47-57.

Donaldson, B. (1973) An empirical investigation into the concept of sectorial bias in the mental maps, search spaces and migration patterns of intra-urban migrants, *Geografiska Annaler* **55**B, 13-33.

Hartshorn, T.A. (1980) *Interpreting the City: An Urban Geography*, John Wiley & Sons, New York, 498p.

Hashimoto, Y. (1992) Transformation of the central place system in Miura Peninsula, *Geographical Review of Japan* **65**A, 665-686. (J with E)

Gober, P., McHugh, K. and Reid, N. (1991) Phoenix in flux: household instability, residential mobility, and neighborhood change, *Annals of the Association of American Geographers* **81**, 80-88.

McCarthy, K. (1976) The household life cycle and housing choices, *Papers and Proceedings of the Regional Science Association* **37**, 55-80.

Murayama,Y. and Inoue, T. (1989) Intra-urban migration patterns in Yokkaichi City, Japan, *Annual Report of the Institute of Geoscience, the University of Tsukuba* **15**, 10-16.

Lewis, G.J. (1982) *Human Migration: A Geographical Perspective*, Croom Helm, London, 200p.

Wolpert, J. (1965) Behavioral aspects of the decision to migrate, *Papers and Proceedings of the Regional Science Association* **15**, 159-169.

(J = in Japanese; J with E = in Japanese with English abstract)

CHAPTER 11

LAND USE CHANGE IN TOKYO

11.1. Introduction

The period of *High Economic Growth* which began in the latter half of the 1950s led to a massive migration of population from rural to urban areas in Japan. The population concentration into Tokyo was particularly noteworthy. Accompanying this economic growth was a rapid expansion of urban zones into the surrounding area. However, the first oil crisis of 1973 severely weakened this process, and the pace of subsequent land use changes declined sharply.

This chapter will examine land use conversion in the Tokyo Metropolitan Area. This study will consider changes in land use patterns and attempt to estimate the structure of land use by types of uses during the ten-year period beginning 1984. To achieve these goals, this study employs Markov chain model.

The study area consists of the region of restricted urbanization and suburban development zones (6324.5 km^2) governed by the National Capital Region Development Act (as of 1974). In this study, this area is defined as the Tokyo Metropolitan Area. The following 15 types of land use is subjected to the analysis: 1) Forests and wasteland, 2) paddy fields, 3) upland and other fields, 4) land undergoing preparation or reclamation, 5) open land, 6) industrial land, 7) low-density housing, 8) dense, low-rise housing, 9) medium- and high-rise housing, 10) commercial and office land, 11) roads (existing and proposed), 12) parks, 13) other public facilities, 14) lakes and streams, and 15) miscellaneous land uses. To expedite the analysis, these 15 categories are sorted into the following five groups: rural land uses (1-3), transitional land uses (4 and 5), urban land uses (6-10), public land uses (11-13), and miscellaneous land uses (14 and 15). The data used here consists of a detailed 100 meter/mesh numerical database (632,450 meshes) compiled by the Geographical Survey Institute.

Analysis is conducted according to the following procedure: 1) Seven zones of 10 km radial increments from the central point, Tokyo Station, will be drawn, then a transition probability matrix (15 rows × 15 columns) for the period 1974-84 will be created for each of these zones. Next, Markov chain model is applied to each of these seven matrices to predict the land use structure for the seven zones in 1994. 2) A comparison will be made between the 1994 land use structure derived from this analysis and actual land use patterns in 1984 to gain a quantitative understanding of the changes in land use during this ten-year period.

11.2. Changes in Land Use Patterns from 1974 to 1984

During this period, nearly 70% of the land in the Tokyo Metropolitan Area was classified under the following four land-use categories (figures in parentheses refer to category number): forests and wasteland, paddy fields, upland and other fields, and low-density housing. In 1974, forests and wasteland occupied 22.49% of the land, making this the largest single land use category. Next came upland and other fields at 16.79%, low-density residential housing at 15.58%, and paddy fields at 14.25%. However, by 1984, the combined share of forests-wasteland and paddy fields to total land area was 20.53%, or a decrease of 12.82% during the ten-year period. On the other hand, low-density housing increased to 18.91%, indicating the relative progress of housing construction.

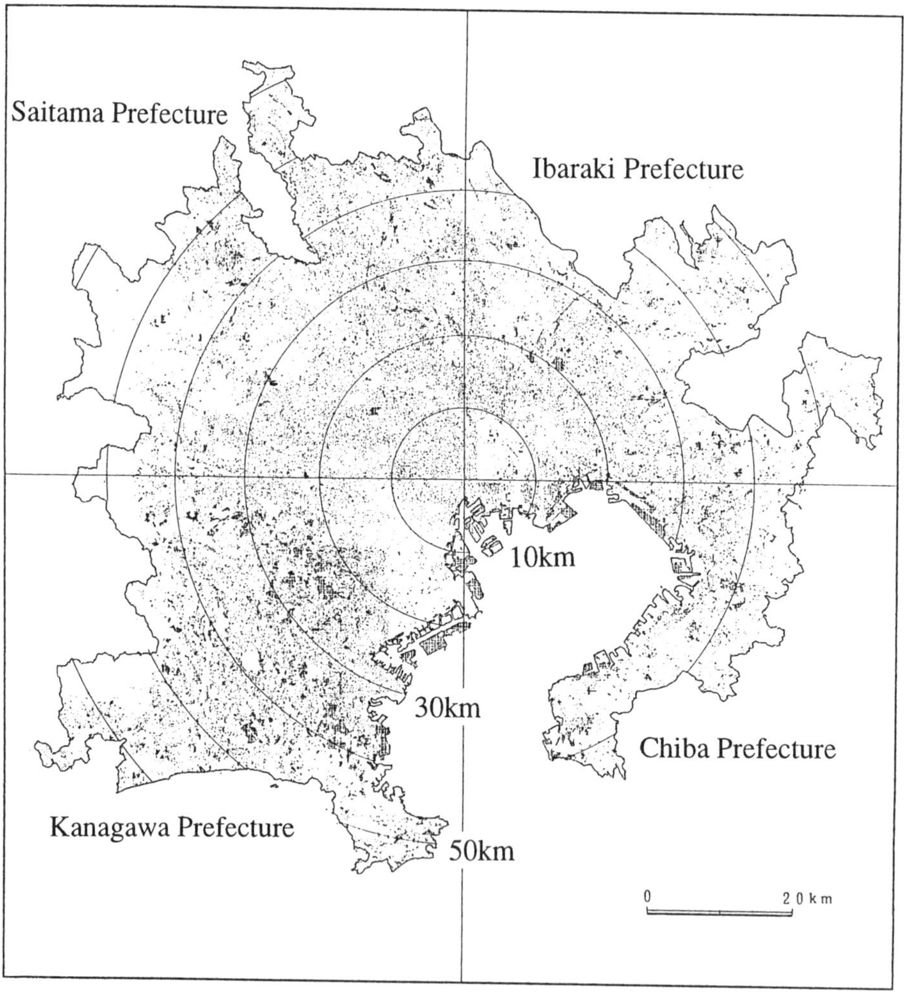

Figure 11.1. The areas which changed the land use from 1974 to 1984

As we can see in Figure 11.1, large-scale changes in land use patterns occurred in the Tokyo Bay area, the Tama Hills, and southern Yokohama. A substantial change was also evident in central Chiba Prefecture due to golf course development. Little change occurred in the central wards of Suginami, Meguro, Setagaya, etc., as well as in the peripheral areas of the metropolitan area.

Using these changes in land use patterns, a model was created to land use conversions by category for the 10-year period beginning 1974 (see Figure 11.2). The legend within the circle indicates the proportion of land use patterns which did not change during this period. Therefore, the lower the value, the greater the activity of land use conversion is.

There was little conversion of urban land uses (nos. 6, 7, 8, 9, 10) and public land uses (nos.11,12,13) into other land use categories. On the other hand, there was considerable conversion of transitional land use areas (4 and 5). Finally, there was little change one way or the other with miscellaneous land uses (14 and 15), indicating that this land is difficult to convert to other uses.

In Figure 11.2, the largest arrow indicates the conversion of prepared/reclaimed land to open land. Its value is 0.199. In other words, 19.9% of the land undergoing preparation or reclamation in 1974 had become open space during the ten-year period. The second largest arrow shows the conversion of open land into low-density housing, with a value of 0.193. Reclaimed/prepared land was most often converted to other uses, as shown by the large concentration of arrows originating from there; a mere 15.6% of such land remained unchanged during the ten-year period. On the other hand, a large concentration of the arrows points toward low-density housing, showing particularly high conversion from prepared reclaimed land, open land, and roads.

The overall view indicates that there is a considerable proportion of transitional and rural land which is being converted to urban and public uses, while little if any urban land is being transformed into rural, transitional or public land; moreover, there is almost no public land which is reverting to agricultural or transitional land.

11.3. Estimation of 1994 Land Use Patterns Using Markov Chain Model

What sort of ratio of land use structure will be revealed by extending the changes in land use patterns from 1974 to 1984 shown in Figure 11.3 to cover the ten-year period beginning from 1984? Which types of land use will be account for the highest proportions? The lowest? To answer these questions, this chapter will use Markov chain model to estimate ratios of land use categories for 1994. The Markov chain model used for this experiment was

$$ {}_{t+1}L = {}_{t}L \cdot P, $$

where, ${}_{t+1}L$ is the vector for land use structure ratio for 1994, and ${}_{t}L$ is the vector for 1984. The total element value is 1.0, as is the sum of the matrix rows for each category.

P represents the transition probability matrix indicating changes in land use from 1974 to 1984.

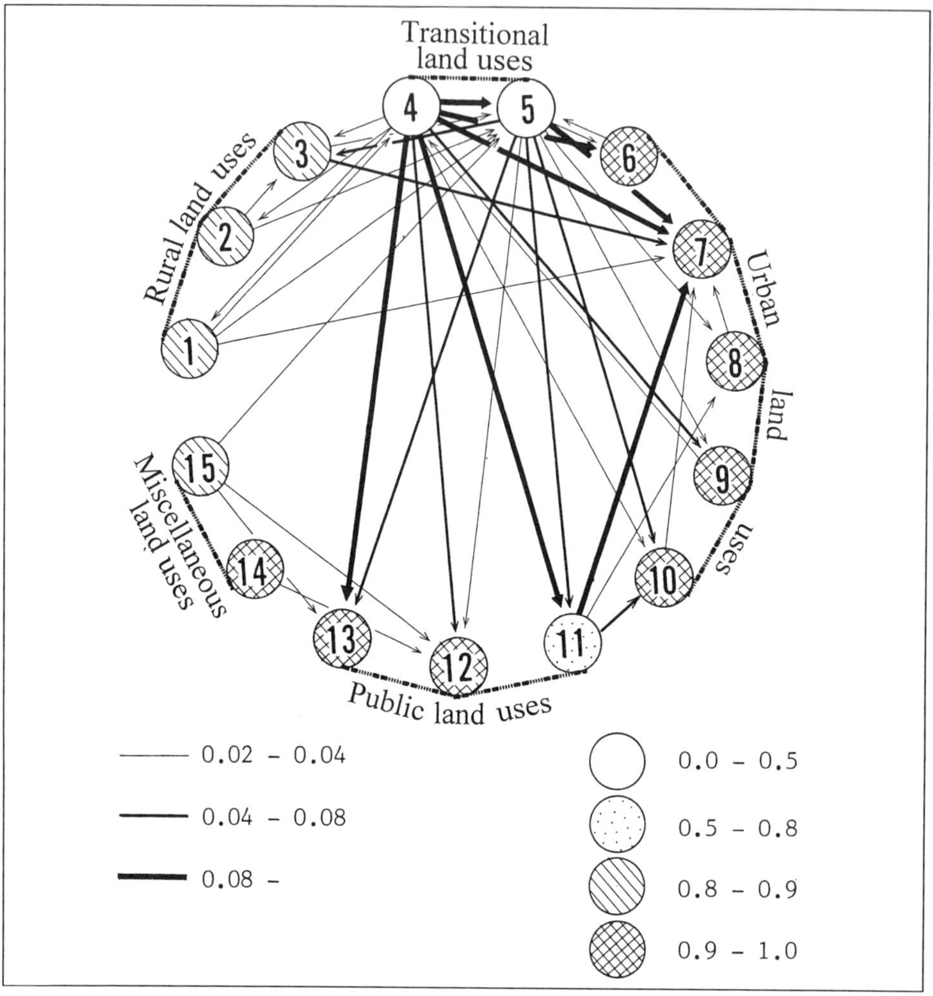

Figure 11.2. Changes in land use associated proportions by category (1974-84)

Note: 1) The numbers in the diagrams represent the following land uses: 1) forests and wasteland, 2) paddy fields, 3) upland and other fields, 4) land undergoing preparation or reclamation, 5) open land, 6) industrial land, 7) low-density housing, 8) dense, low-rise housing, 9) medium- and high-rise housing, 10) commercial and office land, 11) roads (existing and proposed), 12) parks, 13) other public facilities, 14) lakes and streams, and 15) miscellaneous land uses.

2) The arrows in the diagram indicate changes in land use from 1974 to 1984, with the values representing the areal ratios of 1984 land use categories to the areal ratios for 1974. The cross-hatching inside the circles represents the areal ratio of a given land use which did not change during the ten-year period (1974-84).

The results of applying the Markov chain model to the entire Tokyo Metropolitan Area are shown in Table 11.1. Comparing the 1994 structural ratio with that of 1984, we can see that five categories experienced decline: forests and wasteland, paddy fields, upland and other fields, lakes and streams, and miscellaneous land uses. All of these are non-urban uses. The land use showing the greatest increase was low-density housing, which was estimated to have grown by 1.77%. The maximum rate of change was seen in medium- and high-rise housing, which increased 1.285 times over the value of 1984. Next came parks and commercial and office land, in that order. The rate of increase in high-density urban uses was greater than that of low-density housing.

Table 11.1. Actual ratios of land uses (1974, 1984) and ratios estimated from Markov chain model (1994) for all regions in the study area

Type of land use	1974	1984	1994	84/74[1)]	94/84[2)]
1) Forest and wasteland	22.49%	20.53%	18.34%	0.913	0.893
2) Paddy fields	14.25	12.82	11.82	0.900	0.923
3) Upland and other fields	16.79	15.88	14.70	0.946	0.926
4) Land undergoing preparation or reclamation	2.23	1.00	1.06	0.448	1.060
5) Open land	5.48	3.82	3.94	0.697	1.031
6) Industrial land	3.61	3.96	4.03	1.097	1.018
7) Low-density housing	15.58	18.91	20.68	1.214	1.094
8) Dense, low-rise housing	1.80	2.34	2.36	1.300	1.009
9) Medium- and high-rise housing	1.08	1.58	2.03	1.463	1.285
10) Commercial and office land	2.60	3.36	4.01	1.292	1.193
11) Roads (existing and proposed)	3.30	3.71	3.90	1.124	1.051
12) Parks	1.97	2.46	2.98	1.249	1.211
13) Other public facilities	4.03	5.10	6.00	1.266	1.176
14) Lakes and streams	3.75	3.53	3.30	0.941	0.935
15) Miscellaneous land uses	1.04	1.00	0.85	0.962	0.850
Total	100.00%	100.00%	100.00%		

Note: 1) The 1984 land use ratios divided by the 1974 land use ratios.
2) The estimated 1994 land use ratios divided by the 1984 land use ratios.

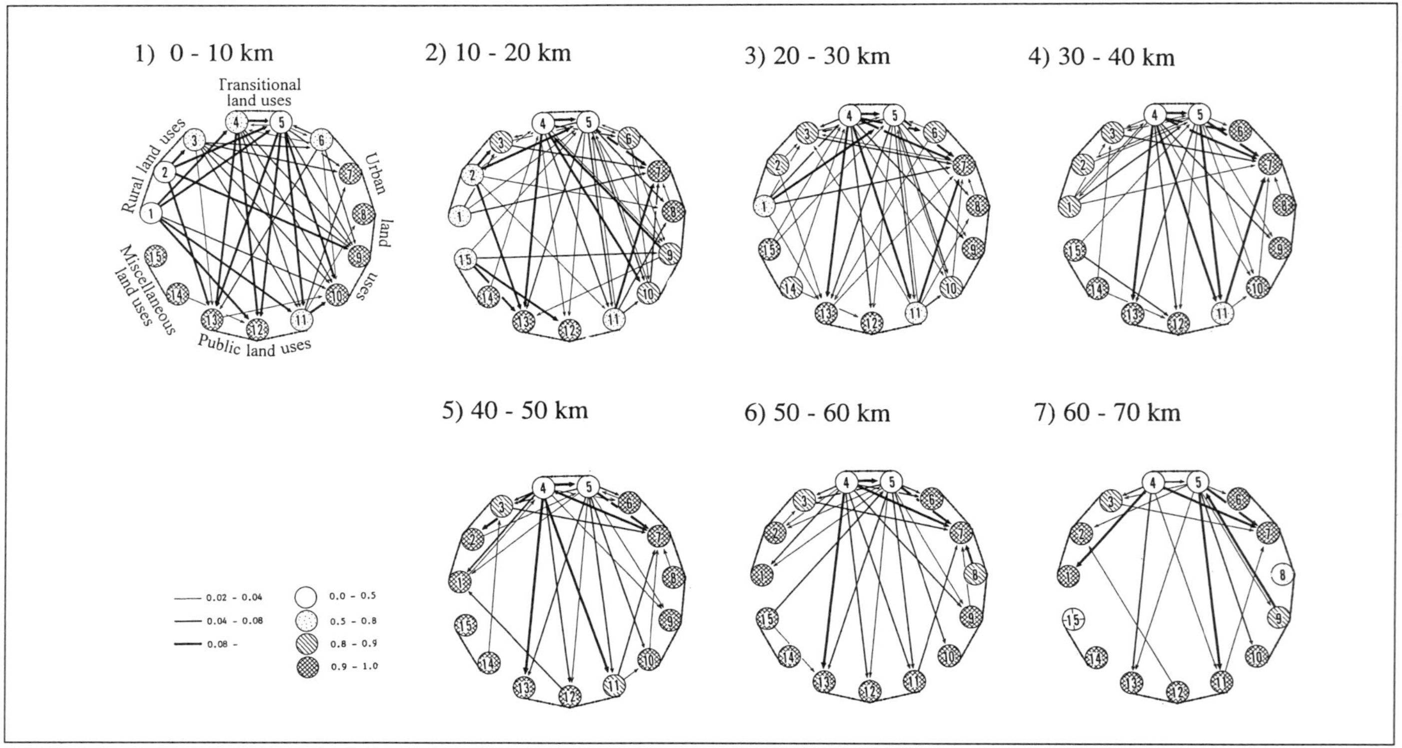

Figure 11.3. Changes in land use and associated proportions by category and distance zone (1974-84)

Note: 1) The numbers represent the following land uses: 1) forests and wasteland, 2) paddy fields, 3) upland and other fields, 4) land undergoing preparation or reclamation, 5) open land, 6) industrial land, 7) low-density housing, 8) dense, low-rise housing, 9) medium- and high-rise housing, 10) commercial and office land, 11) roads, 12) parks, 13) other public facilities, 14) lakes and streams, and 15) miscellaneous land uses.

2) The arrows indicate changes in land use from 1974 to 1984, with the values representing the areal ratios of 1984 land use categories to the areal ratios for 1974. The cross-hatching inside the circles represents the areal ratio of a given land use which did not change during the ten-year period (1974-84).

Table 11.2. Differentials between actual ratios of land uses (1984) and land use ratios estimated from Markov chain model (1994) by distance zone

Type of land use	0 – 10 km		10 – 20 km		20 – 30 km		30 – 40 km		40 – 50 km		50 – 60 km		60 – 70 km	
	LUR	D	LUR	D	LUR	D	LUR	D	LUR	D	LUR	D	LUR	D
	%		%		%		%		%		%		%	
1) Forest and wasteland	0.45	-0.28	2.06	-0.38	11.53	-2.98	15.90	-2.64	31.20	-2.16	47.59	-1.85	40.00	-0.28
2) Paddy fields	0.02	-0.01	1.60	-0.52	10.17	-1.17	16.12	-1.19	17.76	-0.97	14.45	-0.96	12.56	0.13
3) Upland and other fields	0.17	-0.01	10.20	-1.20	20.15	-1.75	16.75	-1.32	17.41	-1.20	14.48	-0.46	20.58	-0.76
4) Land undergoing preparation or reclamation	1.96	-0.37	1.52	-0.47	1.27	-0.09	1.11	-0.16	0.68	0.39	0.32	0.63	0.24	-0.07
5) Open land	3.56	-0.97	6.84	-1.79	6.17	0.12	3.91	0.65	2.02	0.35	0.90	0.32	1.63	-0.34
6) Industrial land	4.81	-0.80	6.46	-0.27	3.43	-0.02	5.10	0.27	2.76	0.22	1.87	0.15	2.23	0.00
7) Low-density housing	19.86	-0.89	33.40	1.16	21.02	2.47	18.85	2.40	14.61	1.81	10.39	1.05	12.68	0.85
8) Dense, low-rise housing	10.79	-0.25	6.42	-0.14	4.14	0.12	1.06	0.04	0.23	0.02	0.05	0.00	0.01	-0.01
9) Medium- and high-rise housing	3.28	1.37	2.53	0.76	2.53	0.67	1.70	0.36	0.66	0.22	0.34	0.10	0.14	0.13
10) Commercial and office land	21.40	1.01	6.14	1.33	3.17	0.79	2.25	0.62	1.37	0.35	1.07	0.20	1.23	0.21
11) Roads (existing and proposed)	8.94	-0.08	4.93	0.42	2.96	0.47	4.47	0.01	2.85	0.11	2.10	0.12	2.72	-0.12
12) Parks	6.09	0.77	4.67	0.51	3.26	0.63	2.22	0.72	1.36	0.24	0.75	0.31	0.64	0.15
13) Other public facilities	12.08	1.00	8.33	1.04	6.64	1.18	4.69	0.90	2.80	0.77	3.24	0.48	1.52	0.26
14) Lakes and streams	5.53	-0.35	4.48	-0.29	2.49	-0.27	3.81	-0.31	3.80	-0.16	2.04	-0.04	3.82	-0.13
15) Miscellaneous land uses	1.06	-0.15	0.42	-0.16	1.07	-0.17	2.06	-0.35	0.49	0.01	0.41	-0.05	0.00	0.00
Total	100.00		100.00		100.00		100.00		100.00		100.00		100.00	

Note: 1) "Land use ratio" (LUR) represents the proportion of each type of land use in 1984.
2) "Differential" (D) represents the values obtained by subtracting the actual 1984 land use ratios from the estimated 1994 ratios derived from the Markov chain model.

Looking at the proportion of land uses for the three years 1974, 1984, and 1994, we can see that the differentials are lower for 1984-94 than that for 1974-84. For example, the rate of change in medium- and high-rise housing stood at 1.463 for 1974-84, but was 1.285 for the next ten-year period; even more striking was the drop in dense, low-rise housing, which fell from 1.300 in 1974-84 to a mere 1.009 in 1984-94 (see Table 11.1). Therefore, we can conclude that changes in land use patterns for 1984-94 were much more subdued than in the previous ten years. Finally, the transitional states of prepared/reclaimed land and open land showed enormous decline as a share of 1974-84 land use patterns, but increased slightly from 1984 to 1994.

11.4. Land Use Patterns by Distance-Zone

Table 11.2 shows the actual proportions of land use patterns in 1974 and 1984. Here we can see that as distance from central Tokyo increases, the proportion of rural land uses (1, 2, 3) increases, while the share of high-density urban uses (8, 9, 10) and public facilities (11, 12, 13) decreases. Finally, transitional land uses (4 and 5) tend to decrease toward the urban fringe.

This section considers the changes in land use patterns by distance-zone over the ten-year period 1974-84. The largest number of arrows are located in the 0-10km zone, indicating a zone of multifarious land use changes. The most notable feature of this zone was the rapid conversion to high-density urban uses such as medium- and high-rise housing and commercial and office land. On the other hand, there was little conversion to low-density uses such as industrial, dense, low-rise housing, and, in particular, low-density housing, which was much less active that in any other zone. The 0-10 km zone was also characterized by the conversion of a considerable proportion of rural land to other uses, with farmland dwindling rapidly. The proportion of prepared/reclaimed land converted to other uses was 0.515, in marked contrast to the other zones (0.1-0.3). In the 0-10 km zone, there was a considerable amount of land that was abandoned as it was undergoing preparation, suggesting that such land was not making a ready transition to other uses. On the other hand, there was a large-scale transition of open land which, having been created from forests and wasteland and paddy field, was transformed into commercial and office land, road, parks and other public facilities. Though there was little conversion of open land to low-rise housing, there was a considerable amount of land which was used for medium- and high-rise housing. In the 10-20 km zone, there was not as noticeable conversion of open land to urban uses and public facilities as there was in the 0-10 km zone.

Land use changes in the next five zones (20-70 km) can be summarized as exhibiting the following three general characteristics: 1) The rate of rural land conversion decreased toward the peripheral areas. In other words, conversion of rural land decreased as distance from the city center increased. 2) The number of arrows in a distance-zone decreased as distance from the city center increased, signifying that land use transformation stagnated toward the peripheral areas. 3) Conversion to high-density urban land uses such as commercial and office land and medium- and high-rise housing was significant up to a 20km radius, beyond which it leveled off rapidly.

Data for the 15 × 15 transition probability matrix were derived from the results of Markov chain model application to each of the seven distance-zones, and are shown as the "land use ratio" by distance-zone in Table 11.2. The "differential" shown in the table represents the value obtained by subtracting the actual 1984 data of land use ratio from 1994 estimated land use ratio obtained from the Markov chain model. Hence, estimated values for 1994 can be obtained by adding the differential value to the land use ratio values in the table. Negative differentials indicate an areal reduction for 1994, while positive differentials signify an increase. Since the three types of rural land uses (forests and wasteland, paddy field, and upland and other fields) showed negative differentials for all zones except 60-70 km, we can predict that the areal extent of rural land will decline in 1994, especially in the 20-50 km zones. In the 0-20 km zones, transitional land uses (prepared/reclaimed land and open land) show a decrease in area for 1994, while in the 20-40 km zones, prepared land decreases while open land can be predicted to increase. The 1994 outlook in the 40-60 km zones is for both types of land uses to increase.

With the exception of certain categories which declined within the 20 km radius, urban land use ratio increased away from the city center. Industrial land can be expected to decline in area up to a 30km radius and increase beyond that. Commercial and office land and medium- and high-rise housing will likely increase in all zones.

Among public land uses, only roads can be expected to decline in the 0-10 km and 60-70 km zones, with the area of all other such uses increasing in all zones. With the exception of a slight increase in the 40-50 km zone, the remaining categories of land use are predicted to decrease in all regions.

Viewed as a whole, the land use transition of 1984-94 in the 20-30 km and 30-40 km zones is much more rampant than in the other zones.

CHAPTER 12

TOKYO AND PARIS: COMPARISON OF URBAN SPATIAL STRUCTURE

12.1. Introduction

After World War II, Tokyo and Paris grew into world-class "megacities" as their economic infrastructure and working populations expanded with dazzling speed, and the central position they play in both national and international urban systems has been steadily increasing. For example, according to statistics of United Nations, the population of the Paris Metropolitan Area grew from 8.5 million persons in 1970 to 9.5 million in 1995, while during the same time period the Tokyo Metropolitan Area grew from 16.5 million to 26.8 million in population. This process has wrought drastic changes on the spatial structure of Tokyo and Paris. The urbanization and intensification of urban functions led to spatial divergence in the 1970s; in the 1980s, interdependencies of cities on both the domestic and international levels helped to enhance the core managerial functions of the central cities.

Given this background, this chapter attempts to examine the present similarities and differences in urban spatial structure between Tokyo and Paris and identify the types of changes that have been occurring in these two cities. This will be done by conducting distance-zone analyses to the fullest extent that the analytical framework and various investigative indices will allow. Indices here include population, household information, age structure, employment, and residential data.

12.2. Framework of Analysis and Study Areas

When we attempt to statistically compare the major megacities of the world, we face many troublesome problems. First of all, the time frame of statistical surveys differs from country to country. The five- to ten-year intervals of censuses and other such surveys are quite long, so it is very difficult to obtain data for the same years for different countries.

Secondly, the framework and systems for conducting statistical surveys, and the items which are being surveyed, also differ. Whether it is a complete count or random sampling survey, differences arise in accuracy and margin of error; this problem is compounded if the definitions of survey items such as income and unemployment rate are not identical.

The third problem we encounter concerns the spatial extent of the city. Depending on how the boundary of the metropolitan area is defined, there may be large differences in the analytical results. Many studies in the past simply used administrative boundaries to delineate the boundaries of metropolitan areas, because it was relatively easy to obtain statistical data. However, administrative boundaries are determined based on the conditions of the respective countries and do not necessarily represent the actual state of the city. In geographical research at least, the objects of our analysis should not be administrative areas which are little more than artificially delineated regions; rather, they should be substantial regions at a common scale.

The fourth problem is the extent and scale of unit regions, which differ from country to country. The lack of uniformity in unit regions is a major obstacle to the study of spatial distribution. Even within the same country, statistical regions may differ from one survey to the next due to annexations, consolidations, and separations of political units.

Fortunately, this study, which uses population census data, is able to avoid the first problem. In France, censuses were taken in 1975, 1982, and 1990; in Japan, they were taken in 1970, 1975, 1980, 1985, and 1990. Therefore, there are two years, 1975 and 1990, for which census data is available for both countries. This will allow us to track the significant changes in urban structure that occurred in the 1980s, which are wedged between these two years.

The second problem is handled in the following manner. First, the definitions of statistical items are scrutinized, then the standards common to both countries' censuses are selected as indices. Those items which are not identical are processed to make them as identical as possible.

The third problem is rather bothersome. Most of the comparative studies conducted so far have included 4 prefectures in the Tokyo Metropolitan Area: Tokyo-to, Chiba-ken, Saitama-ken, and Kanagawa-ken. The Paris Metropolitan Area has consisted of the Ile-de-France area. However, these boundaries have not been determined by geographical standards. One solution to this problem has been to consider homogeneous regions based on population density, land use, etc., or functional or nodal regions which use commuter or traffic volume as indices. However, this study attempts to make comparisons based on distance-zones from the city center, a method which the author believes to be ideal for examining the spatial extent of metropolitan areas. Therefore, this study considers population density and distribution and commuting conditions within a 50 km radius of both central Paris and central Tokyo. Here, the city centers are considered to be the Cathedral of Notre Dame in Paris, and the old prefectural capitol building in Tokyo. Hereafter in this paper, the areas encompassed within a 50 km radius of these landmarks are called metropolitan areas (Figure 12.1).

The fourth problem is somewhat perplexing. The basic unit for statistical areas in France is the commune, while in Japan, it is the "shi-cho-son", or municipality (literally, "city, town, and village"). The problem arises in the fact that communes are much smaller than Japanese municipalities. Since it is important for the number of unit areas in each city to be as close as possible, the analytical unit for Paris is the canton (voting district), which is larger than a commune and somewhat closer in size to the municipalities of Tokyo. As a result of using cantons, there was a total of 273 unit areas in the Paris analysis; there were 182 such areas in the Tokyo analysis.

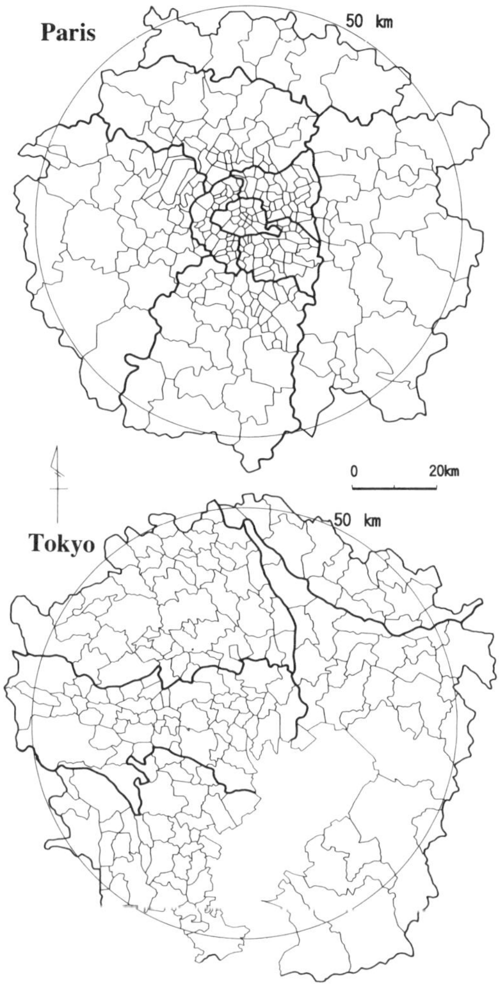

Figure 12.1. Study area

12.3. Spatial Distribution of Population Characteristics

First, let us examine the distribution of population characteristics in Paris and Tokyo in 1990.

12.3.1. POPULATION DESITY

Tokyo is much more densely populated than Paris (see Figure 12.2). According to the 1990 censuses, there were 10,558,302 people living within 50 km of central Paris while there were 29,199,756 people living within the same radius from central Tokyo. The population density of Paris was 1,168 people per square kilometer, while in Tokyo, it was 3,831 people/ km^2. If we limit consideration of population density to 10,000 people or more per km^2, we can see that in Paris, the area so populated extends to 10 km from the city center while in Tokyo, it extends to 20 km. Therefore, there are some differences between Paris and Tokyo in distribution of population density. Furthermore, in contrast to the clear concentricity of Paris, the population density of Tokyo is slightly higher in the western regions than in the east, indicating a clockwise spiral structure. The development of road and rail networks in Tokyo began in the west, thus urbanization in that city spread more quickly in that direction.

In the Tokyo Metropolitan Area, there is a great difference between the population densities along the boundary between Tokyo Metropolis on the one hand and Saitama and Chiba prefectures on the other. Furthermore, as will be discussed later, managerial functions predominate in the Tokyo city center, so the nighttime population density there is somewhat lower than in surrounding areas.

12.3.2. RATIO OF DAYTIME TO NIGHTTIME POPULATION (15 YEARS OLD AND OVER)

In Tokyo, there are 280,000 persons employed within 5 km from of the city center; in Paris, this number is 180,000. Figure 12.3 shows the spatial distribution of the ratio of daytime to nighttime population of employed persons of 15 years of age and over. Although the ratio of daytime to nighttime population of the centers of both cities are quite high, it is especially so in Tokyo (see Table 12.1). In Paris, this ratio is 1.68:1 within 5 km from the city center, but in Tokyo, it is 8.23:1. Nearly all of the people working within 5 km of central Tokyo are commuters living outside of this area. The ratio of daytime population rapidly declines outside of the 5 km radius; the only other place it exceeds 2.0 in the Tokyo area is in the Nishi and Naka districts ("ku") of Yokohama.

In contrast, in the Paris Metropolitan Area, there are four areas excluding Paris proper where the daytime population is at least twice as high as the nighttime population, namely, the cantons of Gonesse (Val d'Oise), Vélizy-Villacoublay (Yvelines), Orly (Val-de-Marne), and Chivelly-Larue (Val-de-Marne). Furthermore, there are only 3 other places in the Tokyo area where this ratio is at least 1.5 to 1: Shinagawa-ku (in Tokyo proper), Kawasaki, and Narita. This is in contrast to the numerous locations in Paris where this ratio is at least 1.5:1. From this information we can see that workplaces are extremely concentrated in central Tokyo while in Paris, that they are expanding out from the city center and dispersion of employment and diversification is progressing. In addition, the daytime-to-nighttime population ratio within the 5 km radius is not as high in Paris as in Tokyo because the resident population in the former is much higher than

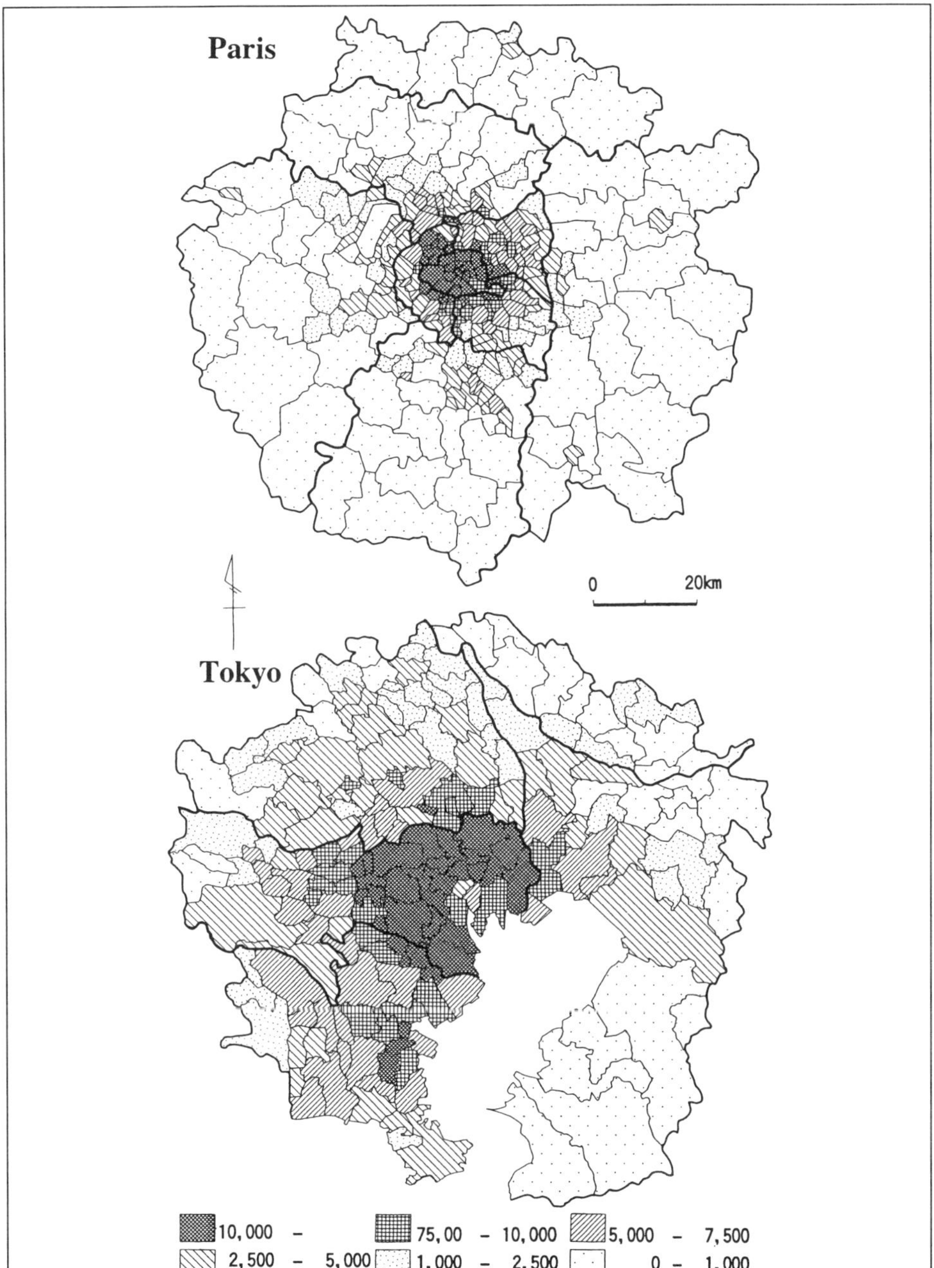

Figure 12.2. Population density in Paris and Tokyo in 1990 (persons / km^2)

Source: Paris: Recensement Général de la Population de 1990.
Tokyo: 1990 Population Census of Japan.

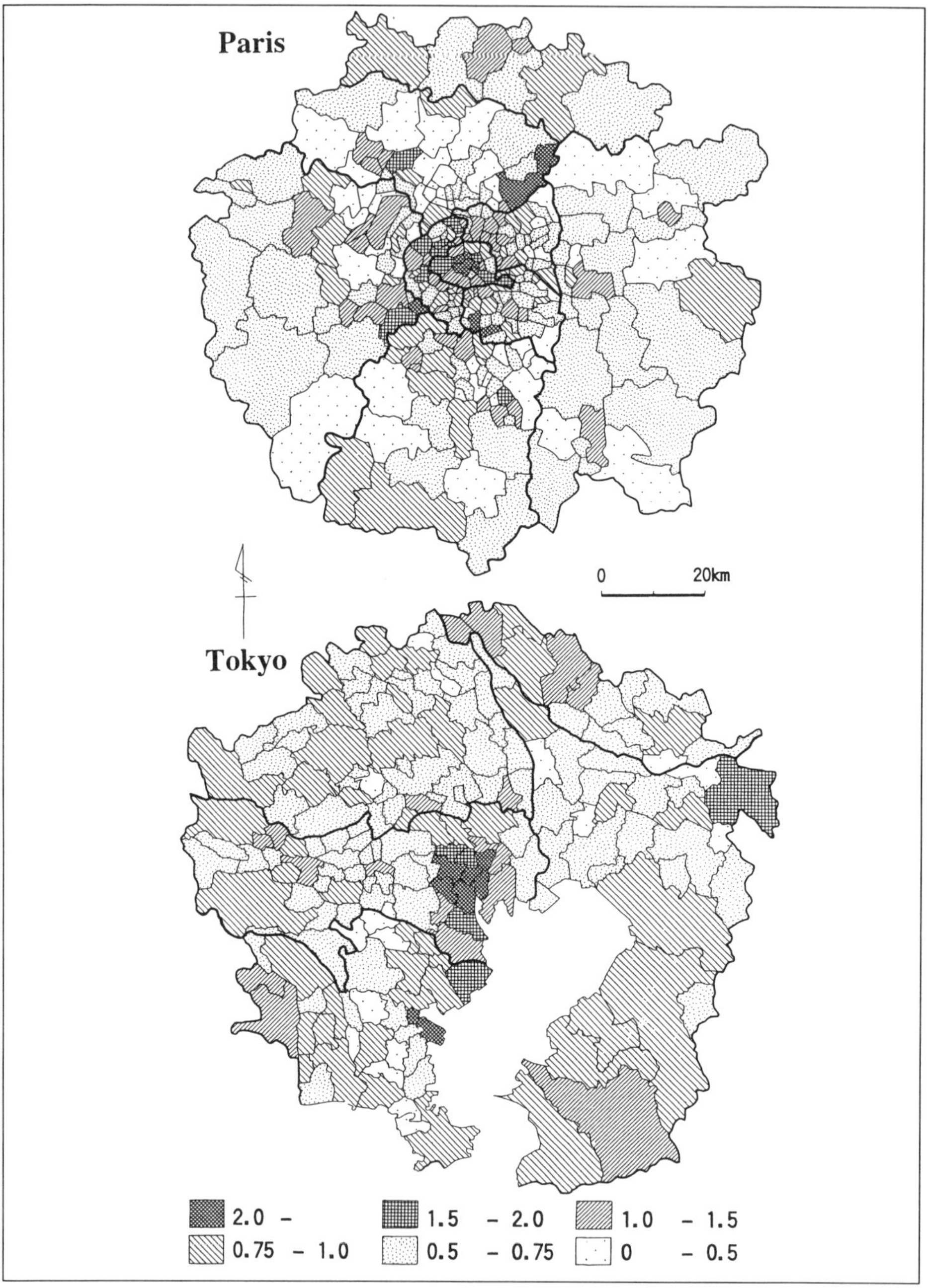

Figure 12.3. Ratio of daytime to nighttime population in 1990 (15 years old and over)

Source: Paris: Recensement Général de la Population de 1990.
Tokyo: 1990 Population Census of Japan.

in the latter; that is, many more workers in Paris live close to their jobs.

Table 12.1. Population structure by distance (1990) *(continued).*

Paris

Distance zone	Population ('000)	Population density	Ratio of daytime to nighttime pop.*
0-50 km	10,558	1,168	1.04
0 - 5 km	2,244	20,534	1.68
5 -10 km	2,200	8,487	1.14
10 - 20 km	3,403	3,009	0.77
20 - 30 km	1,439	792	0.77
30 - 40 km	645	319	0.73
40 - 50 km	592	160	0.71

Tokyo

Distance zone	Population ('000)	Population density	Ratio of daytime to nighttime pop.*
0-50 km	29,200	3,831	1.02
0 - 5 km	610	9,615	8.23
5 -10 km	2,827	15,453	1.59
10 - 20 km	8,399	10,035	0.76
20 - 30 km	6,331	5,232	0.77
30 - 40 km	6,752	2,952	0.71
40 - 50 km	4,280	1,408	0.81

Note: * Ratio of the daytime working population to the nighttime population of employed persons (15 years old and over).

Source: Paris: Recensement Général de la Population de 1990.
Tokyo: 1990 Population Census of Japan.

Figure 12.3 shows that in the regions west of Paris, the ratio of daytime-to-nighttime population is increasing. This may be explained by a wave of new office construction in the area, as many office functions are moving there from the city center. This mass movement has led to a relative decline in the ratio of daytime-to-nighttime population in central Paris.

12.3.3. RESIDENTIAL CHARACTERISTICS

Figure 12.4 shows the ratio of single-family (detached) homes. Within a radius of 20 km from the city center, the ratio of such homes in Tokyo is greater than in Paris, but the opposite is true beyond the 20 km radius. In Paris, there is a rapid rise in the proportion of single-family homes as one moves out of the city center toward the suburbs. In Tokyo, the ratio is greater within a 5 km radius than in the 5-10 km zone, indicating that mostly owners of housing land can live there because land prices are rising and rental housing is being converted into office space.

Figure 12.5 shows that the ratio of owner-occupancy in the 0-5 km zone is quite high in Tokyo. The distance-zone differential of the ratio of owner-occupied housing is not as striking as the ratio of single-family homes (Figure 12.4).

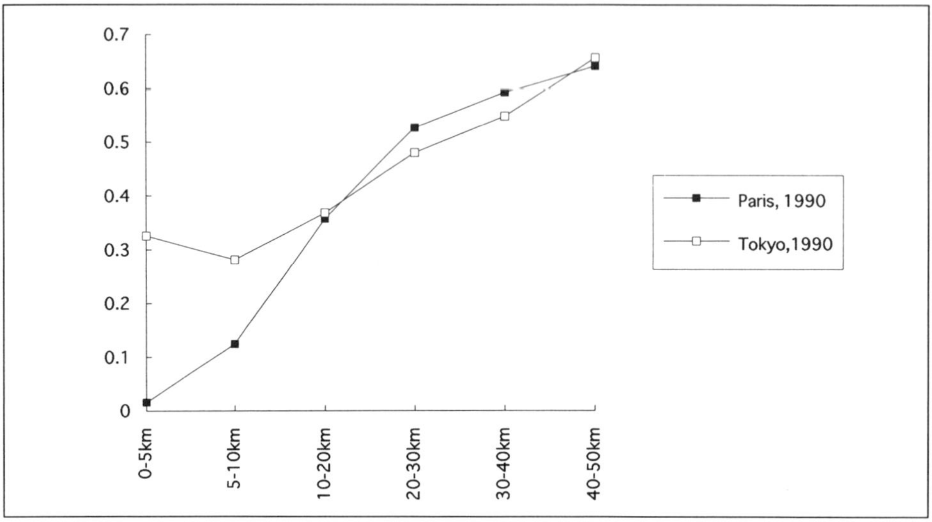

Figure 12.4. Ratio of single-family (detached) home (single-family homes / total homes)

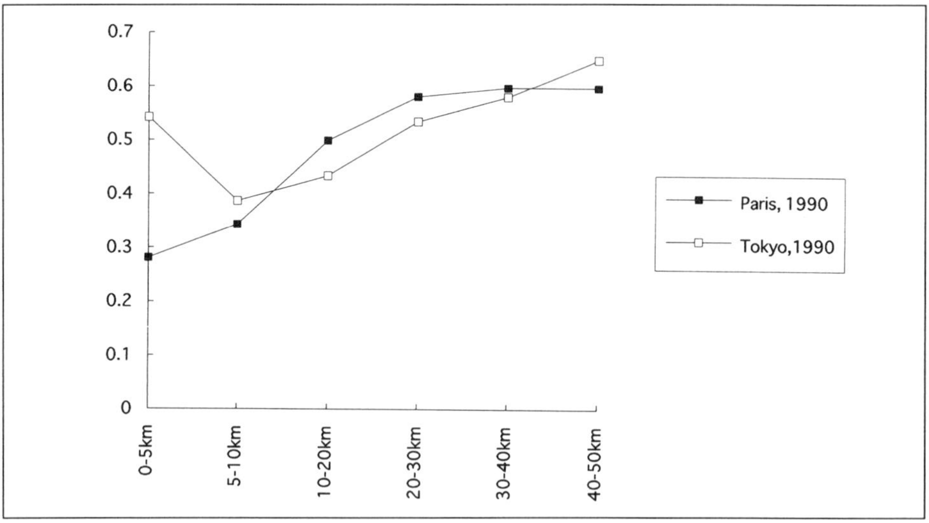

Figure 12.5. Ratio of owner-occupancy (owned properties / total properties)

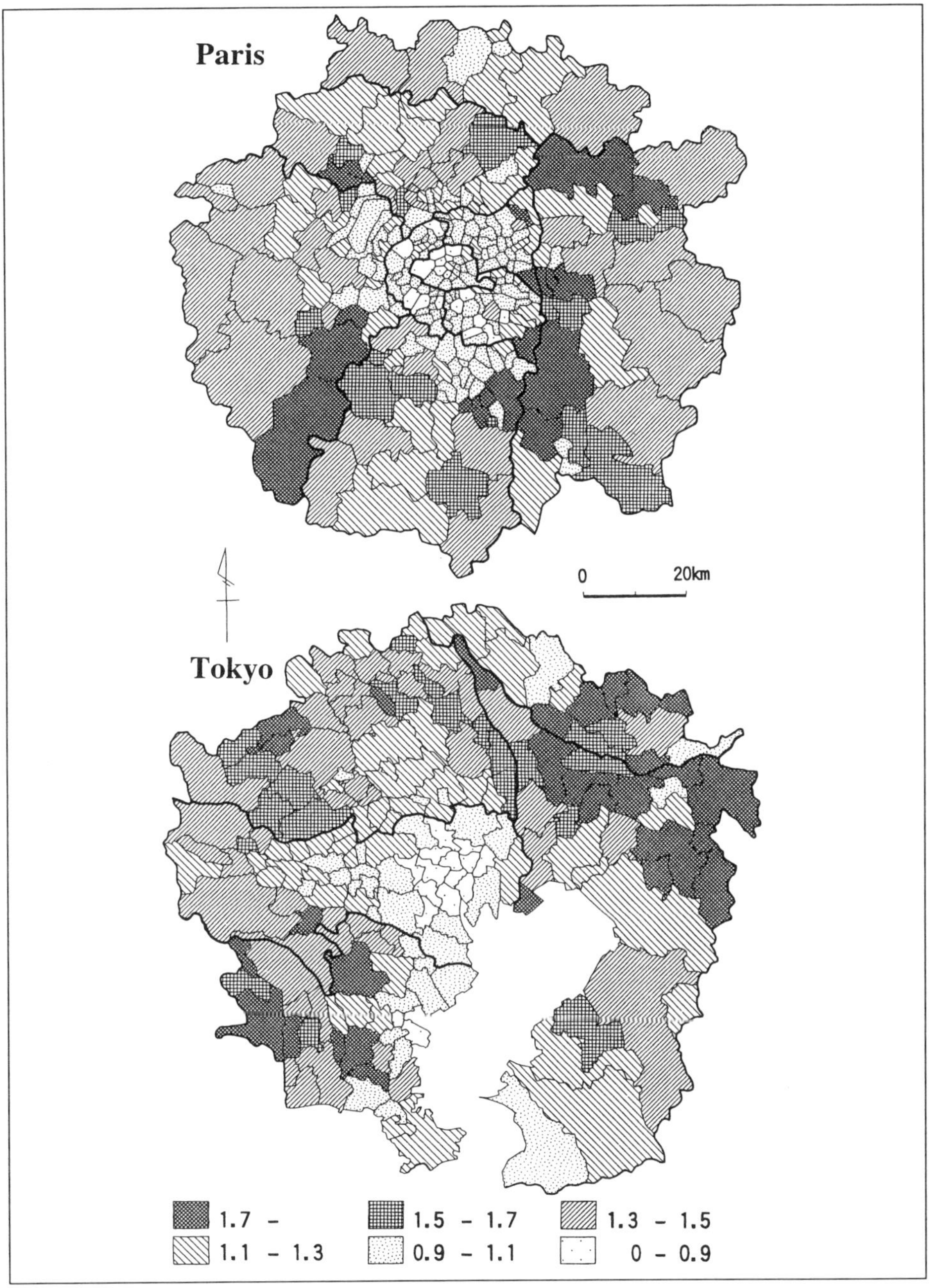

Figure 12.6. Population changes in Paris and Tokyo (1975 – 1990)

Source: Paris: Recensement Général de la Population de 1990.
Tokyo: 1990 Population Census of Japan.

12.4. Changes in Spatial Patterns per Distance-Zone

12.4.1. POPULATION CHANGE

Figure 12.6 shows the distribution of population changes from 1975 to 1990. Although population in the city center declined in both cities, this trend was especially notable in Tokyo. With the exception of two districts, or "ku," in Yokohama, all of the districts in the Tokyo area losing 10% or more of their population were adjacent to the city center. In Paris, on the other hand, this decline was widely dispersed within a 10 km radius. In addition, there were many cantons in the 0-5 km zone in Paris where population decline did not reach 10%. As we can see, population decline in Paris covered a wider area and its pace was less dramatic.

Population increase in the suburbs in both cities was quite remarkable, especially in the new towns of Paris. Rapid increases in population in the Tokyo area were especially noticeable in the northeastern part. Nevertheless, since road and rail access to Tokyo was not convenient, urbanization was very slow in this region. Since land prices were and still are relatively inexpensive here, housing areas have been rapidly developing since the 1980s.

Figure 12.7 is a graphic representation of population density per distance-zone in 1975 and 1990. In both cities, population within a 10 km radius of the city center declined, but increased outside of this zone. From the figure we can see that the differential for distance-zones in both cities decreased, suggesting that there was a movement of population away from the city centers toward the suburbs. The decrease in population density is especially noteworthy within a 10 km radius of central Tokyo.

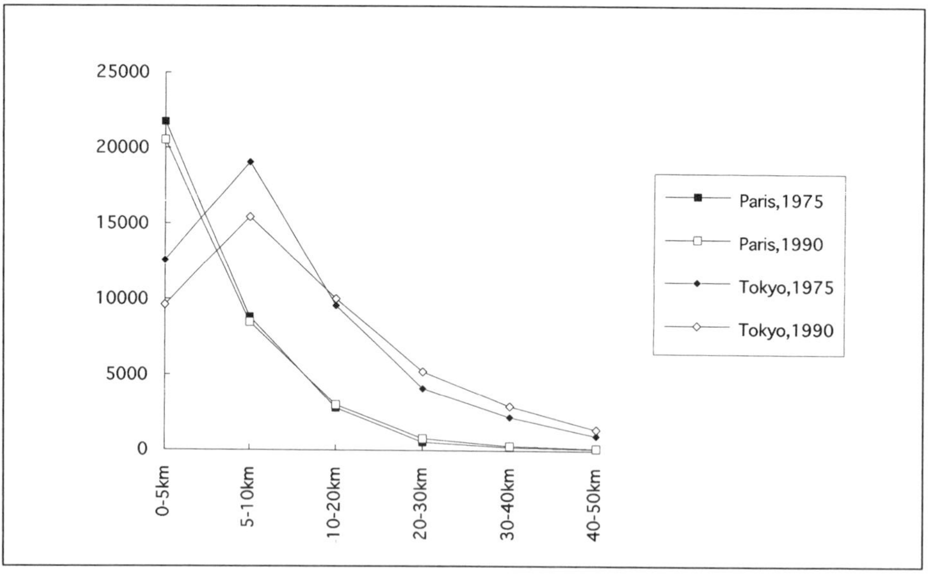

Figure 12.7. Population density (persons/km^2)

12.4.2. RATIO OF INFANT-TODDLER POPULATION

The ratio of infant-toddler population (i.e., the ratio of children under 5 years to the total population) in 1990 was 6.1% in Paris and 4.9% in Tokyo (see Figure 12.8). Decline in the infant-toddler ratio was especially noticeable in Tokyo, which by 1990 had fallen to only about 51% what it had been in 1975. In Paris, on the other hand, there was only a 5% decline. The rapid decline in Tokyo is attributable to four factors: 1) The number of families with infants and toddlers had been declining since the baby boomer generation, 2) families were having fewer children, 3) the population was aging, and 4) the number of divorced and unmarried people was increasing. In both Paris and Tokyo, the infant-toddler ratio was low in the city center and high in the surrounding areas, exhibiting a spatial pattern which had not changed in the 15-year period of the surveys, and the distance-zone differential of both cities tended to decrease. It should be noted that the infant-toddler ratio in the 0-5 km zone of Paris increased during the 15-year period, and this can be explained by a wave of young people moving into the area from the provinces and by a large influx of immigrants from overseas. Generally speaking, immigrants are usually young people, and they tend to have more children than the general population.

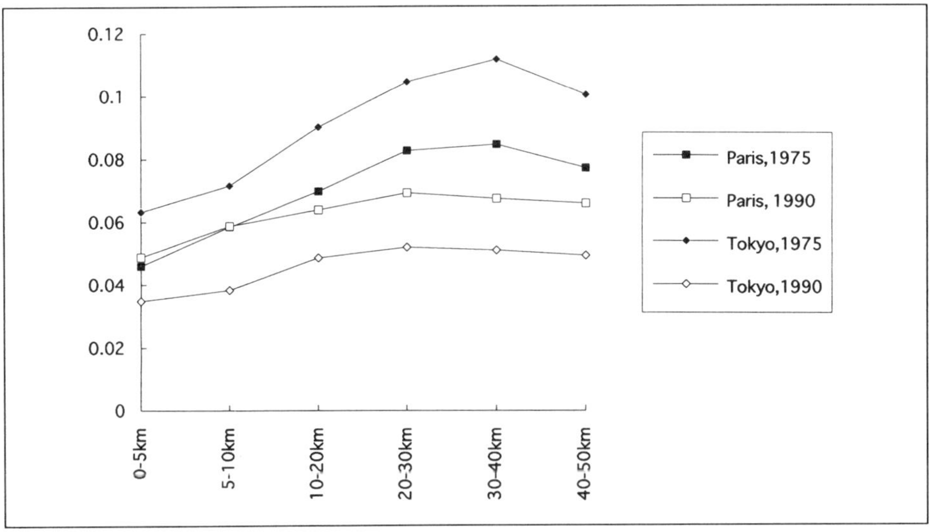

Figure 12.8. Ratio of infant-toddler (0 – 4 years old) population (infant-toddler population / total population)

12.4.3. RATIO OF ELDERLY POPULATION

In both years of the surveys, the ratio of elderly (75 years or older) population to total population in Paris was higher than in Tokyo, although the rate at which the respective populations were aging was higher in Tokyo. As we can see in Figure 12.9, the curve for the elderly ratio is the opposite of the infant-toddler ratio, that is, it is highest in both city centers and decreases as distance from the city centers increases. It should be noted, however, that this decline ends at 30 km in Paris and 40 km in Tokyo; beyond the 40 km zone, the elderly ratio starts to increase. In the survey, the farther an area was from the city center, the higher was the proportion of farmers, many if not most of whom are elderly. The distance-zone differential of the elderly ratio for both cities increased during the 15-year period, while the ratio of elderly population increased toward the city center. This suggests that people with families were moving out to the suburbs so that they could buy their own home. Furthermore, in the case of Paris, quite a number of elderly people were moving from the suburbs to the city center, which may have helped to increase the elderly ratio there.

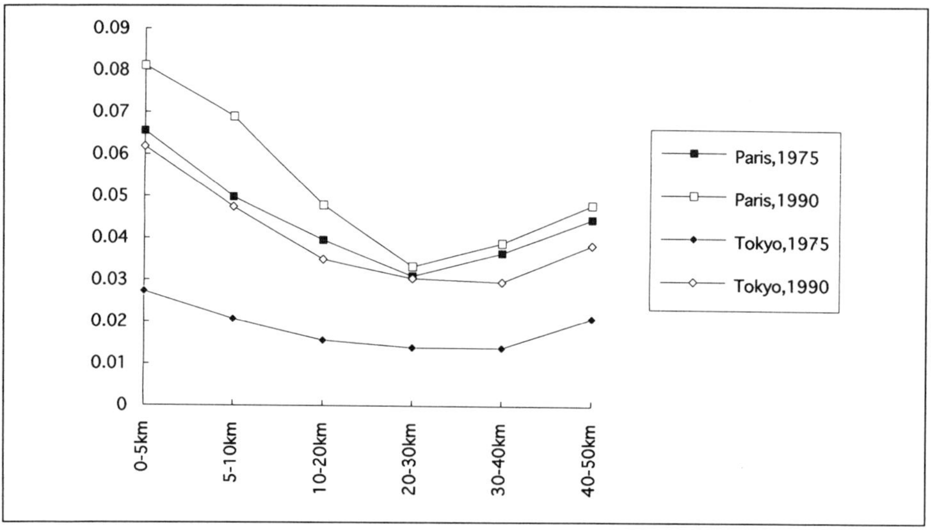

Figure 12.9. Ratio of elderly (75 years or older) population (elderly population / total properties)

In Paris, the elderly ratio shows a positive correlation with population density (0.566) for 1990 (see Figure 12.10). The densely inhabited districts of Paris offer a wealth of medical, cultural, and welfare services, making this a very convenient and thus attractive place for the elderly. In contrast, Tokyo shows no correlation between these two phenomena (see Figure 12.11). The correlation coefficient is lowered in Tokyo by two factors: first, the ratio of elderly population in the city center, which is not so densely populated, is quite high; second, there are many elderly people working in agriculture and part-time jobs in outlying areas where the population density is low.

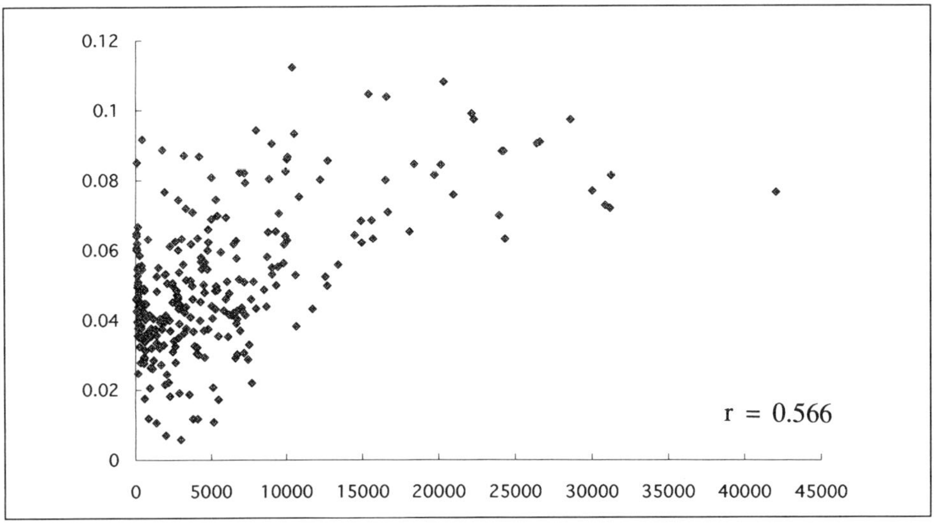

Figure 12.10. Correlation between population density (x-axis) and ratio of elderly population (y-axis) in Paris (1990)

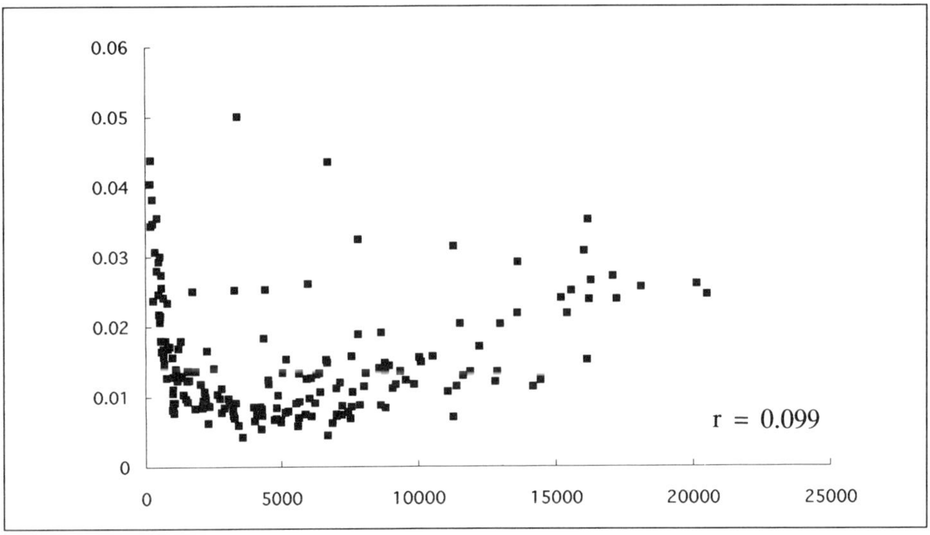

Figure 12.11. Correlation between population density (x-axis) and ratio of elderly population (y-axis) in Tokyo (1990)

12.4.4. RATIO OF FOREIGN POPULATION

As we can see in Figure 12.12, the ratio of foreigners to total population was 14 times higher in Paris than in Tokyo in 1990. In both cities, the number of foreigners living in the 0-5 km and 5-10 km zones increased. In Paris, the distance-zone differential increased dramatically over the 15-year period, that is, the population of foreigners increased up to the 30 km boundary, then decreased beyond that. The increase in foreigners was especially remarkable within the 10 km zone, probably because there were many immigrants and other non-natives originally living in the suburbs who moved to the central city in search of employment. In addition, temporary workers and immigrants newly arriving from overseas likely preferred the city center or adjacent areas rather than suburban areas. The rate of increase in Tokyo, especially in the 0-5 and 5-10 km zones, was high during the 15-year period.

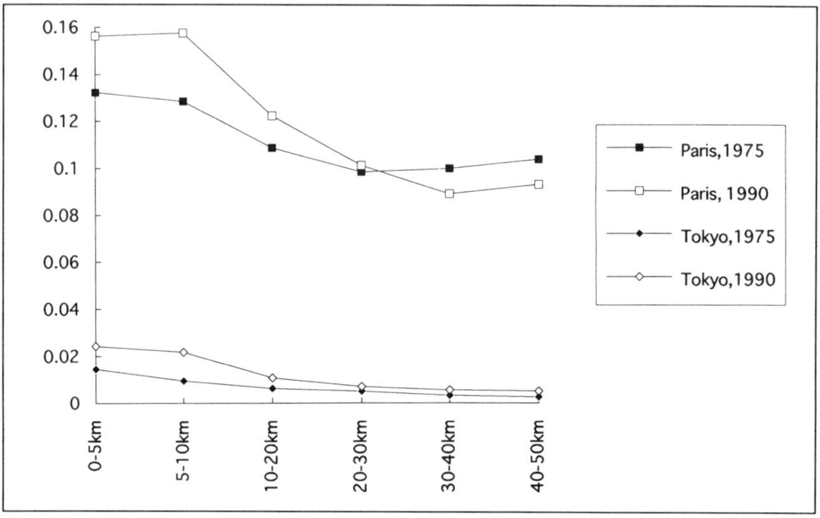

Figure 12.12. Ratio of foreign population (foreign population / total population)

12.4.5. EMPLOYMENT RATIO

In both years, the employment ratio (number of employed workers to total population) in Tokyo was slightly higher than in Paris. Over the 15-year period (1975-90), the employment ratio decreased by 0.5% in Paris but increased by 5% in Tokyo (see Figure 12.13). This increase in Tokyo, where the population is aging, can be explained by the following factors: 1) Both the mandatory retirement age and the number of elderly people seeking employment increased, 2) the population of persons under 15 years old, that is, those who have not reached working age, decreased, and 3) the employment of women increased significantly.

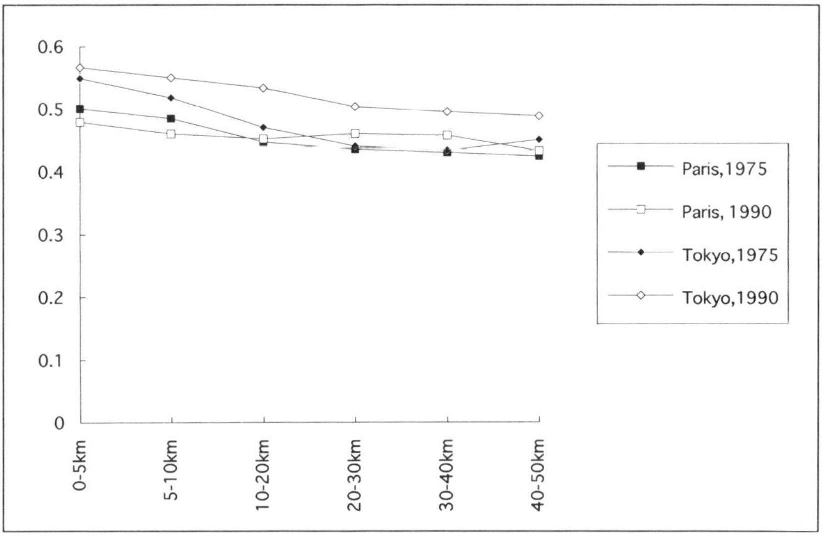

Figure 12.13. Employment ratio (employment population / total population)

In Paris, the employment ratio in the 0-10 km zone declined, mostly because of large numbers of basically unemployed pensioners and college students who were moving to the city center. On the other hand, the employment ratio increased outside of the 10 km radius, especially beyond the 20 km zone, likely because the construction of factories and office facilities there provided new employment opportunities and housewives began to enter the job market. Since unemployment is strongly tied to the employment ratio, let us now examine the unemployment rate.

12.4.6. UNEMPLOYMENT RATE

In both Paris and Tokyo, the unemployment rate (proportion of unemployed persons to total working age population) increased during the 15-year period (see Figure 12.14). However, the rate of this increase was overwhelmingly higher in Paris, 8.5% versus 4.3% in Tokyo. The unemployment rate was highest within 5 km of the Paris city center and lowest in the 20-30 km zone, likely because new towns in the latter zone were providing a large number of new employment opportunities. The reason the unemployment rate was low within 5 km of central Tokyo was probably because this is where the elite of society live, i.e., high income professionals such as managers, lawyers, and diplomats who are not so susceptible to unemployment pressures. In addition, a large number of unemployed Tokyo residents, unable to bear the rising land prices and rents in the central area, are believed to have moved out to suburban areas.

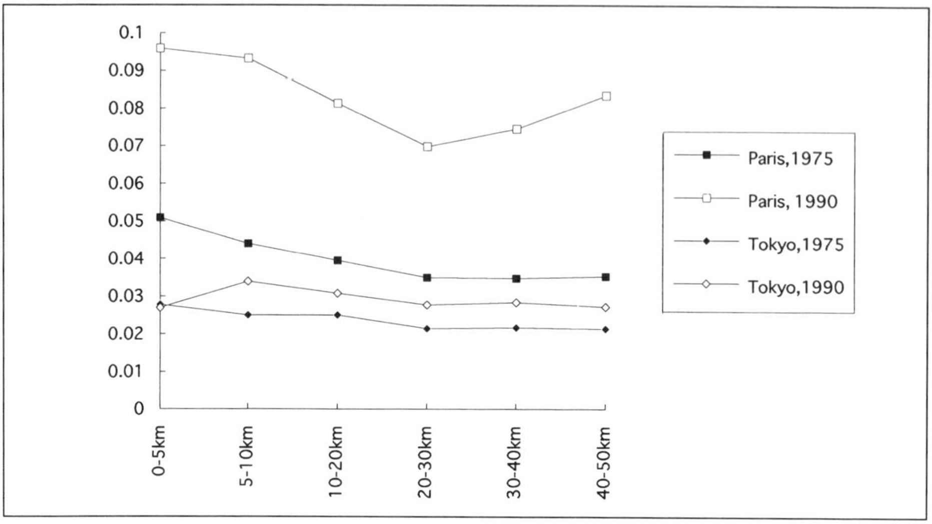

Figure 12.14. Unemployment rate
(unemployed population / (employed + unemployed population))

12.4.7. NUMBER OF PERSONS PER HOUSEHOLD

The number of persons per household in Paris decreased from 2.7 in 1975 to 2.5 in 1990; in Tokyo, this figure decreased from 3.3 in 1975 to 2.8 in 1990 (see Figure 12.15). In both cities, the trend was toward nuclear families, although the rate at which this progressed was higher in Tokyo. In 1990, there were more people per household in Paris in the 5-30 km zone, but there were more in the 0-5 km zone in Tokyo. As reported earlier, the proportion of single-family homes in the 0-5 km zone was higher in Tokyo, providing more space for more family members.

In Paris, the structure of the family has been changing dramatically since around the end of 1960. As the marriage rate declined, the average age of marriage increased. At the same time, the divorce rate was also increasing, reducing the length of the average marriage. The ratio of divorces to marriages was about 1 to 3, and the number of unmarried people of 50 years or over increased by 18%. The long-term trend shows fewer children staying at home after reaching adulthood. This is because, among other reasons, 1) more children are seeking independence at an earlier age, 2) parents want their children to live on their own, and 3) elderly people want to live independent lives. However, due to the rising unemployment rate in recent years, it has been getting harder for the younger generation to become independent. Household size is especially small in central Paris, as the number of young and elderly people there is unusually high. As a general rule, we can say that young people tend to live in the city center while they still have no children, then move to the suburbs to raise a family. After they reach old age, some of them return to the city center.

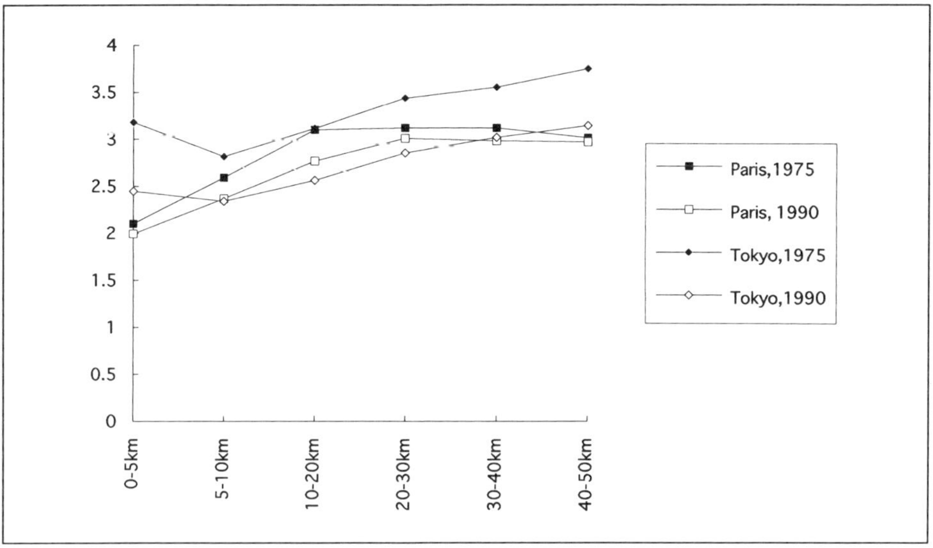

Figure 12.15. Number of persons per household

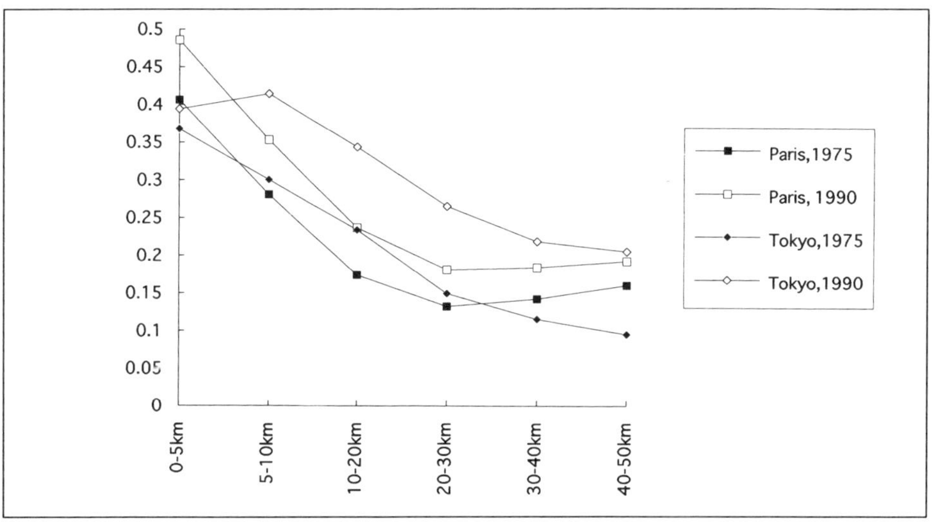

Figure 12.16. Ratio of single-person households (single-person households / total households)

12.4.8. RATIO OF SINGLE-PERSON HOUSEHOLDS

In both cities, there were a higher ratios of single-person households in 1990 than in 1975. Overall, this ratio tended to decrease away from the city center (see Figure 12.16). However, in Paris, it tended to increase beyond the 30 km zone, probably due to the large numbers of elderly people and farm workers living by themselves in this area. In Tokyo, the rate of increase in single-person households in the 0-5 km zone was not as high as in other zones, probably because in central Tokyo, residential areas were being converted into offices, and rental housing geared toward college students and single office workers decreased dramatically.

12.5. Changes in Urban Spatial Structure

The changes in the urban spatial structure of Paris and Tokyo are summarized in the following four points.

1) Both Paris and Tokyo are primate cities, somewhat unusual for major industrialized nations. They serve as the core cities for the economic, political, and cultural activities of their respective countries. About one-fifth of the French population lives in the Paris Metropolitan Area, while a full one-fourth of all Japanese live in the Tokyo Metropolitan Area. Both of these proportions are growing each year. The primacy of both cities in their respective national urban systems is likely the result of their excellent geographical locations, intensive transportation networks, and centralized political systems. Since the 1980s, primacy has been further entrenched in Paris and Tokyo due to the concentration of multi-national corporations and foreign-owned financial and insurance institutions there.

2) The Tokyo Metropolitan Area has about 3 times the population of the Paris Metropolitan Area. As a result, population density of 1,500 persons or more per km^2 extends to 50 km from the center of Tokyo, while it ends at about 25 km in the Paris area. Beyond the 30 km zone in Paris, which is characterized by agricultural areas, population density falls below 500 people/km^2. The distribution of population densities of Paris and Tokyo are concentric and spiral shaped, respectively, but in both cities, population density gradually decreases away from the city center toward the suburbs. These population distribution patterns have developed because a) large expanses of homogeneous flat land extend in all four directions, b) decisions were made to concentrate important functions in the city centers, enhancing the economic centripetal force there, and c) road and rail systems in both cities form radial patterns which converge on the respective city centers.

3) While both cities have been experiencing population declines in their respective centers, population has been increasing in the suburbs, and the distance-zone differential of population density distribution has been shrinking, especially in Tokyo (see Figure 12.17). Between 1975 and 1990, population in the 0-5 km zone in Tokyo decreased by 22.3%, while it increased by 38.1% in the 40-50 km zone. Paris showed a

decrease of 5.6% and increase of 30.3%, respectively, for these two zones. As rents skyrocketed and living environments deteriorated in central Tokyo, young families with small children were moving in large numbers to the suburbs. This was especially conspicuous during the so-called "bubble economy" of the late 1980s, in which the population of the CBD of Tokyo fell into rapid decline.

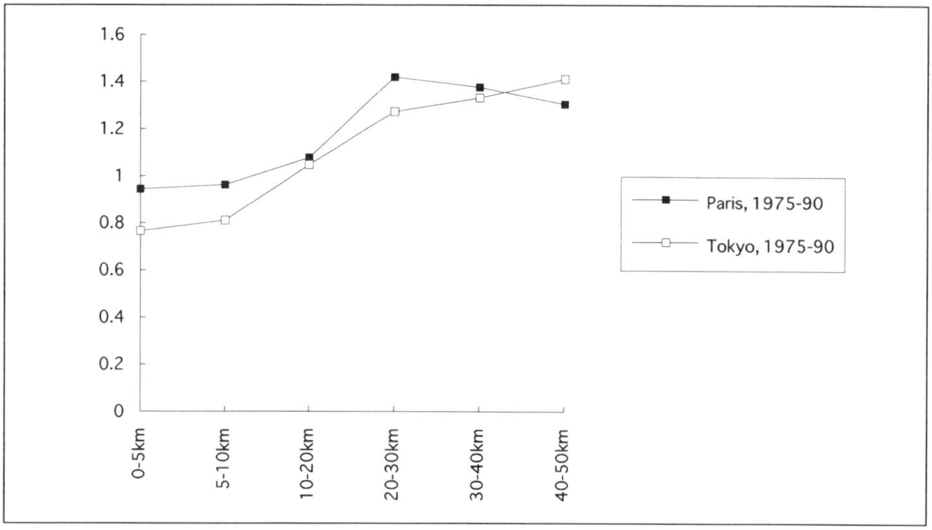

Figure 12.17. Population change, 1975-1990
(population in 1990 / population in 1975)

Although the population distribution of the Tokyo Metropolitan Area had been moving toward regional equilibrium, the void left by lost housing areas was being filled by offices, and the ratio of daytime-to-nighttime population grew at a dramatic pace, especially in the three high-rise districts of Chuo-ku, Chiyoda-ku, and Minato-ku, where the daytime population was at least 10 times the nighttime population. The exodus from the city center to the suburbs and the concentration of office functions which caused this, have resulted in workers living farther and farther from their jobs. This in turn helps to aggravate congestion in the transportation systems. Japanese geographers refer to this situation, in which the ratio of daytime population is extremely high in the central city and extremely low in the suburbs, as the "doughnut phenomenon." The radius of the "doughnut hole" is increasing each year. On the other hand, population decline in central Paris is more subdued than that in Tokyo, and more office functions are moving out to the suburban areas, so the "doughnut phenomenon" is not in evidence there; that is, the creation of employment opportunities in new towns is helping to suppress the concentration of jobs in central Paris.

4) Distance differentials are becoming apparent in the distribution of demographics such as age, household, and social structure of both cities. In densely populated central Paris, where housing is near the workplace, there is a wide variety of social classes. Noteworthy characteristics of central Paris are a large number of foreigners, small

number of single-family homes, large number of single (unmarried and divorced) residents, large number of unemployed workers, high proportion of elderly and young residents, low number of young people with children, and a high proportion of both high-income residents and low-income laborers who support them. Although many different social classes are mixed together in the Paris Metropolitan Area, the spatial segregation of social classes in the city center, while not the same as London or New York, is increasing year by year. Although the traditional 7-story apartment buildings which make up the Parisian landscape are difficult to distinguish from the outside, we can clearly see that the lower classes and immigrants from the Middle-East and North Africa predominate in the northeastern section of the city while the upper classes and international businessmen make up much of the population on the city's western side.
Most Tokyoites during the survey period were middle class residents who were seeking single-family homes with a garden in the suburbs and who worked in the city center. Central Tokyo, which was losing population, was being comprised more and more of two groups: high income groups such as company executives, lawyers, international consultants, international businessmen, and diplomats, and local people (especially elderly residents) who had owned single-family homes for many years in the city. However, among the latter group were many heirs who, being unable to pay the extremely high inheritance taxes resulting from the so-called bubble economy of the 1980s, sold their houses and moved to the suburbs. Although this vacated residential land was converted into office buildings at the height of the bubble, the economy fell into recession after the bubble burst. As a result, construction of office buildings, as one would expect, declined, and many of these lots are now being used for parking lots or are left vacant. In the three districts of central Tokyo, many elementary schools are closing for lack of students, and the collapse of the community is becoming a major social problem. Since the cost of living in central Paris is not so high, the elderly are returning to the city center, while this is not happening in Tokyo due to the extraordinarily high rents there.

In summary, a wide variety of social classes, from the poorest to the richest, live in central Paris, while residents of central Tokyo are mostly limited to high-income groups and people who have inherited land from their parents and grandparents. Middle- and upper-class residents of central Paris own villas or holiday homes outside the metropolitan area in an effort to compensate for the inconveniences of urban life. The Tokyo middle class, on the other hand, is commuting farther and farther to work as a result of their moving out of the city center in search of affordable single-family homes.

12.6. Conclusion

The spatial relationship between employment and housing in the metropolitan areas of Paris and Tokyo can be illustrated by the model shown in Figure 12.18.

In the 1970s, the spatial scale of the metropolitan area of Tokyo was larger than that of Paris, but Parisians lived closer to their jobs than did Tokyoites. Nevertheless, both cities had a polarized internal structure at that time. However, different developments subsequently occurred in the two cities. In brief, Paris moved toward a diversified, dispersed spatial structure which expanded its metropolitan area, while the development of suburban business centers was slow in Tokyo, whose metropolitan area expanded even as its polarized structure became further entrenched.

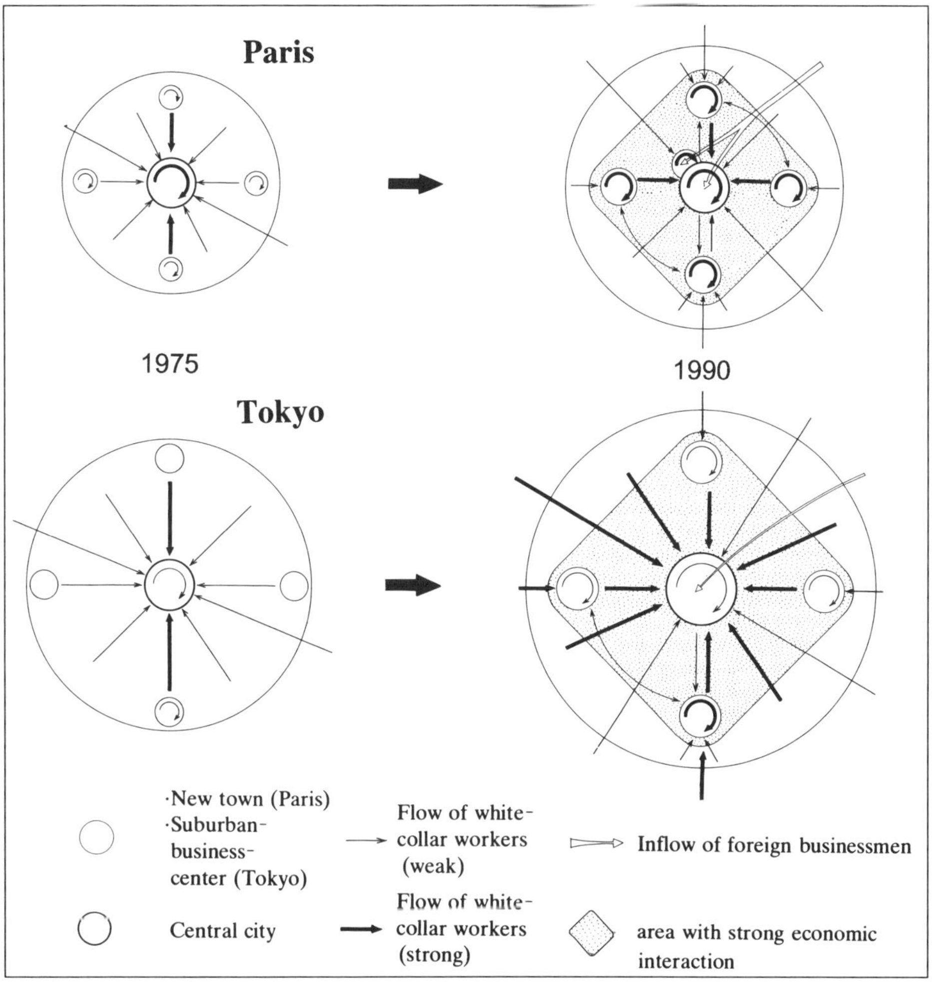

Figure 12.18. A model of changes in urban spatial structure, the Paris and Tokyo metropolitan areas (1975-90)

In Paris, the spatial dispersion of the workplace continued as offices and factories moved from the city center to the suburbs and new research centers and offices were being built in new towns and La Défense districts. In the new towns, offices and residential developments were being built as one unit to meet the demand for proximity between home and workplace, thus attracting families to move from the city center. Hence, both workplaces and housing areas were being dispersed, with a relative spatial equilibrium being reached between daytime and nighttime population. The growth of new towns was supported by the construction of roads and railways directly connecting them to the city center. In recent years, ring or loop roads linking these new towns with each other have been planned and partly built, with the result that there are even a few people who now commute between new towns. Although the number of people commuting from new towns and other suburban areas to the city center has been growing, there are also commuters who are going the other way.

In Tokyo as in Paris, the daytime and nighttime populations were becoming more spatially dispersed during the 15-year period. However, since the spatial dispersion of the nighttime population was much more dramatic, the home became farther distanced from the workplace. One reason is that the ratio of daytime-to-nighttime population grew at a rapid pace in central Tokyo. Furthermore, the residential population of the suburban business centers did not increase rapidly in surrounding areas. Unlike the new towns of Paris, the creation of new jobs in these Japanese business centers was not accompanied by new housing development.

Various types of statistical data show that offices are definitely moving into these new Japanese business centers and that the population of workers is growing. Therefore, numerous government reports and academic papers have indicated that the Tokyo Metropolitan Area is moving toward diversification and dispersion.

However, looking at the overall structure, we must say that the suburban business centers are still not very attractive. Although there are some signs that the lower order business functions are becoming more diverse, it must be pointed out that, at least as far as concentration of core managerial functions and commuting of elite office workers is concerned, the differential between the city center and suburban business centers is growing. Since the driving force of the economy is core managerial functions, polarization will likely continue in the Tokyo Metropolitan Area. Suburban business centers provide locations for branch and back offices, data processing centers, and routine work. Yet, high order business functions such as decision-making are remaining in central Tokyo, which is still attracting elite commuters from a 50 km radius for its ever-increasing core managerial functions. High-income groups almost always want to work in the city center, and the number of commuters in the 0-20 km zone willing to make the opposite commute, that is, from central Tokyo to suburban business centers, is not increasing. At the same time, many office workers who live in these suburban business centers commute to central Tokyo. There is also no indication that there will be any increase in the number of people commuting between suburban business centers.

This author believes that the spatial diversification and dispersion of high-order business functions will make little progress in Tokyo. One reason is that suburban business centers are 30 km or farther away from not only central Tokyo, but from each other. They are too far apart, and their transportation connection with the other suburban business centers is not well developed, making spatial interactions very difficult. In contrast, the new towns of Paris are all located about 25 km from the city center, and it is possible to travel between any two points in the Paris Metropolitan Area within an hour. This compact spatial situation enhances new towns' interactions with both the central city and each other. This is likely responsible for the diversification in the spatial structure of Paris.

On a final note, it should be noted that both Paris and Tokyo were experiencing an influx of foreigners and foreign capital in the 1980s. Headquarters and branches of large foreign-owned financial and insurance companies and multi-national corporations began locating in city centers (including the La Défense district of Paris). As a result, large numbers of foreigners such as international consultants, international businessmen, lawyers, simple laborers, technicians, and college students, began coming to the central city or its environs.

This phenomenon was also happening at the same time in New York, London, and other international megacities, and may have led to the growth of all these cities, including Paris and Tokyo, into "world cities," that is, central nodes in the global urban system which has fostered "global control" capacities. The globalization of the economy and capital flow and the development of exchange and communications technologies has led to the increased flow of people, goods, and information across international boundaries, which is rapidly intensifying the interdependence of the world's major cities where this is mainly taking place. Whether the rapid concentration of international functions since the 1980s has had a great impact on the urban spatial structure of Paris and/or Tokyo is a topic for future research.[1]

Note

1. This study was carried out with the assistance of the Grant-in-Aid for Scientific Research of the Science and Culture Division of the Japanese Ministry of Education (International Scientific Research Program: Regional Changes in the Paris Metropolitan Area). The author would like to thank Professors Nobuo Takahashi, Akihisa Sakurai, Akira Tezuka, Toshio Kikuchi, Komei Matsumura, J. R. Pitte and J. Robert for their important contributions. This paper was read at the French-Japanese Symposium on the Paris and Tokyo Metropolitan Areas, held at the University of Tsukuba on November 30, 1995.

PART VI

CONCLUSION

CHAPTER 13

CHANGING JAPANESE URBAN SYSTEM

13.1. Development of the Japanese Urban System

The formation of the modern Japanese urban system likely began in the Meiji Era. The fall of the old Edo government in 1867 brought about the end of the feudal system and paved the way for the modernization of Japan. The old policy of national seclusion, which had severely restricted free interaction with other countries, had existed since the early 17th century. Thus, Japan was still a backward nation in the 1860s. However, the new Meiji government launched an ambitious campaign to acquire and introduce the best of Western culture and technology. Trade with other countries was opened up, and seaports, especially Kobe and Yokohama, began to attract businessmen and merchants from both Japan and overseas, and residential areas for foreigners quickly came into existence. At the same time, these ports became centers for commerce, distribution facilities, wholesalers, warehouses, stores and shops, offices, and other economic functions. In this way, Kobe and Yokohama quickly grew as the gateways to Japan.

In the latter half of the Meiji Era, the Japanese textile industry began to grow; more than 100 years after its start in England, the Industrial Revolution had finally come to Japan. During this time (ca. 1887- ca. 1907), cities along the coast became industrial centers, while inland cities became industrial and/or mining centers.

The Meiji government made a special effort to develop the railway system, which helped to greatly facilitate the transport of goods and people over long distances. Communities that had stations on major railway lines developed as nodes for the collection and distribution of commercial products. As railway transport expanded, the regional urban systems, which up until then had consisted merely of weak linkages and very small economic hinterlands, began to strengthen ties with one another. The increase in traffic between cities complementarily linked these cities, and functional divisions and a hierarchy began to take shape. The development of the railway network promoted spatial organization. While cities with railway stations grew, ports at the confluences and mouths of rivers, which had flourished as important transport nodes during the Edo Period, fell into decline as the relative importance of river transport waned.

Thus developed the modern national urban system, which saw its framework completed around 1920. During this period, Tokyo emerged as the dominant city of central and northeast Japan, while Osaka became the most important city in the western part of the country. Together these two cities were at the top of the national urban system, giving it a bipolar structure. Just below Tokyo and Osaka in this hierarchy were Kyoto, Nagoya,

Yokohama and Kobe. Even as cities were forming a hierarchy, though, their interdependence during this period was not very pronounced.

In the latter half of the 1950s, Japan entered its period of high economic growth, and the Japanese economy gradually shifted emphasis away from primary industries toward secondary and tertiary industries. During this period there was a population exodus from rural communities toward the major cities, which had a tremendous impact on the urban system. Population tended to concentrate especially in Tokyo, Osaka and Nagoya, forming the so-called "three major metropolitan areas". As the cores of the "Tokaido megalopolis", each of these cities was attracting not only manufacturing, but also a variety of other enterprises to support this industrial development, including the distribution, commercial, and finance and securities industries. The new locations of these enterprises created a vast number of employment opportunities, providing the spur for population concentration. The opening of the Tokaido Shinkansen ("bullet train") line in 1964 and the Tomei expressway in 1969 greatly enhanced the locational potential of Tokyo, Osaka and Nagoya. Outside of this megalopolis, the cities of Sapporo, Sendai, Hiroshima, Fukuoka, etc., were also expanding their respective hinterlands, establishing themselves as regional core cities. During the period of high economic growth, which lasted until the early 1970s, the dominant-subordinate relationships among cities were strengthened, and a Christaller-type of urban hierarchy took shape.

Amidst such economic transition, the capital city of Tokyo underwent some dramatic changes. It grew into a major node both in the national and international urban systems, becoming a world city in the same league with London and New York. In the 1980s, foreign-owned and multinational corporations came one after another to Tokyo, and there was constant competition to build regional coordinating headquarters, branches, and offices. These enterprises, which were conducting the business of the global corporations, were engaged in economic activities day and night. Tokyo came to be known as a "24-hour city". The dramatic growth of Tokyo effectively reduced the economic role of its main rival, Osaka, and companies began relocating their headquarters out of Osaka and into Tokyo. The "straw" effect came into play, in which Tokyo appeared to be "sucking" core management functions out of other areas. This trend marked the destruction of the bipolar urban system in Japan.

In the 1990s, the development of the high-speed transportation system was well underway, and Shinkansen lines, expressways and air networks became organically linked. The shortening of transport time and reduction of associated fares, costs, etc., led to greater movement of goods and people between cities. The high-speed communications network expanded, and information transmission became smoother.

What should be emphasized here is the point at which cities that have come to function as relay points for goods distribution and information transmission, start the gradual loss of their influence as nodes for economic activity. Let us examine the case of how companies transmit information internally. Large corporations employ the territory system and establish foci of operations within the districts they control. For example, in the prefectural capitals of Akita and Yamagata cities, there are offices of major Tokyo-based corporations, but at one time they had been subordinate to Sendai, which had been in charge of affairs for the expansive Tohoku district, of which Akita and Yamagata are a part. However, orders are now coming straight from Tokyo. At the same time, Akita and Yamagata's interaction with Sendai is gradually decreasing, while their links with Tokyo are growing stronger. This is also apparent when viewed in the context of the flow of goods. The rationalization of distribution systems has formed direct links between areas of production and consumption, resulting in a relative decline of the influence of cities that had flourished as wholesale centers and as relay nodes for the delivery of goods.

13.2. Characteristics of the Japanese Urban System and Future Trends

The national urban system of Japan is slowly being transformed from a pyramid-shaped hierarchical system to a non-hierarchical network system, but for all practical purposes the hierarchical system is still firmly entrenched. Centralized governmental administrative systems, as well as the corporate territoriality that these systems have brought about, have supported this hierarchical structure. This support is impeding the transition to the network system. Let us reflect on the "Regional structure of information flow" in Chapter 4. Telephone communications from Tokyo to major regional cities and prefectural capitals, and from major regional cities to prefectural capitals, are mostly in the form of transmissions, rather than receptions. This reflects Tokyo's transmission to major regional cities, and directly to prefectural capitals; in other words, from the upper to the lower echelons, the transmission of operational information from corporations and government organizations is a structurally unilateral expression of Japan's urban system. In order to form the type of network system found in North America and Europe, it may become necessary for the central government to share power with local governments.

As the unification of Europe progresses, the economic significance of national borders is diminishing. It is becoming more appropriate to look upon the major cities of Europe as being nodes in the European international urban system, rather than just being nodes of their respective national systems. In North America as well, American and Canadian urban systems are strengthening their linkages with one another, and year by year the economic friction imposed by the US-Canada border is diminishing.

In contrast is Japan, which is on the eastern rim of East Asia. Its status as an island nation has had a major effect, and historically it has had less openness in its national urban system than the advanced countries of Europe and North America. The degree of openness is still not high by most standards. Tokyo is still the only major center for economic and social interaction with other countries, as it has been since the early days of the Meiji Era (1868-1912).

However, the author should point out that even in this somewhat antiquated structure, there are still places where change can be seen. For example, regional cities are bypassing Tokyo in establishing their own linkages with cities in other countries; Fukuoka, for instance, is strengthening ties with Pusan, Hong Kong and Singapore; Niigata is becoming closer to Nakhodka and Vladivostok; and Naha is forming a stronger relationship with Taipei.

In the past when branch offices in regional cities engaged in business activities, the Tokyo headquarters, after being informed of the intentions of the branch office, would generally initiate negotiations with the overseas company, but today the opportunities have been increasing for branches to become directly involved in such dealings. As technology transfer and factory establishment from Japan to the countries of mainland Asia takes firmer root, the role of regional cities is becoming more important. This is because most Japanese manufacturers are located in the smaller, provincial cities. The strengthening of ties between the smaller cities in Japan and their counterparts overseas does not stop at economic activities; for example, the number of smaller cities forming sister-city relationships and strengthening cultural interaction, is growing.

At the global level, Tokyo, as a core city of Asia, has become a peer to major cities like New York, London, Frankfurt and Paris. However, this status is being threatened by Asian competitors like Singapore and Hong Kong. While there are multi-national corporations from Europe and the United States that are setting up new bases of operations in Tokyo, there are some foreign-owned companies that are moving their regional coordinating headquarters out of Tokyo and into Singapore and Hong Kong.

In the 21st century, the Japanese national urban system will become more open, and many if not most of the provincial cities, not just Tokyo and the other major cities, will undoubtedly be incorporated as nodes into the international urban system. Naturally, when this happens, the prosperity or decline of these cities will be determined by their interdependency and competition with cities overseas.

13.3. Characteristics of the Japanese Intra-Urban System

Unlike the cities of Europe and North America, the central areas of Japanese cities maintain a strong centripetality. The modern cities of Japan grew around railway stations. These stations were focal points for the distribution of goods and functioned as nodes for passenger transportation, while the "ekimae" (downtown) area grew as the core of office and commercial functions. Even today, railways are a major form of transportation in Japan, helping to maintain the centripetality of the downtown area. This is a nationwide feature of Japanese cities in the 300,000-class and above. Naturally, as is pointed out in Chapter 9, smaller cities where commercial businesses are moving to outlying areas, do not fall into this category.

In Japan, the single-pole structure is dominant not only in medium-size cities, but also in the major cities, and the conclusion of the author is that multi-polar dispersion is not happening to any great extent. The largest city in Japan, Tokyo, is like this. Although it's been said that Kawasaki-Yokohama, Hachioji, Omiya-Urawa, Tsuchiura-Tsukuba,

Chiba, etc., are growing as outlying cores, relatively few functions are being transferred away from the central Tokyo to the outlying areas. On the contrary, the accumulation of high-level economic core functions into the central core of Tokyo is continuing. There are many businessmen and office workers who spend two hours commuting to central Tokyo, and their numbers are not decreasing.

In contrast, the decline of rail transportation in North American cities since the mid-20th century has led to a rapid loss of centripetal force of the city center. As automobile transportation became more entrenched, the population spread out into suburban and exurban areas, suburban cores having business functions grew, and urban areas began to form a multi-polar structure. In the 1970s, when this phenomenon was still very unusual in most of the Western world, counter urbanization in North America became quite remarkable. Since the 1990s, advancements in information and communication technologies have taken this multi-polar dispersion to the next level. Numerous "satellite offices" are being established in suburban cores and so-called home-based "teleworking" is becoming increasingly common; so much so that a growing number of American workers are living near their workplaces in suburban cores. This phenomenon has only now begun to appear in Japanese cities.

INDEX

The GeoJournal Library

1. B. Currey and G. Hugo (eds.): *Famine as Geographical Phenomenon.* 1984 ISBN 90-277-1762-1
2. S.H.U. Bowie, F.R.S. and I. Thornton (eds.): *Environmental Geochemistry and Health.* Report of the Royal Society's British National Committee for Problems of the Environment. 1985 ISBN 90-277-1879-2
3. L.A. Kosiński and K.M. Elahi (eds.): *Population Redistribution and Development in South Asia.* 1985 ISBN 90-277-1938-1
4. Y. Gradus (ed.): *Desert Development.* Man and Technology in Sparselands. 1985 ISBN 90-277-2043-6
5. F.J. Calzonetti and B.D. Solomon (eds.): *Geographical Dimensions of Energy.* 1985 ISBN 90-277-2061-4
6. J. Lundqvist, U. Lohm and M. Falkenmark (eds.): *Strategies for River Basin Management.* Environmental Integration of Land and Water in River Basin. 1985 ISBN 90-277-2111-4
7. A. Rogers and F.J. Willekens (eds.): *Migration and Settlement.* A Multiregional Comparative Study. 1986 ISBN 90-277-2119-X
8. R. Laulajainen: *Spatial Strategies in Retailing.* 1987 ISBN 90-277-2595-0
9. T.H. Lee, H.R. Linden, D.A. Dreyfus and T. Vasko (eds.): *The Methane Age.* 1988 ISBN 90-277-2745-7
10. H.J. Walker (ed.): *Artificial Structures and Shorelines.* 1988 ISBN 90-277-2746-5
11. A. Kellerman: *Time, Space, and Society.* Geographical Societal Perspectives. 1989 ISBN 0-7923-0123-4
12. P. Fabbri (ed.): *Recreational Uses of Coastal Areas.* A Research Project of the Commission on the Coastal Environment, International Geographical Union. 1990 ISBN 0-7923-0279-6
13. L.M. Brush, M.G. Wolman and Huang Bing-Wei (eds.): *Taming the Yellow River: Silt and Floods.* Proceedings of a Bilateral Seminar on Problems in the Lower Reaches of the Yellow River, China. 1989 ISBN 0-7923-0416-0
14. J. Stillwell and H.J. Scholten (eds.): *Contemporary Research in Population Geography.* A Comparison of the United Kingdom and the Netherlands. 1990 ISBN 0-7923-0431-4
15. M.S. Kenzer (ed.): *Applied Geography.* Issues, Questions, and Concerns. 1989 ISBN 0-7923-0438-1
16. D. Nir: *Region as a Socio-environmental System.* An Introduction to a Systemic Regional Geography. 1990 ISBN 0-7923-0516-7
17. H.J. Scholten and J.C.H. Stillwell (eds.): *Geographical Information Systems for Urban and Regional Planning.* 1990 ISBN 0-7923-0793-3
18. F.M. Brouwer, A.J. Thomas and M.J. Chadwick (eds.): *Land Use Changes in Europe.* Processes of Change, Environmental Transformations and Future Patterns. 1991 ISBN 0-7923-1099-3

The GeoJournal Library

---

19. C.J. Campbell: *The Golden Century of Oil 1950–2050.* The Depletion of a Resource. 1991 ISBN 0-7923-1442-5
20. F.M. Dieleman and S. Musterd (eds.): *The Randstad: A Research and Policy Laboratory.* 1992 ISBN 0-7923-1649-5
21. V.I. Ilyichev and V.V. Anikiev (eds.): *Oceanic and Anthropogenic Controls of Life in the Pacific Ocean.* 1992 ISBN 0-7923-1854-4
22. A.K. Dutt and F.J. Costa (eds.): *Perspectives on Planning and Urban Development in Belgium.* 1992 ISBN 0-7923-1885-4
23. J. Portugali: *Implicate Relations.* Society and Space in the Israeli-Palestinian Conflict. 1993 ISBN 0-7923-1886-2
24. M.J.C. de Lepper, H.J. Scholten and R.M. Stern (eds.): *The Added Value of Geographical Information Systems in Public and Environmental Health.* 1995 ISBN 0-7923-1887-0
25. J.P. Dorian, P.A. Minakir and V.T. Borisovich (eds.): *CIS Energy and Minerals Development.* Prospects, Problems and Opportunities for International Cooperation. 1993 ISBN 0-7923-2323-8
26. P.P. Wong (ed.): *Tourism vs Environment: The Case for Coastal Areas.* 1993 ISBN 0-7923-2404-8
27. G.B. Benko and U. Strohmayer (eds.): *Geography, History and Social Sciences.* 1995 ISBN 0-7923-2543-5
28. A. Faludi and A. der Valk: *Rule and Order. Dutch Planning Doctrine in the Twentieth Century.* 1994 ISBN 0-7923-2619-9
29. B.C. Hewitson and R.G. Crane (eds.): *Neural Nets: Applications in Geography.* 1994 ISBN 0-7923-2746-2
30. A.K. Dutt, F.J. Costa, S. Aggarwal and A.G. Noble (eds.): *The Asian City: Processes of Development, Characteristics and Planning.* 1994 ISBN 0-7923-3135-4
31. R. Laulajainen and H.A. Stafford: *Corporate Geography.* Business Location Principles and Cases. 1995 ISBN 0-7923-3326-8
32. J. Portugali (ed.): *The Construction of Cognitive Maps.* 1996 ISBN 0-7923-3949-5
33. E. Biagini: *Northern Ireland and Beyond.* Social and Geographical Issues. 1996 ISBN 0-7923-4046-9
34. A.K. Dutt (ed.): *Southeast Asia: A Ten Nation Region.* 1996 ISBN 0-7923-4171-6
35. J. Settele, C. Margules, P. Poschlod and K. Henle (eds.): *Species Survival in Fragmented Landscapes.* 1996 ISBN 0-7923-4239-9
36. M. Yoshino, M. Domrös, A. Douguédroit, J. Paszynski and L.D. Nkemdirim (eds.): *Climates and Societies – A Climatological Perspective.* A Contribution on Global Change and Related Problems Prepared by the Commission on Climatology of the International Geographical Union. 1997 ISBN 0-7923-4324-7
37. D. Borri, A. Khakee and C. Lacirignola (eds.): *Evaluating Theory-Practice and Urban-Rural Interplay in Planning.* 1997 ISBN 0-7923-4326-3

The GeoJournal Library

38. J.A.A. Jones, C. Liu, M-K. Woo and H-T. Kung (eds.): *Regional Hydrological Response to Climate Change.* 1996 ISBN 0-7923-4329-8
39. R. Lloyd: *Spatial Cognition.* Geographic Environments. 1997 ISBN 0-7923-4375-1
40. I. Lyons Murphy: *The Danube: A River Basin in Transition.* 1997 ISBN 0-7923-4558-4
41. H.J. Bruins and H. Lithwick (eds.): *The Arid Frontier.* Interactive Management of Environment and Development. 1998 ISBN 0-7923-4227-5
42. G. Lipshitz: *Country on the Move: Migration to and within Israel, 1948–1995.* 1998 ISBN 0-7923-4850-8
43. S. Musterd, W. Ostendorf and M. Breebaart: *Multi-Ethnic Metropolis: Patterns and Policies.* 1998 ISBN 0-7923-4854-0
44. B.K. Maloney (ed.): *Human Activities and the Tropical Rainforest.* Past, Present and Possible Future. 1998 ISBN 0-7923-4858-3
45. H. van der Wusten (ed.): *The Urban University and its Identity.* Roots, Location, Roles. 1998 ISBN 0-7923-4870-2
46. J. Kalvoda and C.L. Rosenfeld (eds.): *Geomorphological Hazards in High Mountain Areas.* 1998 ISBN 0-7923-4961-X
47. N. Lichfield, A. Barbanente, D. Borri, A. Khakee and A. Prat (eds.): *Evaluation in Planning.* Facing the Challenge of Complexity. 1998 ISBN 0-7923-4870-2
48. A. Buttimer and L. Wallin (eds.): *Nature and Identity in Cross-Cultural Perspective.* 1999 ISBN 0-7923-5651-9
49. A. Vallega: *Fundamentals of Integrated Coastal Management.* 1999 ISBN 0-7923-5875-9
50. D. Rumley: *The Geopolitics of Australia's Regional Relations.* 1999 ISBN 0-7923-5916-X
51. H. Stevens: *The Institutional Position of Seaports.* An International Comparison. 1999 ISBN 0-7923-5979-8
52. H. Lithwick and Y. Gradus (eds.): *Developing Frontier Cities.* Global Perspectives – Regional Contexts. 2000 ISBN 0-7923-6061-3
53. H. Knippenberg and J. Markusse (eds.): *Nationalising and Denationalising European Border Regions, 1800–2000.* Views from Geography and History. 2000 ISBN 0-7923-6066-4
54. R. Gerber and G.K. Chuan (eds.): *Fieldwork in Geography: Reflections, Perspectives and Actions.* 2000 ISBN 0-7923-6329-9
55. M. Dobry (ed.): *Democratic and Capitalist Transitions in Eastern Europe.* Lessons for the Social Sciences. 2000 ISBN 0-7923-6331-0
56. Y. Murayama: *Japanese Urban System.* 2000 ISBN 0-7923-6600-X

KLUWER ACADEMIC PUBLISHERS – DORDRECHT / BOSTON / LONDON